Introduction to Environmental Impact Assessment

2nd edition

The Natural and Built Environment series

Editors: Professor Michael J. Bruton, The Residuary Body for Wales
        Professor John Glasson, Oxford Brookes University

Methods of environmental impact assessment
Peter Morris & Riki Therivel (editors)

Public transport
Peter White

Planning, the market and private housebuilding
Glen Bramley, Will Bartlett, Christine Lambert

Housing policy in Britain and Europe
Gavin McCrone & Mark Stephens

Partnership agencies in British urban policy
Nick Bailey (with Alison Barker and Kelvin MacDonald)

British planning policy in transition
Mark Tewdwr-Jones (editor)

Urban planning and real estate development
John Ratcliffe & Michael Stubbs

Controlling development
Philip Booth

Development control
Keith Thomas

Landscape planning and environmental impact design, 2nd edition
Tom Turner

Introduction to environmental impact assessment, 2nd edition
John Glasson, Riki Therivel, Andrew Chadwick

Introduction to
Environmental
Impact
Assessment

Principles and procedures, process, practice and prospects

2nd edition

John Glasson
Riki Therivel
Andrew Chadwick

First published in 1994 by UCL Press

Reprinted 1999 (twice)

UCL Press Limited is an imprint of the Taylor & Francis Group

UCL Press Limited
11 New Fetter Lane
London EC4P 4EE

and

325 Chestnut Street, 8th Floor
Philadelphia
PA 19106–1598
USA

The name of University College London (UCL) is a registered
trade mark used by UCL Press with the consent of the owner.

British Library Cataloguing in Publication Data
A CIP catalogue record for this book is available from the British Library.

Library of Congress Cataloging-in-Publication Data are available.

ISBNS: 1–84142–002–6 HB
       1–85728–945–5 PB

Set by Graphicraft Limited, Hong Kong
Printed and bound in Great Britain by T. J. International Ltd, Padstow.

Contents

Preface to the first edition

There has been a remarkable and refreshing interest in environmental issues over the past few years. A major impetus was provided by the 1987 Report of the World Commission on the Environment and Development (the Brundtland Report); the Rio Summit in 1992 sought to accelerate the impetus. Much of the discussion on environmental issues and on sustainable development is about the better management of current activity in harmony with the environment. However, there will always be pressure for new development. How much better it would be to avoid or mitigate the potential harmful effects of future development on the environment at the planning stage. Environmental impact assessment (EIA) assesses the impacts of planned activity on the environment in advance, thereby allowing avoidance measures to be taken: prevention is better than cure.

Environmental impact assessment was first formally established in the USA in 1969. It has spread worldwide and received a significant boost in Europe with the introduction of an EC Directive on EIA in 1985. This was implemented in the UK in 1988. Subsequently there has been a rapid growth in EIA activity, and over three hundred environmental impact statements (EISs) are now produced in the UK each year. EIA is an approach in good currency. It is also an area where many of the practitioners have limited experience. This text provides a comprehensive introduction to the various dimensions of EIA. It has been written with the requirements of both undergraduate and postgraduate students in mind. It should also be of considerable value to those in practice – planners, developers and various interest groups. EIA is on a rapid "learning curve"; this text is offered as a point on the curve.

The book is structured into four parts. The first provides an introduction to the principles of EIA and an overview of its development and agency and legislative context. Part 2 provides a step-by-step discussion and critique of the EIA process. Part 3 examines current practice, broadly in the UK and in several other countries, and in more detail through selected UK case studies. Part 4 considers possible future developments. It is likely that much more of the EIA iceberg will become visible in the 1990s and beyond. An outline of important and associated developments in environmental auditing and in strategic environmental assessment concludes the text.

Although the book has a clear UK orientation, it does draw extensively on EIA experience worldwide, and it should be of interest to readers from many countries. The book seeks to highlight best practice and to offer enough insight to methods, and to supporting references, to provide valuable guidance to the practitioner. For information on detailed methods for assessment of impacts in particular topic areas (e.g. landscape, air quality, traffic impacts), the reader is referred to the complementary volume, *Methods of environmental impact assessment* (Morris & Therivel, 1995, London, UCL Press).

JOHN GLASSON   RIKI THERIVEL   ANDREW CHADWICK

Oxford Brookes University

Preface

The aims and scope of this second edition are unchanged from those of the first edition. But, as noted in the first preface, EIA is on a rapid learning curve, and any commentary on the subject must be seen as part of an ongoing discussion. The worldwide spread of EIA procedures and practice is becoming increasingly comprehensive. In the European Union, there is now ten years' experience of the implementation of the pioneering EIA Directive, and an amended directive will become operational in 1999. There has also been considerable interest in the development of the EIA process, in extending the scope of activity, and also in assessing effectiveness. Reflecting such changes, this revised edition updates the commentary by introducing and developing a number of issues which are seen as of growing importance to both the student and practitioner of EIA.

The structure of the first edition has been retained, plus much of the original material, but variations and additions have been made to specific sections. In Part 1 (principles and procedures), a significant addition has been the incorporation of the amendment to the EC EIA Directive and consideration of the implications for EIA practice. In Part 2 (discussion of the EIA process), many elements have been updated, including screening, assessment of significance, participation, presentation, review and the overall management of the process.

We have made very substantial changes to Part 3 (overview of practice), drawing on the findings of several major international and UK reviews of EIA effectiveness. While there is general consensus on the utility of EIA, there is also concern about some weaknesses in the procedures and practice to date. The more detailed studies of UK practice for new settlements, roads and electricity supply have been updated, and the important area of waste disposal projects has been added. Major changes have also been made to the chapter on comparative practice, with more discussion of emerging EIA systems and the role of international funding institutions, such as the World Bank.

Part 4 of the book (prospects) has also been substantially revised to reflect some of the changing prospects for EIA including, for example, more consideration of cumulative impacts, socio-economic impacts, and public participation, plus possible shifts towards more integrated environmental assessment. Similarly, in the final chapter, there is a substantial update of the developing principles, procedures and practice of Strategic Environmental Assessment. Additions to the Appendices include the amended EC Directive, World Bank EIA procedures, environment impact statement review pro-formas. There is an expanded bibliography of key references.

JOHN GLASSON   RIKI THERIVEL   ANDREW CHADWICK

Oxford 1998

Dedicated to our families

Acknowledgements

Our grateful thanks are due to many people without whose help this book would not have been produced. We are particularly grateful to Carol Glasson, who typed and retyped several drafts to tight deadlines and to high quality, and who provided invaluable assistance in bringing together the disparate contributions of the three authors. Our thanks also go to Rob Woodward for his production of many of the illustrations. We are very grateful to our consultancy clients and research sponsors, who have underpinned the work of the Impact Assessment Unit in the School of Planning at Oxford Brookes University (formerly Oxford Polytechnic). Michael Gammon provided the initiative and constant support; Phil Saunders and Andrew Hammond maintained the positive link with the electricity supply industry. Other valuable support has been provided by the Royal Society for the Protection of Birds, the Economic and Social Research Council and PCFC.

Our students at Oxford Brookes University on both undergraduate and post-graduate programmes have critically tested many of our ideas. In this respect we should like to acknowledge, in particular, the students on the MSc course in Environmental Assessment and Management. The editorial and presentation support for the second edition by staff at Taylor and Francis is very gratefully acknowledged. We have benefited from the support of colleagues in the Schools of Planning and Biological and Molecular Sciences, and from the wider community of EIA academics, researchers and consultants, who have helped to keep us on our toes.

We are also grateful for permission to use material from the following sources:

- Environmental Data Service (Figs 3.2, 3.3)
- British Association of Nature Conservationists (cartoons: Parts 2 and 3)
- Rendel Planning (Fig. 4.3)
- UNEP Industry and Environment Office (Fig. 4.5 and Table 4.3)
- University of Manchester, Department of Planning and Landscape, EIA Centre (Tables 5.8, 8.1, Figs. 8.4, 8.7, Appendix 3)
- John Wiley & Sons (Tables 6.1, 6.2)
- Baseline Environmental Consulting, West Berkeley, California (Fig. 7.2)
- David Tyldesley and Associates (Fig. 9.1 and Table 9.5)
- UK Department of Environment (Table 6.4)

- UK Department of Transport (Table 10.1)
- *Planning* newspaper (cartoon: Part 4)
- Kent County Council Planning Department (Fig. 13.3).
- Hertfordshire County Council Planning Department (Table 13.1, Fig. 13.4)

Abbreviations

AEE	assessment of environmental effects
AONB	Area of Outstanding Natural Beauty
BATNEEC	best available technique not entailing excessive costs
BPEO	best practicable environmental option
CBA	cost-benefit analysis
CC	county council
CEAA	Canadian Environmental Assessment Agency
CEC	Commission of the European Communities
CEGB	Central Electricity Generating Board
CEPA	Commonwealth Environmental Protection Agency (Australia)
CEQ	Council on Environmental Quality (US)
CEQA	California Environmental Quality Act
CHP	combined heat and power
CIE	community impact evaluation
CPRE	Council for the Protection of Rural England
CVM	contingent valuation method
DOEn	Department of Energy
DC	district council
DETR	Department of Environment, Transport and the Regions
DG	Directorate General (CEC)
DMRB	Design Manual for Roads and Bridges
DOE	Department of the Environment
DOT	Department of Transport
DTI	Department of Trade and Industry
EA	environmental assessment
EBRD	European Bank for Reconstruction and Development
EC	European Community
EES	environmental evaluation system
EIA	environmental impact assessment
EIR	environmental impact report
EIS	environmental impact statement
EMAS	eco-management and audit scheme (CEC)
EMS	environmental management system

EN	English Nature
EPA	Environmental Protection Act
ES	environmental statement
ESI	electricity supply industry
ESRC	Economic and Social Research Council
EU	European Union
FGD	flue gas desulphurization
FOE	Friends of the Earth
FONSI	finding of no significant impact
GAM	goals achievement matrix
GIS	geographical information system
GNP	gross national product
ha	hectares
HMIP	Her Majesty's Inspectorate of Pollution
HMSO	Her Majesty's Stationery Office
IAIA	International Association for Impact Assessment
IAU	Impact Assessment Unit (Oxford Brookes)
IEA	Institute of Environmental Assessment
IFI	International Funding Institution
IPC	integrated pollution control
km	kilometre
LCP	large combustion plant
LI	Landscape Institute
LPA	local planning authority
LULU	locally unacceptable land use
MAFF	Ministry of Agriculture, Fisheries and Food
MAUT	multi-attribute utility theory
MEA	Manual of environmental appraisal
MW	megawatts
NEPA	National Environmental Policy Act (US)
NEPP	National Environmental Policy Plan (Netherlands)
NGO	non-government organization
NIMBY	not in my back yard
NRA	National Rivers Authority
PADC	project appraisal for development control
PBS	planning balance sheet
PPG	Planning Policy Guidance note
PPPS	policies, plans and programmes
PWR	pressurised water reactor
RA	risk assessment
RSPB	Royal Society for the Protection of Birds
RTPI	Royal Town Planning Institute
SACTRA	Standing Advisory Committee on Trunk Road Assessment
SDD	Scottish Development Department

SEA	strategic environmental assessment
SIA	social impact assessment
SOS	Secretary of State
SSSI	Site of Special Scientific Interest
T&CP	town and country planning
UK	United Kingdom
UNCED	United Nations Conference on Environment and Development
UNECE	United Nations Economic Commission for Europe
UNEP	United Nations Environment Programme
US	United States
WRAM	Water Resources Assessment Method

Part 1

Principles and procedures

CHAPTER 1
Introduction and principles

1.1 Introduction

In recent years there has been a remarkable growth of interest in environmental issues – in sustainability and the better management of development in harmony with the environment. Associated with this growth of interest has been the introduction of new legislation, emanating from national and international sources, such as the European Commission, that seeks to influence the relationship between development and the environment. Environmental impact assessment (EIA) is an important example. EIA legislation was introduced in the USA over 25 years ago. A European Community directive in 1985 accelerated its application in EU Member States and, since its introduction in the UK in 1988, it has been a major growth area for planning practice. The originally anticipated 20 environmental impact statements (EIS) per year in the UK quickly escalated to over 300, and this is only the tip of the iceberg. The scope of EIA will widen greatly in the coming years.

It is therefore perhaps surprising that the introduction of EIA met with strong resistance from many quarters, particularly in the UK. Planners argued, with partial justification, that they were already making such assessments. Many developers saw it as yet another costly and time-consuming constraint on development, and central government was also unenthusiastic. Interestingly, current UK legislation refers to environmental assessment (EA), leaving out the apparently politically sensitive, negative sounding reference to impacts. Much of the terminology is still at the formative stage. This first chapter therefore introduces EIA as a process, the purposes of this process, types of development, environment and impacts and current issues in EIA.

1.2 The nature of environmental impact assessment

Definitions

Definitions of environmental impact assessment abound. They range from the oft-quoted and broad definition of Munn (1979), which refers to the need "to identify

3

and predict the impact on the environment and on man's health and well-being of legislative proposals, policies, programmes, projects and operational procedures, and to interpret and communicate information about the impacts", to the narrow UK DOE (1989) operational definition: "The term 'environmental assessment' describes a technique and a process by which information about the environmental effects of a project is collected, both by the developer and from other sources, and taken into account by the planning authority in forming their judgements on whether the development should go ahead." The United Nations Economic Commission for Europe (1991) has an altogether more succinct and pithy definition: "an assessment of the impact of a planned activity on the environment".

Environmental impact assessment: a process

In essence, EIA is a *process*, a systematic process that examines the environmental consequences of development actions, in advance. The emphasis, compared with many other mechanisms for environmental protection, is on prevention. Of course planners have traditionally assessed the impacts of developments on the environment, but invariably not in the systematic, holistic and multidisciplinary way required by EIA. The process involves a number of steps, as outlined in Figure 1.1. These are briefly described below, pending a much fuller discussion in Chapters 4–7. It should be noted at this stage that, although the steps are outlined in linear fashion, EIA should be a cyclical activity, with feedback and interaction between the various steps. It should also be noted that practice can and does vary considerably from the process illustrated in Figure 1.1. For example, until recently UK EIA legislation did not require some of the steps, including the consideration of alternatives, and still does not require post-decision monitoring (DOE 1989). The order of the steps in the process may also vary.

- *Project screening* narrows the application of EIA to those projects that may have significant environmental impacts. Screening may be partly determined by the EIA regulations operating in a country at the time of assessment.
- *Scoping* seeks to identify at an early stage, from all of a project's possible impacts and from all the alternatives that could be addressed, those that are the crucial, significant issues.
- *The consideration of alternatives* seeks to ensure that the proponent has considered other feasible approaches, including alternative project locations, scales, processes, layouts, operating conditions and the "no action" option.
- *The description of the project/development action* includes a clarification of the purpose and rationale of the project, and an understanding of its various characteristics – including stages of development, location and processes.
- *The description of the environmental baseline* includes the establishment of both the present and future state of the environment, in the absence of the project, taking into account changes resulting from natural events and from other human activities.

4

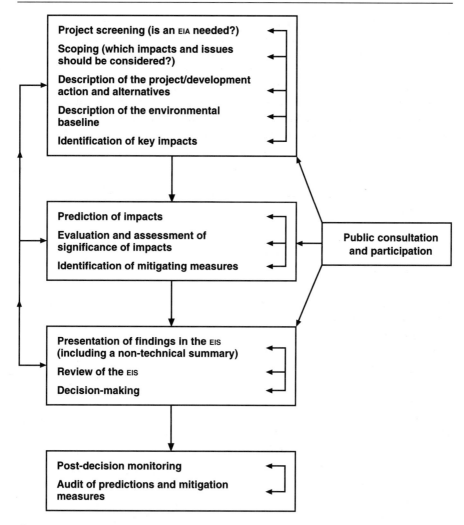

Figure 1.1 Important steps in the EIA process. Note: EIA should be a cyclical process with considerable interaction between the various steps. For example, public participation can be useful at most stages of the process; monitoring systems should relate to parameters established in the initial project and baseline descriptions.

- *The identification of the main impacts* brings together the previous steps with the aims of ensuring that all potentially significant environmental impacts (adverse and beneficial) are identified and taken into account in the process.
- *The prediction of impacts* aims to identify the magnitude and other dimensions of identified change in the environment with a project/action, by comparison with the situation without that project/action.

- *The evaluation and assessment of significance* assesses the relative significance of the predicted impacts to allow a focus on the main adverse impacts.
- *Mitigation* involves the introduction of measures to avoid, reduce, remedy or compensate for any significant adverse impacts.
- *Public consultation and participation* aim to ensure the quality, comprehensiveness and effectiveness of the EIA, and that the public's views are adequately taken into consideration in the decision-making process.
- *EIS presentation* is a vital step in the process. If done badly, much good work in the EIA may be negated.
- *Review* involves a systematic appraisal of the quality of the EIS, as a contribution to the decision-making process.
- *Decision-making* on the project involves a consideration by the relevant authority of the EIS (including consultation responses) together with other material considerations.
- *Post-decision monitoring* involves the recording of outcomes associated with development impacts, after a decision to proceed. It can contribute to effective project management.
- *Auditing* follows from monitoring. It can involve comparing actual outcomes with predicted outcomes, and can be used to assess the quality of predictions and the effectiveness of mitigation. It provides a vital step in the EIA learning process.

Environmental impact statements: the documentation

The environmental impact statement documents the information and estimates of impacts derived from the various steps in the process. Prevention is better than cure; an EIS revealing many significant unavoidable adverse impacts would provide valuable information that could contribute to the abandonment or substantial modification of a proposed development action. Where adverse impacts can be successfully reduced through mitigation measures, there may be a different decision. Table 1.1 provides an example of the content of an EIS for a project.

The *non-technical summary* is an important element in the documentation; EIA can be complex, and the summary can help to improve communication with the various parties involved. Reflecting the potential complexity of the process, a *methods statement*, at the beginning, provides an opportunity to clarify some basic information (e.g. who the developer is, who has produced the EIS, who has been consulted and how, what methods have been used, what difficulties have been encountered and what the limitations of the EIA are). A *summary statement of key issues*, up-front, can also help to improve communications. A more enlightened EIS would also include a *monitoring programme*, either here or at the end of the document. The *background to the proposed development* covers the early steps in the EIA process, including clear descriptions of a project, and baseline conditions (including relevant planning policies and plans). Within each of the *topic areas* of an EIS there

Table 1.1 An EIS for a project – example or contents.

Non-technical summary

Part 1: Methods and key issues

1 Methods statement
2 Summary of key issues; monitoring programme statement

Part 2: Background to the proposed development

3 Preliminary studies: need, planning, alternatives and site selection
4 Site description, baseline conditions
5 Description of proposed development
6 Construction activities and programme

Part 3: Environmental impact assessment – topic areas

7 Land use, landscape and visual quality
8 Geology, topography and soils
9 Hydrology and water quality
10 Air quality and climate
11 Ecology: terrestrial and aquatic
12 Noise
13 Transport
14 Socio-economic impact
15 Interrelationships between effects

would normally be a discussion of existing conditions, predicted impacts, scope for mitigation and residual impacts.

EIA and EIS practices vary from study to study, from country to country, and best practice is constantly evolving. A recent UN study of EIA practice in several countries advocated changes in the process and documentation (United Nations Economic Commission for Europe 1991). These included giving a greater emphasis to the socio-economic dimension, to public participation, and to "after the decision" activity, such as monitoring.

Other relevant definitions

Development actions may have impacts not only on the physical environment but also on the social and economic environment. Typically, employment opportunities, services (e.g. health, education) and community structures, life-styles and values may be affected. *Socio-economic impact assessment* or *social impact assessment* (SIA), is regarded here as an integral part of EIA. However, in some countries it is (or has been) regarded as a separate process, sometimes parallel to EIA, and the reader should be aware of its existence (Carley & Bustelo 1984, Finsterbusch 1985, International Association for Impact Assessment, 1994).

Strategic environmental assessment (SEA) expands EIA from projects to policies, plans and programmes. Development actions may be for a project (e.g. a nuclear power station), for a programme (e.g. a number of pressurized water reactor (PWR) nuclear power stations), for a plan (e.g. in the town and country planning system in England and Wales, for local plans and structure plans), or for a policy (e.g. the development of renewable energy). EIA to date has generally been used for individual projects, and that role is the primary focus of this book. But EIA for programmes, plans and policies, otherwise known as strategic environmental assessment, is currently generating much interest in the EU and beyond (Therivel et al. 1992). SEA informs a higher, earlier, more strategic tier of decision-making. In theory, EIA should be carried out first for policies, then for plans, programmes, and finally for projects.

Risk assessment (RA) is another term sometimes found associated with EIA. Partly in response to events such as the chemicals factory explosion at Flixborough (UK), and nuclear power station accidents at Three Mile Island (USA) and Chernobyl (Ukraine), risk assessment has developed as an approach to the analysis of risks associated with various types of development. The major study of the array of petrochemicals and other industrial developments at Canvey Island in the UK provides an example of this approach (Health and Safety Commission 1978). See Calow (1997) for a recent overview of the growing area of environmental risk assessment and management.

Vanclay and Bronstein (1995) and others note several other relevant definitions, based largely on particular foci of specialization and including: demographic impact assessment, health impact assessment, climate impact assessment, gender impact assessment, psychological impact assessment and noise impact assessment. Other more encompassing definitions include policy assessment, technology assessment and economic assessment. There is a semantic explosion which requires some clarification. As a contribution to the latter, Sadler (1996) suggests that we should view "EA as the generic process that includes EIA of specific projects, SEA of policies plans and programmes, and their relationships to a larger set of impact assessment and planning-related tools".

1.3   The purposes of environmental impact assessment

An aid to decision-making

Environmental impact assessment is a process with several important purposes. It is an aid to decision-making. For the decision-maker, for example a local authority, it provides a systematic examination of the environmental implications of a proposed action, and sometimes alternatives, before a decision is taken. The EIS can be considered by the decision-maker along with other documentation related to the planned activity. EIA is normally wider in scope and less quantitative than other techniques, such as cost–benefit analysis. It is not a substitute for decision-making, but it does help to clarify some of the trade-offs associated with a proposed development action, which should lead to more rational and structured decision-making. The EIA process has the potential, not always taken up, to be a basis for negotiation between the developer, public interest groups and the planning regulator. This can lead to an outcome that balances the interests of the development action and the environment.

An aid to the formulation of development actions

Many developers no doubt see EIA as another set of hurdles to jump before they can proceed with their various activities; the process can be seen as yet another costly and time-consuming activity in the permission process. However, EIA can be of great benefit to them, since it can provide a framework for considering location and design issues and environmental issues in parallel. It can be an aid to the formulation of development actions, indicating areas where a project can be modified to minimize or eliminate altogether its adverse impacts on the environment. The consideration of environmental impacts early in the planning life of a development can lead to environmentally sensitive development; to improved relations between the developer, the planning authority and the local communities; to a smoother planning permission process; and sometimes, as argued by developers such as British Gas, to a worthwhile financial return on the extra expenditure incurred (Breakell & Glasson 1981). O'Riordan (1990) links such concepts of negotiation and redesign to the current dominant environmental themes of "green consumerism" and "green capitalism". The emergence of a growing demand by consumers for goods that do no environmental damage, plus a growing market for clean technologies, is generating a response from developers. EIA can be the signal to the developer of potential conflict; wise developers may use the process to negotiate "green gain" solutions, which may eliminate or offset negative environmental impacts, reduce local opposition and avoid costly public inquiries.

An instrument for sustainable development

Underlying such immediate purposes is of course the central and ultimate role of EIA as one of the instruments to achieve sustainable development: development that does not cost the Earth! Existing environmentally harmful developments have to be managed as best they can. In extreme cases, they may be closed down, but they can still leave residual environmental problems for decades to come. How much better it would be to mitigate the harmful effects in advance, at the planning stage, or in some cases avoid the particular development altogether. Prevention is better than cure.

Economic development and social development must be placed in their environmental contexts. Boulding (1966) vividly portrays the dichotomy between the

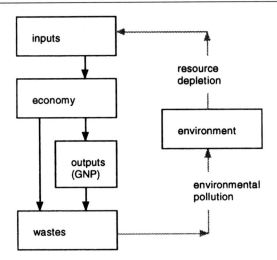

Figure 1.2 The economic development process in its environmental context. (Adapted from: Boulding 1966)

"throughput economy" and the "spaceship economy" (Fig. 1.2). The economic goal of increased GNP, using more inputs to produce more goods and services, contains the seeds of its own destruction. Increased output brings with it not only goods and services but also more waste products. Increased inputs demand more resources. The natural environment is the "sink" for the wastes and the "source" for the resources. Environmental pollution and the depletion of resources are invariably the ancillaries to economic development.

The interaction of economic and social development with the natural environment and the reciprocal impacts between human actions and the biophysical world have been recognized by governments from local to international levels. Attempts have been made to manage the interaction better, but a recent European Community report, *Towards sustainability* (CEC 1992), reveals disquieting trends that could have devastating consequences for the quality of the environment. Such EU trends include: a 25 per cent increase in energy consumption by 2010 if there is no change in current energy demand growth rates; a 25 per cent increase in car ownership and a 17 per cent increase in miles driven by 2000; a 13 per cent increase in municipal waste between 1987 and 1992, despite increased recycling; a 35 per cent increase in the EU's average rate of water withdrawal between 1970 and 1985; and a 60 per cent projected increase in Mediterranean tourism between 1990 and 2000. These trends are likely to be even more pronounced in developing countries, where, because population growth is greater and current living standards lower, there will be more pressure on environmental resources. The revelation of the state of the environment in many central and eastern European countries, and worldwide, adds weight to the assertion in the same EC report that "the great environmental struggles will be won or lost during this decade; by the next century it could be too late".

The 1987 Report of the World Commission on Environment and Development (usually referred to as the Brundtland Report, after its chairwoman) defined sustainable development as "development which meets the needs of the present generation without compromising the ability of future generations to meet their own needs" (UN World Commission on Environment and Development 1987). Sustainable development means handing down to future generations not only "man-made capital", such as roads, schools and historic buildings, and "human capital", such as knowledge and skills, but also "natural/environmental capital", such as clean air, fresh water, rain forests, the ozone layer and biological diversity. The Brundtland Report identified the following chief characteristics of sustainable development: it maintains the quality of life, it maintains continuing access to natural resources, and it avoids lasting environmental damage. It means living on the Earth's income rather than eroding its capital (DOE 1990). In addition to a concern for the environment and the future, Brundtland also emphasizes participation and equity, thus highlighting both inter- and intra-generational equity.

There is, however, a danger that "sustainable development" may become a weak catch-all phrase; there are already many alternative definitions. Holmberg and Sandbrook (1992) found over 70 definitions of sustainable development. Redclift (1987) saw it as "moral convictions as a substitute for thought"; to O'Riordan (1988) it was "a good idea which cannot sensibly be put into practice". But to Skolimowski (1995), sustainable development

> struck a middle ground between more radical approaches which denounced all development, and the idea of development conceived as business as usual. The idea of sustainable development, although broad, loose and tinged with ambiguity around its edges, turned out to be palatable to everybody. This may have been its greatest virtue. It is radical and yet not offensive.

Readers are referred to Reid (1995) and Kirkby et al. (1995) for an overview of the concept, debate and responses.

Turner & Pearce (1992) and Pearce (1992) have drawn attention to alternative interpretations of maintaining the capital stock. A policy of conserving the whole capital stock (man-made, human and natural) is consistent with running down any part of it, as long as there is substitutability between capital degradation in one area and investment in another. This can be interpreted as a "weak sustainability" position. In contrast, a "strong sustainability" position would argue that it is not acceptable to run down environmental assets, for several reasons: uncertainty (we do not know the full consequences for human beings), irreversibility (lost species cannot be replaced), life-support (some ecological assets serve life-support functions), and loss aversion (people are highly averse to environmental losses). The "strong sustainability" position has much to commend it, but institutional responses have varied.

Institutional responses to meet the goal of sustainable development are required at several levels. Issues of global concern, such as ozone-layer depletion, climate change, deforestation and biodiversity loss, require global political commitments to action. The United Nations Conference on Environment and Development (UNCED)

11

held in Rio de Janeiro in 1992 was an example of international concern, but also of the problems of securing concerted action to deal with such issues. Agenda 21, an 800-page action plan for the international community into the twenty-first century, sets out what nations should do to achieve sustainable development. It includes topics such as biodiversity, desertification, deforestation, toxic wastes, sewage, oceans and the atmosphere. For each of 115 programmes, the need for action, the objectives and targets to be achieved, the activities to be undertaken, and the means of implementation are all outlined. Agenda 21 offers policies and programmes to achieve a sustainable balance between consumption, population and the Earth's life-supporting capacity. Unfortunately it is not legally binding. It relies on national governments, local governments and others to implement most of the programmes. The Rio Conference called for a Sustainable Development Commission to be established to progress the implementation of Agenda 21. The Commission met for the first time in 1993 and reached agreement on a thematic programme of work for 1993–7. This provided the basis for an appraisal of Agenda 21 in preparation for a special session of the UN in 1997.

Within the EU, four Community Action Programmes on the Environment were implemented between 1972 and 1992. These gave rise to specific legislation on a wide range of topics, including waste management, the pollution of the atmosphere, the protection of nature and environmental impact assessment. The Fifth Programme, "Towards sustainability", is set in the context of the completion of the Single European Market. The latter, with its emphasis on major changes in economic development resulting from the removal of all remaining fiscal, material and technological barriers between Member States, could pose additional threats to the environment. The Fifth Programme recognizes the need for the clear integration of performance targets – in relation to environmental protection – for several sectors, including manufacturing, energy, transport and tourism. EU policy on the environment will be based on the "precautionary principle", that preventive action should be taken, that environmental damage should be rectified at source, and that the polluter should pay. Whereas previous EU programmes relied almost exclusively on legislative instruments, the Fifth Programme advocates a broader mixture, including "market-based instruments", such as the internalization of environmental costs through the application of fiscal measures, and "horizontal, supporting instruments", such as improved baseline and statistical data and improved spatial and sectoral planning. Figure 1.3 illustrates the interdependence of resources, sectors and policy areas. EIA has a clear role to play.

In the UK, the publication of *This common inheritance: Britain's environmental strategy* (DOE 1990) provided the country's first comprehensive White Paper on the environment. The report includes a discussion of the greenhouse effect, town and country, pollution control, and awareness and organization with regard to environmental issues. Throughout it emphasizes that responsibility for our environment should be shared between the government, business and the public. The range of policy instruments advocated includes legislation, standards, planning and economic measures. The last, building on work by Pearce et al. (1989), include charges,

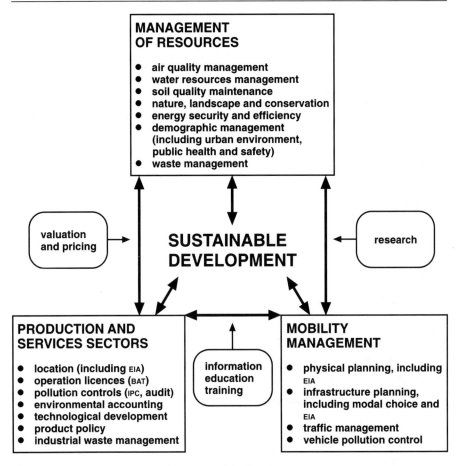

Figure 1.3 An EC framework for sustainable development. (*Source:* CEC 1992)

subsidies, market creation and enforcement incentives. The report also notes, cautiously, the recent addition of EIA to the "toolbox" of instruments. Subsequent UK government reports, such as *Sustainable development: the UK strategy* (HMG 1994), recognize the role of EIA in contributing to sustainable development and raise the EIA profile among key user groups.

Changing perspectives on EIA roles

The arguments for EIA vary in time, in space and according to the perspective of those involved. From a minimalist defensive perspective, developers, and possibly also some parts of government, might see EIA as a necessary evil, an administrative exercise, something to be gone through that might result in some minor, often

13

cosmetic, changes to a development that would probably have happened anyway. For the "deep ecologists" or "deep Greens", EIA cannot provide total certainty about the environmental consequences of development proposals; they feel that any projects carried out under uncertain or risky circumstances should be abandoned.

EIA and its methods must straddle such perspectives, partly reflecting the previous discussion on weak and strong sustainability. EIA can be, and is now often, seen as a positive process that seeks a harmonious relationship between development and the environment. The nature and use of EIA will change as relative values and perspectives also change. O'Riordan (1990) provides an appropriate conclusion to this subsection:

> One can see that EIA is moving away from being a defensive tool of the kind that dominated the 1970s to a potentially exciting environmental and social betterment technique that may well come to take over the 1990s . . . If one sees EIA not so much as a technique, rather as a process that is constantly changing in the face of shifting environmental politics and managerial capabilities, one can visualize it as a sensitive barometer of environmental values in a complex environmental society. Long may EIA thrive.

1.4 Projects, environment and impacts

The nature of major projects

As noted in Section 1.2, EIA is relevant to a broad spectrum of development actions, including policies, plans, programmes and projects. The focus here is on projects, reflecting the dominant role of project EIA in practice. The strategic environmental assessment of the "upper tiers" of development actions is considered further in Chapter 13. The scope of projects covered by EIA is widening, and is discussed further in Chapter 4. Traditionally, project EIA has applied to major projects; but what are major projects, and what criteria can be used to identify them? One could take Lord Morley's approach to defining an elephant: it's difficult, but you easily recognize one when you see it. In a similar vein, the acronym LULU (locally unacceptable land-uses) has been applied in the USA to many major projects, such as in energy, transport and manufacturing, clearly reflecting the public perception of the negative impacts associated with such developments. There is no easy definition, but it is possible to highlight some important characteristics (Table 1.2).

Most large projects involve considerable investment. In the UK context, "mega-projects" such as the Sizewell B PWR nuclear power station (budgeted to cost about £2 billion), the Channel Tunnel (about £6 billion) and the proposed Severn Barrage (about £8 billion) constitute one end of the spectrum. At the other end may be industrial estate developments, small stretches of road, various waste-disposal

Table 1.2   Characteristics of major projects.

- Substantial capital investment
- Cover large areas; employ large numbers (construction and/or operation)
- Complex array of organizational links
- Wide-ranging impacts (geographical and by type)
- Significant environmental impacts
- Require special procedures
- Extractive and primary (including agriculture); services; infrastructure and utilities
- Band, point

facilities, with considerably smaller, but still substantial, price tags. Such projects often cover large areas and employ many workers, usually in construction, but also in operation for some projects. They also invariably generate a complex array of inter- and intra-organizational activity during the various stages of their lives. The developments may have wide-ranging, long-term and often very significant impacts on the environment. The definition of significance with regard to environmental effects is an important issue in EIA. It may relate, *inter alia*, to scale of development, to sensitivity of location and to the nature of adverse effects; it will be discussed further in later chapters. Like a large stone thrown into a pond, a major project can create major ripples with impacts spreading far and wide. In many respects such projects tend to be regarded as exceptional, requiring special procedures. In the UK, there procedures have included public inquiries, hybrid bills that have to be passed through parliament (for example for the Channel Tunnel) and EIA procedures.

Major projects can also be defined according to type of activity. They include: manufacturing and extractive projects, such as petrochemicals plants, steelworks, mines and quarries; services projects, such as leisure developments, out-of-town shopping centres, new settlements and education and health facilities; and utilities and infrastructure, such as power stations, roads, reservoirs, pipelines and barrages. An EC study adopted a further distinction between band and point infrastructures. Point infrastructure would include, for example, power stations, bridges and harbours; band or linear infrastructure would include electricity transmission lines, roads and canals (CEC 1982).

A major project also has a planning and development life-cycle, including a variety of stages. It is important to recognize such stages, because impacts can vary considerably between them. The main stages in a project's life cycle are outlined in Figure 1.4. There may be variations in timing between stages, and internal variations within each stage, but there is a broadly common sequence of events. In EIA, an important distinction is between "before the decision" (stages A and B) and "after the decision" (stages C, D and E). As noted in Section 1.2, the monitoring and auditing of the implementation of a project following approval are often absent from the EIA process.

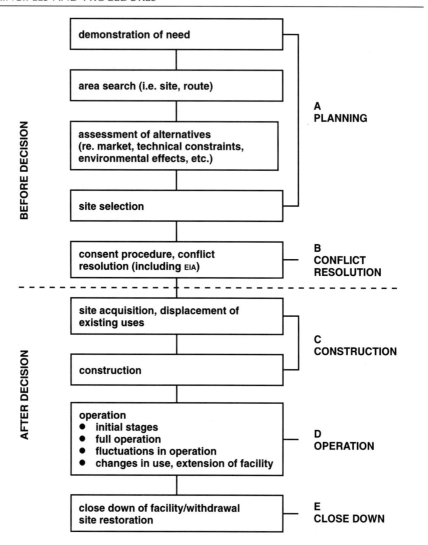

Figure 1.4 Generalized planning and development life-cycle for major projects (with particular reference to impact assessment on host area). (Adapted from Breese et al. 1965)

Projects are initiated in several ways. Many are responses to market opportunities (e.g. a holiday village, a subregional shopping centre, a gas-fired power station); others may be seen as necessities (e.g. the Thames Barrier); others may have an explicit prestige role (e.g. the programme of Grands Travaux in Paris including the Bastille Opera, Musée d'Orsay and Great Arch). Many major projects are public-sector initiatives, but with the move towards privatization in many countries, there

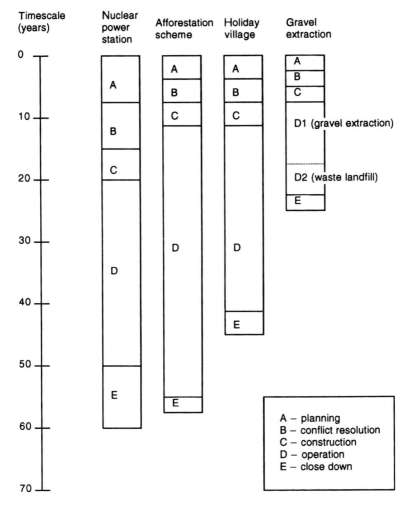

Figure 1.5 Broad variations in life-cycle stages between different types of project.

has been a move towards private sector funding, exemplified by such projects as the Mersey Barrage and the Channel Tunnel. The initial planning stage A may take several years, and lead to a specific proposal for a particular site. It is at stage B that the various control and regulatory procedures, including EIA, normally come into play. The construction stage can be particularly disruptive, and may last up to ten years for some projects. Major projects invariably have long operational lives, although extractive projects can be short compared with infrastructure projects. The environmental impact of the eventual close-down of a facility should not be forgotten; for nuclear power facilities it is a major undertaking. Figure 1.5 shows how the stages in the life-cycles of different kinds of project may vary.

17

Table 1.3 Environmental components.

Physical environment (adapted from DOE 1991)

Air and atmosphere	air quality
Water resources and water bodies	water quality and quantity
Soil and geology	classification, risks (e.g. erosion, contamination)
Flora and fauna	birds, mammals, fish, etc.; aquatic and terrestrial vegetation
Human beings	physical and mental health and wellbeing
Landscape	characteristics and quality of landscape
Cultural heritage	conservation areas; built heritage; historic and archaeological sites
Climate	temperature, rainfall, wind, etc.
Energy	light, noise, vibration, etc.

Socio-economic environment

Economic base – direct	direct employment; labour market characteristics; local and non-local trends
Economic base – indirect	non-basic and services employment; labour supply and demand
Demography	population structure and trends
Housing	supply and demand
Local services	supply and demand of services: health, education, police, etc.
Socio-cultural	lifestyles, quality of life; social problems (e.g. crime); community stress and conflict

Dimensions of the environment

The environment can be structured in several ways, including components, scale/space and time. A narrow definition of environmental components would focus primarily on the biophysical environment. For example, the UK Department of the Environment takes the term to include all media susceptible to pollution, including air, water and soil; flora, fauna and human beings; landscape, urban and rural conservation and the built heritage (DOE 1991). The DOE checklist of environmental components is outlined in Table 1.3. However, as already noted in Section 1.2, the environment has important economic and socio-cultural dimensions. These include economic structure, labour markets, demography, housing, services (education, health, police, fire, etc.), life-styles and values, and these are added to the checklist in Table 1.3. This wider definition is more in tune with an Australian definition, "For the purposes of EIA, the meaning of environment incorporates physical, biological, cultural, economic and social factors" (ANZECC 1991).

The environment can also be analyzed at various scales (Fig. 1.6). Many of the spatial impacts of projects affect the local environment, although the nature of "local" may vary according to the aspect of environment under consideration and to the stage in a project's life. However, some impacts are more than local. Traffic

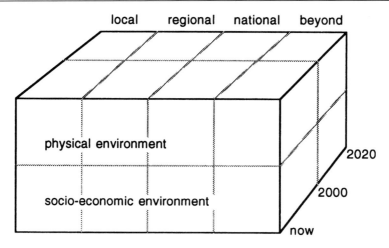

Figure 1.6 Environment: components, scale and time dimensions.

noise, for example, may be a local issue, but changes in traffic flows caused by a project may have a regional impact, and the associated CO_2 pollution contributes to the global greenhouse problem. The environment also has a time dimension. Baseline data on the state of the environment are needed at the time a project is being considered. This in itself may be a daunting request. In the UK, local development plans and national statistical sources, such as the Digest of Environmental Protection and Water Standards, may provide some relevant data. However, tailor-made state-of-the-environment reports and audits are still in limited supply (see Ch. 12 for further information). Even more limited are time-series data highlighting trends in environmental quality. The environmental baseline is constantly changing, irrespective of any development under consideration, and it requires a dynamic rather than a static analysis.

The nature of impacts

The environmental impacts of a project are those resultant changes in environmental parameters, in space and time, compared with what would have happened had the project not been undertaken. The parameters may be any of the type of environmental receptors noted previously: air quality, water quality, noise, levels of local unemployment and crime, for example. Figure 1.7 provides a simple illustration of the concept.

Table 1.4 provides a summary of some of the types of impact that may be encountered in EIA. The biophysical and socio-economic impacts have already been noted. These are often seen as synonymous with adverse and beneficial. Thus, new developments may produce harmful wastes but also produce much needed jobs in

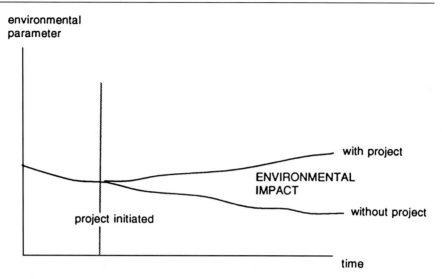

Figure 1.7 The nature of an environmental impact.

areas of high unemployment. However, the correlation does not always apply. A project may bring physical benefits when, for example, previously polluted and derelict land is brought back into productive use; similarly the socio-economic impacts of a major project on a community could include pressure on local health services and on the local housing market, and increases in community conflict and crime. Projects may also have immediate and direct impacts that give rise to secondary and indirect impacts later. A reservoir based on a river system not only takes land for the immediate body of water but also may have severe downstream implications for flora and fauna and for human activities such as fishing and sailing.

The direct and indirect impacts may sometimes correlate with short-run and long-run impacts. For some impacts the distinction between short-run and long-run may also relate to the distinction between a project's construction and its operational

Table 1.4 Types of impact.

- Physical and socio-economic
- Direct and indirect
- Short-run and long-run
- Local and strategic (including regional, national and beyond)
- Adverse and beneficial
- Reversible and irreversible
- Quantitative and qualitative
- Distribution by group and/or area
- Actual and perceived
- Relative to other developments

stage; however, other construction-stage impacts, such as change in land-use, are much more permanent. Impacts also have a spatial dimension. One distinction is between local and strategic, the latter covering impacts on areas beyond the immediate locality. These are often regional, but may sometimes be of national or even international significance.

Environmental resources cannot always be replaced; once destroyed, some may be lost for ever. The distinction between reversible and irreversible impacts is a very important one, and the irreversible impacts, not susceptible to mitigation, can constitute particular significant impacts in an EIA. It may be possible to replace, compensate for or reconstruct a lost resource in some cases, but substitutions are rarely ideal. The loss of a resource may become more serious later, and valuations need to allow for this. Some impacts can be quantified, others are less tangible. The latter should not be ignored. Nor should the distributional impacts of a proposed development be ignored. Impacts do not fall evenly on affected parties and areas. Although a particular project may be assessed as bringing a general benefit, some groups and/or geographical areas may be receiving most of any adverse effects, the main benefits going to others elsewhere. There is also a distinction between actual and perceived impacts. Subjective perceptions of impacts may significantly influence the responses and decisions of people towards a proposed development. They constitute an important source of information, to be considered alongside more objective predictions of impacts. Finally, all impacts should be compared with the "do-nothing" situation, and the state of the environment predicted without the project. This can be widened to include comparisons with anticipated impacts from alternative development scenarios for an area.

We conclude on a semantic point: the words "impact" and "effect" are widely used in the literature and legislation on EIA, but it is not always clear whether they are interchangeable or should be used only for specifically different meanings. In the United States, the regulations for implementing the National Environmental Policy Act expressly state that "effects and impacts as used in these regulations are synonymous". This interpretation is widespread, and is adopted in this text. But there are other interpretations relating to timing and to value judgements. Catlow and Thirlwall (1976) make a distinction between effects which are ". . . the physical and natural changes resulting, directly or indirectly, from development" and impacts which are ". . . the consequences or end products of those effects represented by attributes of the environment on which we can place an objective or subjective value". In contrast, a recent Australian study (CEPA 1994) reverses the arguments, claiming that "there does seem to be greater logic in thinking of an impact resulting in an effect, rather than the other way round". Other commentators have introduced the concept of value judgement into the differentiation. Preston and Bedford (1988) state that "the use of the term 'impacts' connotes a value judgement". This view is supported by Stakhiv (1988), who sees a distinction between "scientific assessment of facts (effects), and the evaluation of the relative importance of these effects by the analyst and the public (impacts)". The debate continues!

21

1.5   Current issues in environmental impact assessment

Although EIA now has almost 30 years of history in the USA, elsewhere the development of concepts and practice is more recent. Development is moving apace in many other countries, including the UK and the other EU Member States. Such progress has not been without its problems, and a number of the current issues in EIA are highlighted here and will be discussed more fully in later chapters.

Scope of the assessment

Whereas legislators may seek to limit coverage, best practice may lead to its widening. For example, project EIA may be mandatory only for a limited set of major projects. In practice many others have been included. But which projects should have assessments? In the UK, case law is now building up, but the criteria for the inclusion or exclusion of a project for EIA are still developing. In a similar vein, there is a case for widening the dimensions of the environment under consideration to include socio-economic impacts more fully. The trade-off between the adverse biophysical impacts of a development and its beneficial socio-economic impacts often constitutes the crucial dilemma for decision-makers. Coverage can also be widened to include other types of impacts only very partially covered to date. Distributional impacts would fall into this category. Lichfield and others are seeking to counter this problem (see Lichfield 1996).

The nature of methods of assessment

As noted in Section 1.2, some of the main steps in the EIA process (e.g. the consideration of alternatives, monitoring) may be missing from many studies. There may also be problems with the steps that are included. The prediction of impacts raises various conceptual and technical problems. The problem of establishing the environmental baseline position has already been noted. It may also be difficult to establish the dimensions and development stages of a project clearly. Further conceptual problems include establishing what would have happened in the relevant environment without a project, clarifying the complexity of interactions of phenomena, and making trade-offs in an integrated way (i.e. assessing the trade-offs between economic apples, social oranges and physical bananas). Other technical problems are the general lack of data and the tendency to focus on the quantitative, and often single, indicators in some areas. There may also be delays and discontinuities between cause and effect, and projects and policies may discontinue. The lack of auditing of predictive techniques limits the feedback on the effectiveness of methods. Nevertheless, innovative methods are being developed to predict impacts, ranging from simple checklists and matrices to complex mathematical models. These methods are not neutral, in the sense that the more complex they are, the more difficult it becomes for the general public to participate in the EIA process.

The relative roles of participants in the process

The various "actors" in the EIA process – the developer, the affected parties, the general public and the regulators at various levels of government – have different accesses to the process, and their influence on the outcome varies. Many would argue that in countries such as the UK, the process is too developer-orientated. The developer or the developer's consultant carries out the EIA and prepares the EIS, and is unlikely to predict that the project will be an environmental disaster. Notwithstanding this, developers themselves are concerned about the potential delays associated with the requirement to submit an EIS. They are also concerned about cost. Details about costs are difficult to obtain. Clark (1984) estimates EIA costs of 0.5–2.0 per cent of a project's value. Hart (1984) and Wathern (1988) suggest figures of a similar order. More recent estimates by Coles et al. (1992) suggest a much wider range, from 0.000025 to 5 per cent, for EISs in the UK.

Procedures for and the practice of public participation in the EIA process vary between, and sometimes within, countries, from the very comprehensive to the very partial and largely cosmetic. An important issue is the stages in the EIA process to which the public should have access. Government roles in the EIA process may be conditioned by caution at extending systems, by limited experience and expertise in this new and rapidly developing area, and by resource considerations. A central government may offer limited guidance on best practice, and make inconsistent decisions. A local government may find it difficult to handle the scope and complexity of the content of EISs.

The quality of assessments

Many EISs fail to meet even minimum standards. For example, a survey by Jones et al. (1991) of the EISs published under UK environmental impact assessment regulations highlighted shortcomings. They found that "one-third of the EISs did not appear to contain the required non-technical summary, that, in a quarter of the cases, they were judged not to contain the data needed to assess the likely environmental effects of the development, and that in the great majority of cases, the more complex, interactive impacts were neglected". An update by Glasson et al. (DOE 1996) suggests that although there has been some learning from experience, many EISs in the UK are still unsatisfactory (see Ch. 8 for further discussion). Quality may vary between types of project. It may also vary between countries supposedly operating under the same legislative framework.

23

Beyond the decision

Many EISs are for one-off projects, and there is little incentive for developers to audit the quality of the assessment predictions and to monitor impacts as an input to a better assessment for the next project. EIA up to and no further than the decision on a project is a very partial linear process, with little opportunity for a cyclical learning process. In some areas of the world (e.g. California, Western Australia), the monitoring of impacts is mandatory, and monitoring procedures must be included in an EIS. The extension of such approaches constitutes another significant current issue in the largely project-based EIA process.

Beyond project assessment

As noted in Section 1.2, the strategic environmental assessment (SEA) of policies, plans and programmes represents a logical extension of project assessment. SEA can cope better with cumulative impacts, alternatives and mitigation measures than project assessment. SEA systems already exist in California and the Netherlands, and to a lesser extent in Canada, Germany and New Zealand. Discussions are in hand to introduce an EU-wide system (Therivel et al. 1992). The Fifth Community Action Programme on the Environment states: "Given the goal of achieving sustainable development, it seems only logical, if not essential, to apply an assessment of the environmental implications of all relevant policies, plans and programmes" (CEC 1992).

1.6 An outline of subsequent parts and chapters

This book is in four parts. The first establishes the context of EIA in the growth of concern about environmental issues and in relevant legislation, with particular reference to the UK. Following from this first chapter, which provides an introduction to EIA and an overview of principles, Chapter 2 focuses on the origins of EIA under the US National Environmental Policy Act (NEPA) of 1969, on interim developments in the UK, and on the subsequent introduction of EC Directive 85/337 and subsequent amendments. The details of the UK legislative framework for EIA, under town and country planning and other legislation, are discussed in Chapter 3.

Part 2 provides a rigorous step-by-step approach to the EIA process. This is the core of the text. Chapter 4 covers the early starting-up stages, establishing a management framework, clarifying the type of developments for EIA, and outlining approaches to scoping, the consideration of alternatives, project description, establishing the baseline and identifying impacts. Chapter 5 explores the central issues of prediction, the assessment of significance and the mitigation of adverse impacts. The approach draws out broad principles affecting prediction exercises, exemplified with reference to particular cases. Chapter 6 provides coverage of an important issue identified above: participation in the EIA process. Communication in the EIA process, EIS presentation and EIS review are also covered in this chapter. Chapter 7 takes the process beyond the decision on a project and examines the importance of, and approaches to, monitoring and auditing in the EIA process.

Part 3 exemplifies the process in practice. Chapter 8 provides an overview of UK practice to date, including quantitative and qualitative analyses of the EISs prepared. Chapters 9 and 10 provide case studies of current practice in particular sectors; Chapter 9 includes analyses of several new settlement proposals, produced under the Town and Country Planning (Assessment of Environmental Effects) Regulations. New settlements include a variety of activities and land-uses and provide some of the most comprehensive projects, akin to development plans, for the new procedures. The important project type of waste disposal facilities is also discussed in this chapter. Chapter 10 includes analyses of major road proposals and power station proposals, which are produced under associated legislation, respectively the Highways (Assessment of Environmental Effects) Regulations and Electricity and Pipe-line Works (Assessment of Environmental Effects) Regulations. Chapter 11 draws on comparative experience from a number of developed countries (the Netherlands, Canada, Australia and Japan) and from a number of countries from the developing and emerging economies (Peru, China and Poland) – presented to highlight some of the strengths and weaknesses of other systems in practice; the important role of international agencies in EIA practice – such as the UN and the World Bank – are also discussed in this chapter.

Part 4 looks to the future. It illuminates many of the issues noted in Section 1.5. Chapter 12 focuses on improving the effectiveness of the current system of project assessment. Particular emphasis is given to the development of environmental auditing to provide better baseline data, to various procedural developments and to achieving compatibility for EIA systems in Europe. Chapter 13 discusses the extension of assessment to policies, plans and programmes, concluding full circle with a further consideration of EIA, SEA and sustainable development.

A set of appendices provide details of legislation and practice not considered appropriate to the main text. A list of further reading is included there.

References

ANZECC (Australian and New Zealand Environment and Conservation Council) 1991. *A national approach to EIA in Australia*. Canberra: ANZECC.

Boulding, K. 1966. The economics of the coming Spaceship Earth. In *Environmental quality in a growing economy*, H. Jarrett (ed.), 3–14. Baltimore: Johns Hopkins University Press.

Breakell, M. & J. Glasson (eds) 1981. *Environmental impact assessment: from theory to practice.* Oxford School of Planning, Oxford Polytechnic.

Breese, G. et al. 1965. *The impact of large installations on nearby urban areas.* Los Angeles: Sage.

Calow, P. (ed.) 1997. *Handbook of environmental risk assessment and management.* Oxford: Blackwell Science.

Carley, M. J. & E. S. Bustelo 1984. *Social impact assessment and monitoring: a guide to the literature.* Boulder: Westview Press.

Catlow, J. & C. G. Thirlwall 1976. *Environmental impact analysis.* London: DOE.

CEC (Commission of the European Communities) 1982. *The contribution of infrastructure to regional development.* Brussels: CEC.

CEC 1992. *Towards sustainability: a European Community programme of policy and action in relation to the environment and sustainable development,* vol. II. Brussels: CEC.

CEPA (Commonwealth Environmental Protection Agency) 1994. *Assessment of cumulative impacts and strategic assessment in EIA.* Canberra: CEPA.

Clark, B. D. 1984. Environmental impact assessment (EIA): scope and objectives. In *Perspectives on environmental impact assessment,* B. D. Clark et al. (eds). Dordrecht: Reidel.

Coles, T., K. Fuller, M. Slater 1992. *Practical experience of environmental assessment in the UK.* East Kirkby, Lincolnshire: Institute of Environmental Assessment.

DOE (Department of the Environment) 1989. *Environmental assessment: a guide to the procedures.* London: HMSO.

DOE 1991. *Policy appraisal and the environment.* London: HMSO.

DOE et al. 1990. *This common inheritance: Britain's environmental strategy* (Cmnd 1200). London: HMSO.

DOE 1996. *Changes in the quality of environmental impact statements.* London: HMSO.

Finsterbusch, K. 1985. State of the art in social impact assessment. *Environment and Behaviour* **17**, 192–221.

Hart, S. L. 1984. The costs of environmental review. In *Improving impact assessment,* S. L. Hart et al. (eds). Boulder: Westview Press.

Health and Safety Commission 1978. *Canvey: an investigation of potential hazards from the operations in the Canvey Island/Thurrock area.* London: HMSO.

HMG, Secretary of State for the Environment 1994. *Sustainable development: the UK strategy.* London: HMSO.

Holmberg, J. & R. Sandbrook 1992. Sustainable development: what is to be done? In *Policies for a small planet* J. Holmberg (ed.), 19–38. London: Earthscan.

International Association for Impact Assessment 1994. Guidelines and principles for social impact assessment. *Impact Assessment* **12**(2).

Jones, C. E., N. Lee, C. M. Wood 1991. *UK environmental statements 1988–1990: an analysis.* Occasional Paper 29, Department of Planning and Landscape, University of Manchester.

Kirkby, J., P. O'Keefe, L. Timberlake 1995. *The earthscan reader in sustainable development.* London: Earthscan.

Lichfield, N. 1996. *Community impact evaluation.* London: UCL Press.

Lovejoy, D. 1992. What happened at Rio? *The Planner* **78**(15), 13–14.

Munn, R. E. 1979. *Environmental impact assessment: principles and procedures,* 2nd edn. New York: Wiley.

O'Riordan, T. 1988. The politics of sustainability. In *Sustainable environmental management: principles and practice,* R. K. Turner (ed.). London: Belhaven.

O'Riordan, T. 1990. EIA from the environmentalist's perspective. *VIA* **4**, March, 13.

Pearce, D. W. 1992. *Towards sustainable development through environment assessment.* Working Paper PA92–11, Centre for Social and Economic Research in the Global Environment, University College London.

Pearce, D., A. Markandya, E. Barbier 1989. *Blueprint for a Green economy.* London: Earthscan.

Preston, D. & B. Bedford 1988. Evaluating cumulative effects on wetland functions: a conceptual overview and generic framework. *Environmental Management* **12**(5).

Redclift, M. 1987. *Sustainable development: exploring the contradictions.* London: Methuen.

Reid, D. 1995. *Sustainable development: an introductory guide.* London: Earthscan.

Sadler, B. 1996. *Environmental assessment in a changing world: evaluating practice to improve performance.* International study on the effectiveness of environmental assessment. Ottawa: Canadian Environmental Assessment Agency.

Skolimowski, P. 1995. Sustainable development – how meaningful? *Environmental Values* **4**.

Stakhiv, E. 1988. An evaluation paradigm for cumulative impact analysis. *Environmental Management* **12**(5).

Therivel, R., E. Wilson, S. Thompson, D. Heaney, D. Pritchard 1992. *Strategic environmental assessment.* London: RSPB/Earthscan.

Turner, R. K. & D. W. Pearce 1992. *Sustainable development: ethics and economics.* Working Paper PA92–09, Centre for Social and Economic Research in the Global Environment, University College London.

United Nations Economic Commission for Europe 1991. *Policies and systems of environmental impact assessment.* Geneva: United Nations.

UN World Commission on Environment and Development 1987. *Our common future.* Oxford: Oxford University Press.

Vanclay, F. & D. Bronstein (eds) 1995. *Environment and social impact assessments.* London: Wiley.

Wathern, P. (ed.) 1988. *Environmental impact assessment: theory and practice.* London: Unwin Hyman.

CHAPTER 2
Origins and development

2.1 Introduction

Environmental impact assessment was first formally established in the USA in 1969 and has since spread, in various forms, to most other countries. In the UK, EIA was initially an *ad hoc* procedure carried out by local planning authorities and developers, primarily for oil- and gas-related developments. A 1985 European Community directive on EIA (Directive 85/337) introduced broadly uniform requirements for EIA to all EU Member States and significantly affected the development of EIA in the UK. However, ten years after the Directive was agreed, Member States were still carrying out widely diverse forms of EIA, contradicting the Directive's aim of "levelling the playing field". Amendments of 1997 aimed to improve this situation. The nature of EIA systems – e.g. mandatory or discretionary, level of public participation, types of action requiring EIA – and their implementation in practice vary widely from country to country. However, the rapid spread of the concept of EIA and its central role in many countries' programmes of environmental protection attest to its universal validity as a proactive planning tool.

This chapter first discusses how the system of EIA evolved in the US. The present status of EIA worldwide is then briefly reviewed (Chapter 11 will consider a number of countries' systems of EIA in greater depth). EIA in the UK and the EU are then discussed. Finally, we review the various systems of EIA in the EU Member States.

2.2 The National Environmental Policy Act and subsequent US systems

The US National Environmental Policy Act of 1969, also known as NEPA, was the first legislation to require EIAS. Consequently it has become an important model for other EIA systems, both because it was a radically new form of environmental policy, and because of the successes and failures of its subsequent development. Since its enactment, NEPA has resulted in the preparation of well over 10,000 full

EISS and many more partial appraisals, which have influenced countless decisions and represent a powerful base of environmental information. On the other hand, NEPA is unique. Other countries have shied away from the form it takes and the procedures it sets out, not least because they are unwilling to face a situation like that in the USA, where there has been extensive litigation over the interpretation and workings of the EIA system.

This section covers NEPA's legislative history, i.e. the early development before it became law, the interpretation of NEPA by the courts and the Council on Environmental Quality (CEQ), the main EIA procedures arising from NEPA, and likely future developments. The reader is referred to Anderson et al. (1984), Bear (1990), Orloff (1980) and the annual reports of the CEQ for further information.

Legislative history

NEPA is in many ways a fluke, strengthened by what should have been amendments weakening it, and interpreted by the courts to have powers that were not originally intended. The legislative history of NEPA is interesting not only in itself but also because it explains many of the anomalies of its operation and touches on some of the major issues involved in designing an EIA system. Several proposals to establish a national environmental policy were discussed in the US Senate and House of Representatives in the early 1960s. These proposals all included some form of unified environmental policy and the establishment of a high-level committee to foster it. In February 1969, Bill S1075 was introduced in the Senate; it proposed a programme of federally funded ecological research and the establishment of a Council on Environmental Quality. A similar Bill, HR6750, introduced in the House of Representatives, proposed the formation of a CEQ and a brief statement on national environmental policy. Subsequent discussions in both chambers of Congress focused on several points:

- the need for a declaration of national environmental policy (now Title I of NEPA);
- a proposed statement that "each person has a fundamental and inalienable right to a healthful environment" (which would put environmental health on a par with, say, free speech). This was later weakened to the statement in §101(c) that "each person should enjoy a healthful environment";
- action-forcing provisions similar to those then being proposed for the Water Quality Improvement Act, which would require federal officials to prepare a detailed statement concerning the probable environmental impacts of any major action; this was to evolve into NEPA's §102(2)(C) which requires EIA. The initial wording of the Bill had required a "finding", which would have been subject to review by those responsible for environmental protection, rather than a "detailed statement" subject to inter-agency review. The Senate had intended to weaken the Bill by requiring only a detailed statement. Instead, the "detailed

29

Table 2.1 Main points of NEPA.

NEPA consists of two titles. Title I establishes a national policy on the protection and restoration of environmental quality. Title II sets up a three-member Council on Environmental Quality (CEQ) to review environmental programmes and progress, and to advise the President on these matters. It also requires the President to submit an annual "Environmental Quality Report" to Congress. The provisions of Title I are the main determinants of EIA in the USA, and they are summarized here.

Section 101 contains requirements of a substantive nature. It states that the Federal Government has a continuing responsibility to "create and maintain conditions under which man and nature can exist in productive harmony, and fulfil the social, economic and other requirements of present and future generations of Americans". As such the government is to use all practicable means, "consistent with other essential considerations of national policy", to minimize adverse environmental impact and to preserve and enhance the environment through federal plans and programmes. Finally, "each person should enjoy a healthful environment", and citizens have a responsibility to preserve the environment.

Section 102 requirements are of a procedural nature. Federal agencies are required to make full analyses of all the environmental effects of implementing their programmes or actions. Section 102(1) directs agencies to interpret and administer policies, regulations and laws in accordance with the policies of NEPA. Section 102(2) requires federal agencies

- to use "a systematic and interdisciplinary approach" to ensure that social, natural and environmental sciences are used in planning and decision-making;
- to identify and develop procedures and methods so that "presently unquantified environmental amenities and values may be given appropriate consideration in decision-making along with traditional economic and technical considerations";
- to "include in every recommendation or report on proposals for legislation and other *major Federal actions significantly affecting the quality of the human environment, a detailed statement* by the responsible official on:
  - the environmental impact of the proposed action;
  - any adverse environmental effects which cannot be avoided should the proposal be implemented;
  - alternatives to the proposed action;
  - the relationship between local short-term uses of man's environment and the maintenance and enhancement of long-term productivity;
  - any irreversible and irretrievable commitments of resources which would be involved in the proposed action should it be implemented (authors' emphases).

Section 103 requires federal agencies to review their regulations and procedures for adherence with NEPA, and to suggest any necessary remedial measures.

---

assessment" became the subject of external review and challenge; the public availability of the detailed statements became a major force shaping the law's implementation in its early years. NEPA became operational on 1 January 1970. Table 2.1 summarizes its main points.

An interpretation of NEPA

NEPA is a generally worded law that required substantial early interpretation. The CEQ, which was set up by NEPA, prepared guidelines to assist in the Act's interpretation.

However, much of the strength of NEPA came from early court rulings. NEPA was immediately seen by environmental activists as a significant vehicle for preventing environmental harm, and the early 1970s saw a series of influential lawsuits and court decisions based on it. These lawsuits were of three broad types, as described by Orloff (1980):

- Challenging an agency's decision not to prepare an EIA. This generally raised issues such as whether a project was major, federal, an "action", or had significant environmental impacts (see NEPA §102(2)(C)). For instance, the issue of whether an action is federal came into question in some lawsuits concerning the federal funding of local government projects.[1]
- Challenging the adequacy of an agency's EIS. This raised issues such as whether an EIS adequately addressed alternatives, and whether it covered the full range of significant environmental impacts. A famous early court case concerned the Chesapeake Environmental Protection Association's claim that the Atomic Energy Commission did not adequately consider the water quality impacts of its proposed nuclear power plants, particularly in the EIA for the Calvert Cliffs power plant.[2] The Commission argued that NEPA merely required the consideration of water quality standards; opponents argued that it required an assessment beyond mere compliance with standards. The courts sided with the opponents.
- Challenging an agency's substantive decision, namely its decision to allow or not to allow a project to proceed in light of the contents of its EIS. Another influential early court ruling[3] laid down guidelines for the judicial review of agency decisions, noting that the court's only function was to ensure that the agency had taken a "hard look" at environmental consequences, not to substitute its judgement for that of the agency.

The early proactive role of the courts greatly strengthened the power of environmental movements and caused many projects to be stopped or substantially amended. In many cases the lawsuits delayed construction for long enough to make them economically infeasible or to allow the areas where projects would have been sited to be designated as national parks or wildlife areas (Turner 1988). More recent decisions have been less clearly pro-environment than the earliest decisions. The flood of early lawsuits, with the delays and costs involved, was a lesson to other countries in how *not* to set up an EIA system. As will be shown later, many countries carefully distanced their EIA systems from the possibility of lawsuits.

The CEQ was also instrumental in establishing guidelines to interpret NEPA, producing interim guidelines in 1970, and guidelines in 1971 and 1973. Generally the courts adhered closely to these guidelines when making their rulings. However, the guidelines were problematic: they were not detailed enough, and were interpreted by the federal agencies as being discretionary rather than binding. To combat these limitations, President Carter issued Executive Order 11992 in 1977, giving the CEQ authority to set enforceable regulations for implementing NEPA. These were issued in 1978 (CEQ 1978) and sought to make the NEPA process more useful for decision-makers and the public, to reduce paperwork and delay and to emphasize real environmental issues and alternatives.

A summary of NEPA procedures

The process of EIA established by NEPA, and developed further in the CEQ regulations, is summarized in Figure 2.1. The following citations are from the CEQ regulations (CEQ 1978).

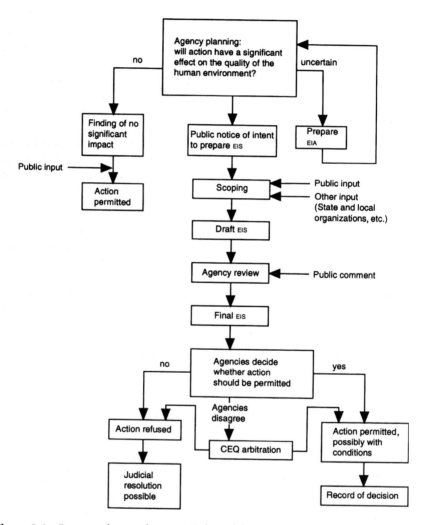

Figure 2.1 Process of EIA under NEPA. (Adapted from Legore 1984)

[The EIA process begins] as close as possible to the time the agency is developing or is presented with a proposal . . . The statement shall be prepared early enough so that it can serve practically as an important contribution to the decision-making process and will not be used to rationalize or justify decisions already made. (§1502.5)

A "lead agency" is designated that co-ordinates the EIA process. The lead agency first determines whether the proposal requires the preparation of a full EIS, no EIS at all, or a "finding of no significant impact" (FONSI). This is done through a series of tests. A first test is whether a federal action is likely to individually or cumulatively have a significant environmental impact. All federal agencies have compiled lists of "categorical exclusions" which are felt not to have such impacts. If an action is on such a list, then no further EIA action is generally needed. If an action is not categorically excluded, an "environmental assessment" is carried out to determine whether a full EIS or a FONSI is needed. A FONSI is a public document which explains why the action is not expected to have a significant environmental impact.

If a FONSI is prepared, then a permit would usually be granted following public discussion. If a full EIS is found to be needed, the lead agency publishes a "Notice of Intent", and the *process of scoping begins*. The aim of the scoping exercise is to determine the issues to be addressed in the EIA: to eliminate insignificant issues, focus on those that are significant and identify alternatives to be addressed. The lead agency invites the participation of the proponent of the action, affected parties and other interested persons.

[The alternatives] section is the heart of the environmental impact statement . . . [It] should present the environmental impacts of the proposal and the alternatives in comparative form, thus sharply defining the issues and providing a clear basis for choice . . . (§1502.14)

A draft EIS is then prepared, and is reviewed and commented on by the relevant agencies and the public. These comments are taken into account in the subsequent preparation of a final EIS. An EIS is normally presented in the format shown in Table 2.2. In an attempt to be comprehensive, early EISs tended to be so bulky as to be virtually unreadable. The CEQ guidelines consequently emphasize the need to concentrate only on important issues and to prepare readable documents:

The text of final environmental impact statements shall normally be less than 150 pages . . . Environmental impact statements shall be written in plain language . . . (§1502.7–8)

The public is involved in this process, both at the scoping stage and after publication of the draft and final EISs:

Agencies shall: (a) Make diligent efforts to involve the public in preparing and implementing NEPA procedures . . . (b) Provide public notice of NEPA-related hearings, public meetings and the availability of environmental documents . . .

33

Table 2.2 Typical format for an EIS under NEPA.

(a) Cover sheet

- list of responsible agencies
- title of proposed action
- contact persons at agencies
- designation of EIS as draft, final or supplement
- abstract of EIS
- date by which comments must be received

(b) Summary (usually 15 pages or less)

- major conclusions
- areas of controversy
- issues to be resolved

(c) Table of contents
(d) Purpose of and need for action
(e) Alternatives, including proposed action
(f) Affected environment
(g) Environmental consequences

- environmental impacts of alternatives, including proposed action
- adverse environmental effects which cannot be avoided if proposal is implemented
- mitigation measures to be used and residual effects of mitigation
- relation between short-term uses of the environment and maintenance and enhancement of long-term productivity
- irreversible or irretrievable commitments of resources if proposal is implemented
- discussion of:
    - direct and indirect effects and their significance
    - possible conflicts between proposed action and objectives of relevant land-use plans, policies and controls
    - effects of alternatives, including proposed action
    - energy requirements and conservation potential of various alternatives and mitigation measures
    - natural or depletable resource requirements and conservation of various alternatives and mitigation measures
    - effects on urban quality, historic and cultural resources, and built environment
    - means to mitigate adverse impacts

(h) List of preparers
(i) List of agencies etc. to which copies of EIS are sent
(j) Index
(k) Appendices, including supporting data

(c) Hold or sponsor public hearings . . . whenever appropriate . . . (d) Solicit appropriate information from the public. (e) Explain in its procedures where interested persons can get information or status reports . . . (f) Make environmental impact statements, the comments received, and any underlying documents available to the public pursuant to the provisions of the Freedom of Information Act . . . (§1506.6)

Finally, a decision is made about whether the proposed action should be permitted:

Agencies shall adopt procedures to ensure that decisions are made in accordance with the policies and purposes of the Act. Such procedures shall include but not be limited to: (a) Implementing procedures under section 102(2) to achieve the requirements of sections 101 and 102(1) . . . (e) Requiring that . . . the decision-maker consider the alternatives described in the environmental impact statement. (§1505.1)

Where all relevant agencies agree that the action should not go ahead, permission is denied, and a judicial resolution may be attempted. Where agencies agree that the action can proceed, permission is given, possibly subject to specified conditions (e.g. monitoring, mitigation). Where the relevant agencies disagree, the CEQ acts as arbiter (§1504). Until a decision is made, "no action concerning the proposals shall be taken which could: (1) have an adverse environmental impact; or (2) limit the choice of reasonable alternatives . . ." (§1506.1)

Recent trends

During the first ten years of NEPA's implementation, about 1,000 EISs were prepared annually. Recently, negotiated improvements to the environmental impacts of proposed actions have become increasingly common during the preparation of "environmental assessments". This has led to many "mitigated findings of no significant impact" (no nice acronym exists for this), reducing the number of EISs prepared: whereas 1,273 EISs were prepared in 1979, only 456 were prepared in 1991 (CEQ 1993). This trend can be viewed positively, since it means that environmental impacts are considered earlier in the decision-making process, and since it reduces the costs of preparing EISs. However, the fact that this abbreviated process allows less public participation causes some concern. Of the 456 EISs prepared in 1991, 145 were filed by the Department of Agriculture (primarily for forestry and range management), and 87 were filed by the Department of Transportation (primarily for road construction). Between 1979 and 1991, the number of EISs filed by the Department of Housing and Urban Development fell from 170 to 7! The number of legal cases filed against federal departments and agencies on the basis of NEPA also fell slightly, from 139 in 1979 to 128 in 1991. The most common complaints were "no EIS when one should have been prepared" (41 cases in 1991), and "inadequate EIS" (26 cases in 1991).

NEPA's twentieth year of operation, 1990, was marked by a series of conferences on the Act and the presentation to Congress of a bill of NEPA amendments. Under the Bill (HR1113), which was not passed, federal actions that take place outside the USA (e.g. projects built in other countries with US federal assistance) would have been subject to EIA, and all EIAS would have been required to consider global climatic change, the depletion of the ozone layer, the loss of biological diversity and trans-boundary pollution. This latter amendment was controversial: although the need to consider the global impacts of programmes was undisputed, it was felt to be infeasible at the level of project EIA. Finally, the Bill would have required all federal agencies to survey a statistically significant sample of EISs to determine whether mitigation measures promised in the EIS had been implemented and, if so, whether they had been effective.

The context of EIA has also become a matter of concern. EIA is only one part of a broader environmental policy (NEPA), but the procedural provisions set out in NEPA's §102(2)(C) have overshadowed the rest of the Act. It has been argued that mere compliance with these procedures is not enough, and that greater emphasis should be given to the environmental goals and policies stated in §101. EIA must also be seen in the light of other environmental legislation. In the USA, many laws dealing with specific aspects of the environment were enacted or strengthened in the 1970s, including the Clean Water Act and the Clean Air Act. These laws have in many ways superseded NEPA's substantive requirements and have complemented and buttressed its procedural requirements. Compliance with these laws does not necessarily imply compliance with NEPA. However, the permit process associated with these other laws has become a primary method for evaluating project impacts, reducing NEPA's importance except for its occasional role as a focus of debate on major projects (Bear 1990).

The scope of EIA, and in particular the recognition of the social dimension of the environment, has been another matter of concern. After long campaigning by black and ethnic groups, particularly about inequalities in the distribution of hazardous waste landfills and incinerators, a working group was set up within the EPA to make recommendations for dealing with environmental injustice (Hall 1994). The outcome was the Clinton "Executive Order on Federal Actions to Address Environmental Justice in Minority Populations and Low-Income Populations" (White House 1994). Under this Order, each federal agency must analyse the environmental effects, including human health, economic and social effects, of federal actions, including effects on minority and low-income communities, when such analysis is required under NEPA. Mitigation measures, wherever feasible, should also address the significant and adverse environmental effects of federal actions on the same communities. In addition, each federal agency must provide opportunities for communities to contribute to the NEPA process, identifying potential effects and mitigation measures in consultation with affected communities and improving the accessibility of meetings and crucial documents.

Other issues remain, and Canter (1996) highlights four areas for which NEPA requirements need further elaboration:

- how much an agency should identify and plan mitigation before issuing an EIS;
- ways to assess the cumulative impacts of proposed developments;
- ways to conduct "reasonable foreseeability" (or worst-case) analyses; and
- the monitoring and auditing of impact predictions.

Little NEPAs and the particular case of California

Many state-level EIA systems have been established in the USA in addition to NEPA. Sixteen of the USA's 50 states[4] have so-called "little NEPAs", which require EIA for state actions (actions that require state funding or permission) and/or projects in sensitive areas. Other states[5] have no specific EIA regulations, but have EIA requirements in addition to those of NEPA.

Of particular interest is the Californian system, established under the California Environmental Quality Act (CEQA) of 1973, and subsequent amendments. This is widely recognized as one of the most advanced EIA systems in the world. The legislation applies not only to government actions but also to the activities of private parties that require the approval of a government agency. It is not merely a procedural approach but one that requires state and local agencies to protect the environment by adopting feasible mitigation measures and alternatives in environmental impact reviews (EIRs). The legislation extends beyond projects to higher levels of actions, and an amendment in 1989 also added mandatory mitigation, monitoring and reporting requirements to CEQA. Annual guidance on the California system is provided in an invaluable publication by the State of California, which sets out the CEQA Statutes and Guidelines in considerable detail (State of California 1992).

2.3 The worldwide spread of EIA

Since the enactment of NEPA, EIA systems have been established in various forms throughout the world, beginning with more developed countries – e.g. Canada in 1973, Australia in 1974, West Germany in 1975, France in 1976 – and later also in the less developed countries. The approval of a European Directive on EIA in 1985 stimulated the enactment of EIA legislation in many European countries in the late 1980s. The formation of new countries after the break-up of the Soviet Union in 1991 led to the enactment of EIA legislation in many of these countries in the early to mid-1990s. The early 1990s also saw a large growth in the number of EIA regulations and guidelines established in Africa and South America. By 1996, more than 100 countries had EIA systems (Sadler 1996). Figure 2.2 summarizes the present state of EIA systems worldwide to the best of the authors' knowledge.

These EIA systems vary greatly. Some are in the form of *mandatory regulations, acts,* or *statutes*; these are generally enforced by the authorities' requiring the preparation of an adequate EIS before permission is given for a project to proceed. In other cases, EIA *guidelines* have been established. These are not enforceable but generally impose obligations on the administering agency. Other legislation allows government officials to require EIAs to be prepared at their *discretion.* Elsewhere, EIAs are prepared in an *ad hoc* manner, often because they are required by funding

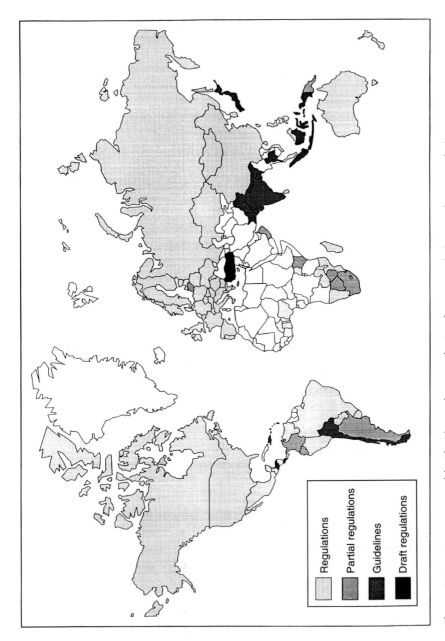

Figure 2.2 EIA systems worldwide (the authors apologize for any omissions or inaccuracies).

bodies (e.g. the World Bank, USAID) as part of a funding approval process. However, these classifications are not necessarily indicative of how thoroughly EIA is carried out. For instance, the EIA regulations of Brazil and the Philippines are not well carried out or enforced in practice (Abracosa & Ortolano 1987, Moreira 1988), whereas Japan's guidelines are thoroughly implemented, and some very good *ad hoc* EIAs have been prepared in the UK.

Another important distinction between types of EIA system is that sometimes the actions that require EIA are given as a *definition* (e.g. the USA's definition of "major federal actions significantly affecting the quality of the human environment"), sometimes as a *list of projects* (e.g. roads of more than 10 kilometres in length). Most countries use a list of projects, in part to avoid legal wrangling such as that surrounding NEPA's definition. Another distinction asks whether EIA is required for *government projects only* (as in NEPA), for *private projects only* or for both.

Finally, some international development and funding agencies have set up EIA guidelines, including the European Bank for Reconstruction and Development (1992), Overseas Development Administration (1996), United Nations Environment Programme (1997), and World Bank (1992, 1995).

2.4 Development in the UK

The UK has enacted formal legislation for EIA quite recently, since 1988, in the form of several laws that implement European Community Directive 85/337/EEC (CEC 1985) and subsequent amendments. It is quite possible that without pressure from the European Commission such legislation would never have been enacted, since the UK government felt that its existing planning system more than adequately controlled environmentally unsuitable developments. However, this does not mean that the UK had no EIA system at all before 1988; many EIAs were prepared voluntarily or at the request of local authorities, and guidelines for EIA preparation were drawn up.

Limitations of the land-use planning system

The UK's statutory land-use planning system has since 1947 required local planning authorities (LPAS) to anticipate likely development pressures, assess their significance, and allocate land, as appropriate, to accommodate them. Environmental factors are a fundamental consideration in this assessment. Most developments require planning consent, so environmentally harmful developments can be prevented by its denial. This system resulted in the accumulation of considerable planning expertise concerning the likely consequences of development proposals.

39

After the mid-1960s, however, the planning system began to seem less effective at controlling the impacts of large developments. The increasing scale and complexity of developments, the consequently greater social and physical environmental impacts and the growing internationalization of developers (e.g. oil and chemicals companies) all outstripped the capability of the development control system to predict and control the impacts of developments. In the late 1960s, public concern about environmental protection also grew considerably, and the relation between statutory planning controls and the development of large projects came under increasing scrutiny. This became particularly obvious in the case of the proposed third London Airport. The Roskill Commission was established to select the most suitable site for an airport in southeast England, with the mandate to prepare a cost–benefit analysis of alternative sites. The resulting analysis (HMSO 1971) focused on socio-economic rather than physical environmental impacts; it led to an understanding of the difficulties of expanding cost–benefit analysis to impacts not easily measured in monetary terms, and to the realization that other assessment methods were needed to achieve a balance between socio-economic and physical environmental objectives.

North Sea oil- and gas-related EIA initiatives

The main impetus towards the further development of EIA, however, was the discovery of oil and gas in the North Sea. The extraction of these resources necessitated the construction of large developments in remote areas renowned for their scenic beauty and distinctive ways of life (e.g. the Shetlands, the Orkneys and the Highlands Region). Planning authorities in these areas lacked the experience and resources needed to assess the impacts of such large developments. In response, the Scottish Development Department (SDD) issued a technical advice note to LPAs (Scottish Development Department 1974). *Appraisal of the impact of oil-related development* noted that these developments and other large and unusual projects need "rigorous appraisal", and suggested that LPAs should commission an impact study of the developments if needed. This was the first government recognition that major developments needed special appraisal. Some EIAS were carried out in the early 1970s, mostly for oil and gas developments. Many of these were sponsored by the SDD and LPAs, and were prepared by environmental consultants, but some (e.g. for the Flotta Oil Terminal and Beatrice Oilfield) were commissioned by the developers. Other early EIAS concerned a coal mine in the Vale of Belvoir, a pumped-storage electricity scheme at Loch Lomond and various motorway and trunk road proposals (Clark & Turnbull 1984).

In 1973, the Scottish Office and Department of the Environment (DOE) commissioned the University of Aberdeen's Project Appraisal for Development Control (PADC) team to develop a systematic procedure for planning authorities to make a balanced appraisal of the environmental, economic and social impacts of large industrial developments. PADC produced an interim report, *The assessment of major industrial applications – a manual* (Clark et al. 1976), which was issued free of charge to all LPAs in the UK and "commended by central government for use by planning authorities, government agencies and developers". The PADC procedure was designed to fit into the existing planning framework, and was used to assess a variety of (primarily private-sector) projects. An extended and updated version of the manual was issued in 1981 (Clark et al. 1981).

In 1974, the Secretaries of State for the Environment, Scotland and Wales commissioned two consultants, J. Catlow and C. G. Thirwall, to investigate the "desirability of introducing a system of impact analysis in Great Britain, the circumstances in which a system should apply, the projects it should cover and the way in which it might be incorporated into the development control system" (Catlow & Thirwall 1976). The resulting report made recommendations about who should be responsible for preparing and paying for EIAs, what legislative changes would be needed to institute an EIA system, and similar issues. The report concluded that about 25–50 EIAs per year would be needed, for both public- and private-sector projects. EIA was given additional support by the Dobry Report on the development control system (Dobry 1975), which advocated that LPAs should require developers to submit impact studies for particularly significant development proposals. The report outlined the main topics such a study should address, and the information that should be required from developers. Government reactions to the Dobry Report were mixed: the Royal Commission on Environmental Pollution endorsed the report, but the Stevens Committee (1976) on Mineral Workings recommended that a comprehensive standard form for mineral applications should be introduced, arguing that such a form would make EIAs for mineral workings unnecessary.

Department of the Environment scepticism

However, overall the DOE remained sceptical about the need, practicality and cost of EIA. In fact, the government's approach to EIA has been described as being "from the outset grudging and minimalist" (CPRE 1991). In response to the Catlow & Thirwall report, the DOE stated: "Consideration of the report by local authorities should not be allowed to delay normal planning procedures and any new procedures involving additional calls on central or local government finance and manpower are unacceptable during the present period of economic restraint" (DOE 1977). A year later, after much deliberation, the DOE was slightly more positive:

> We fully endorse the desirability . . . of ensuring careful evaluation of the possible effects of large developments on the environment . . . The approach suggested by Thirwall/Catlow is already being adopted with many [projects] . . . The sensible use of this approach [should] improve the practice in handling these relatively few large and significant proposals (DOE 1978).

The government's foreword to the PADC manual of 1981 also emphasized the need to minimize the costs of EIA procedures: "It is important that the approach suggested in the report should be used selectively to fit the circumstances of the proposed development and with due economy" (Clark et al. 1981). As will be seen in later chapters, the government still remains sceptical about the value of EIA, and is generally unwilling to extend its remit, as is being suggested by the EC.

By the early 1980s, more than two hundred studies on the environmental impacts of projects in the UK had been prepared on an *ad hoc* basis. These are listed in Petts & Hills (1982). Many of these studies were not full EIAs, but focused on only a few impacts. However, large developers such as British Petroleum, British Gas, the Central Electricity Generating Board and the National Coal Board were preparing a series of increasingly comprehensive statements. In the case of British Gas, these were shown to be a good investment, saving the company £30 million in ten years (House of Lords 1981a).

2.5 EC Directive 85/337

The development and implementation of Directive 85/337 greatly influenced the EIA systems of the UK and other EU Member States. In the UK, central government research on a UK system of EIA virtually stopped after the mid-1970s, and attention focused instead on ensuring that any future Europe-wide system of EIA would fully incorporate the needs of the UK for flexibility and discretion. Other Member States were eager to ensure that the Directive reflected the requirements of their own more rigorous systems of EIA. Since the Directive's implementation, EIA activity in all the EU Member States has increased dramatically.

Legislative history

The EC had two main reasons for wanting to establish a uniform system of EIA in all its Member States. First, it was concerned about the state of the physical environment and eager to prevent further environmental deterioration. The EC's First Action Programme on the Environment of 1973 (CEC 1973) advocated the prevention of environmental harm: "the best environmental policy consists of preventing the creation of pollution or nuisances at source, rather than subsequently trying to counteract their effects", and, to that end, "effects on the environment should be taken into account at the earliest possible stage in all technical planning and decision-making processes". Further Action Programmes of 1977, 1983, 1987 and 1992 have reinforced this emphasis. Land-use planning was seen as an important way of putting these principles into practice, and EIA was viewed as a crucial technique for incorporating environmental considerations into the planning process.

Secondly, the EC was concerned to ensure that no distortion of competition should arise through which one Member State could gain unfair advantage by permitting developments that, for environmental reasons, might be refused by another. In other words, it considered environmental policies necessary to the maintenance of a level economic playing field. Further motivation for EC action included a desire to encourage best practice across Member States. In addition, pollution problems transcend territorial boundaries (witness acid rain and river pollution in Europe), and the EC can contribute at least a subcontinental response framework.

The EC began to commission research on EIA in 1975. Five years later and after more than twenty drafts, the European Commission presented a draft directive to the Council of Ministers (CEC 1980); it was circulated throughout the Member States. The 1980 draft attempted to reconcile several conflicting needs. It sought to benefit from the US experience with NEPA, but to develop policies appropriate to European need. It also sought to make EIA applicable to all actions likely to have a significant environmental impact, but to ensure that procedures would be practicable. Finally, and perhaps most challenging, it sought to make EIA requirements flexible enough to adapt to the needs and institutional arrangements of the various Member States, but uniform enough to prevent problems arising from widely varying interpretations of the procedures. The harmonization of the types of project to be subject to EIA, the main obligations of the developers and the contents of the EIAS were considered particularly important (Lee & Wood 1984, Tomlinson 1986).

As a result, the draft directive incorporated a number of important features. First, planning permission for projects was to be granted only after an adequate EIA had been completed. Secondly, LPAS and developers were to co-operate in providing information on the environmental impacts of proposed developments. Thirdly, statutory bodies responsible for environmental issues, and other Member States in cases of trans-frontier effects, were to be consulted. Finally, the public was to be informed, and allowed to comment on issues related to project development.

In the UK the draft directive was examined by the House of Lords Select Committee on the European Commission, where it received widespread support:

> The present draft Directive strikes the right kind of balance: it provides a framework of common administrative practices which will allow Member States with effective planning controls to continue with their system . . . while containing enough detail to ensure that the intention of the draft cannot be evaded . . . The Directive could be implemented in the United Kingdom in a way which would not lead to undue additional delay and costs in planning procedures and which need not therefore result in economic and other disadvantages. (House of Lords 1981a)

However, the Parliamentary Under-Secretary of State at the DOE dissented. Although accepting the general need for EIA, he was concerned about the bureaucratic hurdles, delaying objections and litigation that would be associated with the proposed directive (House of Lords 1981b). The UK Royal Town Planning Institute also commented on several drafts of the directive. Generally the RTPI favoured it, but was concerned that it might cause the planning system to become too rigid:

> The Institute welcomes the initiative taken by the European Commission to secure more widespread use of EIA as it believes that the appropriate use of EIA could both speed up and improve the quality of decisions on certain types of development proposals. However, it is seriously concerned that the proposed Directive, as presently drafted, would excessively codify and formalize procedures of which there is limited experience and therefore their benefits are not yet proven. Accordingly the Institute recommends the deletion of Article 4 and annexes of the draft. (House of Lords 1981a)

More generally, slow progress in the implementation of EC legislation was symptomatic of the wide range of interest groups involved, of the lack of public support for increasing the scope of town planning and environmental protection procedures, and of the unwillingness of Member States to adapt their widely varying planning systems and environmental protection legislation to those of other countries (Williams 1988). In March 1982, after considering the many views expressed by the Member States, the Commission published proposed amendments to the draft directive (CEC 1982). Approval was expected in November 1983. However, this was delayed by the Danish Government, which was concerned about projects authorized by Acts of Parliament. On 7 March 1985, the Council of Ministers agreed on the proposal; it was formally adopted as a Directive on 27 June 1985 (CEC 1985) and became operational on 3 July 1988.

Subsequently, the EC's Fifth Action Programme, *Towards sustainability* (CEC 1992), stresses the importance of EIA, particularly in helping to achieve sustainable development, and the need to expand the remit of EIA:

> Given the goal of achieving sustainable development it seems only logical, if not essential, to apply an assessment of the environmental implications of all relevant policies, plans and programmes. The integration of environmental assessment within the macro-planning process would not only enhance the protection of the environment and encourage optimization of resource management but would also help to reduce those disparities in the international and inter-regional competition for new development projects which at present arise from disparities in assessment practices in the Member States . . .

In response to a (belated) five-year review of the Directive (CEC 1993), amendments to the Directive were agreed in 1997. Appendix 1 gives the complete consolidated version of the amended Directive.

The reader is referred to Clark & Turnbull (1984), Lee & Wood (1984), O'Riordan & Sewell (1981), Swaffield (1981), Tomlinson (1986), Williams (1988), and Wood (1981, 1988) for further discussions on the development of EIA in the UK and EC.

Summary of EC Directive 85/337 procedures

The Directive differs in important respects from NEPA. It requires EIAS to be pre-pared by both public agencies and private developers, whereas NEPA applies only to federal agencies. It requires EIA for a specified list of projects, whereas NEPA uses the definition "major federal actions . . .". It specifically lists the impacts that are to be addressed in an EIA, whereas NEPA does not. Finally, it includes fewer require-ments for public consultation than does NEPA.

Under the provisions of the European Communities Act of 1972, Directive 85/337 is the controlling document, laying down rules for EIA in Member States. Individual states enact their own regulations to implement the Directive and have considerable discretion. According to the Directive, EIA is required for two classes of project, one mandatory (Annex I) and one discretionary (Annex II):

> . . . projects of the classes listed in Annex I shall be made subject to an assessment . . . for projects listed in Annex II, the Member States shall deter-mine through: (a) a case-by-case examination; or (b) thresholds or criteria set by the Member State whether the project shall be made subject to an assess-ment . . . When [doing so], the relevant selection criteria set out in Annex III shall be taken into account. (Article 4)

Table 2.3 summarizes the projects listed in Annexes I and II. The EC (CEC 1995) has also published guidelines to help Member States determine whether a project requires EIA. Similarly, the information required in an EIA is listed in Annex III of the Directive, but must only be provided

> inasmuch as: (a) The Member States consider that the information is relevant to a given stage of the consent procedure and to the specific characteristics of a particular project . . . and of the environmental features likely to be affected; (b) The Member States consider that a developer may reasonably be required to compile this information having regard *inter alia* to current knowledge and methods of assessment. (Article 5.1)

Table 2.4 summarizes the information required by Annex III (Annex IV, post-amendments). A developer is thus required to prepare an EIS that includes the information specified by the relevant Member State's interpretation of Annex III (Annex IV, post-amendments) and to submit it to the "competent authority". This EIS is then circulated to other relevant public authorities and made publicly available:

> Member States shall take the measures necessary to ensure that the authorities likely to be concerned by the project . . . are given an opportunity to express their opinion (Article 6.1).
> Member States shall ensure that:
>
> - any request for development consent and any information gathered pursu-ant to [the Directive's provisions] are made available to the public,

Table 2.3 Projects requiring EIA under EC Directive 85/337 (*as amended*).

Annex I (mandatory)

1. Crude oil refineries, coal/shale gasification and liquefaction
2. Thermal power stations and other combustion installations; nuclear power stations and other nuclear reactors
3. Radioactive waste processing and/or storage installations
4. Cast-iron and steel smelting works
5. Asbestos extraction, processing, or transformation
6. Integrated chemical installations
7. Construction of motorways, express roads, other large roads, railways, airports
8. Trading ports and inland waterways
9. Installations for incinerating, treating, or disposing of toxic and dangerous wastes
10. *Large-scale installation for incinerating or treating non-hazardous waste*
11. *Large-scale groundwater abstraction or recharge schemes*
12. *Large-scale transfer of water resources*
13. *Large-scale waste water treatment plants*
14. *Large-scale extraction of petroleum and natural gas*
15. *Large dams and reservoirs*
16. *Long pipelines for gas, oil or chemicals*
17. *Large-scale poultry or pig rearing installations*
18. *Pulp, timber or board manufacture*
19. *Large-scale quarries or open-cast mines*
20. *Long overhead electrical power lines*
21. *Large-scale installations for petroleum, petrochemical or chemical products.*

Annex II (discretionary)

1. Agriculture, silviculture and aquaculture
2. Extractive industry
3. Energy industry
4. Production and processing of metals
5. *Minerals industry* (projects not included in Annex I)
6. Chemical industry
7. Food industry
8. Textile, leather, wood and paper industries
9. Rubber industry
10. *Infrastructure projects*
11. *Other projects*
12. *Tourism and leisure*
13. Modification, extension or temporary testing of Annex I projects

Note: Amendments are shown in italic.

- the public concerned is given the opportunity to express an opinion before the project is initiated.

The detailed arrangements for such information and consultation shall be determined by the Member States (Article 6.2 and 6.3) [see Chapter 6, Section 6.2 also].

Table 2.4 Information required in an EIA under EC Directive 85/337 (*as amended*).

Annex III (IV)

1. Description of the project.
2. Where appropriate (*an outline of main alternatives studied and an indication of the main reasons for the final choice.*)
3. Aspects of the environment likely to be significantly affected by the proposed project, including population, fauna, flora, soil, water, air, climatic factors, material assets, architectural and archaeological heritage, landscape, and the interrelationship between them.
4. Likely significant effects of the proposed project on the environment.
5. Measures to prevent, reduce and where possible offset any significant adverse environmental effects.
6. Non-technical summary.
7. Any difficulties encountered in compiling the required information.

Note: Amendment is shown in italic.

The competent authority must consider the information presented in an EIS, the comments of relevant authorities and the public, and the comments of other Member States (where applicable) in its consent procedure (Article 8). (The EC (1994) has published a checklist to help competent authorities to review environmental information.) It must then inform the public of the decision and any conditions attached to it (Article 9).

2.6 EC Directive 85/337, as amended by Directive 97/11/EC

Directive 85/337 included a requirement for a five-year review, and a report was published in 1993 (CEC 1993). Whilst there was general satisfaction that the "basics of the EIA are mostly in place", there has been concern about the incomplete coverage of certain projects, insufficient consultation and public participation, the lack of information about alternatives, weak monitoring and the lack of consistency in Member States' implementation. The review process, as with the original Directive, generated considerable debate between the Commission and the Member States, and the amended Directive went through several versions, with some weakening of the proposed changes. The outcome, finalized in March 1997, and to be implemented within two years, includes the following amendments:

- Annex I (mandatory) – the addition of 12 new classes of project (e.g. dams and reservoirs, pipelines, quarries and open-cast mining) (see Table 2.3).
- Annex II (discretionary) – the addition of eight new sub-classes of project (plus extension to ten others), including shopping and car parks, and particularly tourism and leisure (e.g. caravan sites and theme parks).

47

- New Annex III lists matters which must be considered in EIA including:

  - Characteristics of projects: size, cumulative impacts, the use of natural resources, the production of waste, pollution and nuisance, the risk of accidents.
  - Location of projects: designated areas and their characteristics, existing and previous land-uses.
  - Characteristics of the potential impacts: geographical extent, trans-frontier effects, the magnitude and complexity of impacts, the probability of impact, the duration, frequency and reversibility of impacts.

- Change of previous Annex III to Annex IV: small changes in content.
- Other changes:

  - Article 2(3): There is no exemption from consultation with other Member States on transboundary effects.
  - Article 4: When deciding which Annex II projects will require EIA, Member States can use thresholds, case by case or a combination of the two.
  - Article 5.3: The minimum information provided by the developer *must include* an outline of the main alternatives studied and an indication of the main reasons for the final choice between alternatives.
  - Article 5.2: A developer may request an opinion about the information to be supplied in an ES, and a competent authority must provide that information. Member States may require authorities to give an opinion irrespective of the request from the developer.
  - Article 7: This requires consultation with affected Member States, and other countries, about trans-boundary effects.
  - Article 9: A competent authority must make public the *main* reasons and considerations on which decisions are based, together with a description of the *main* mitigation measures.

A consolidated version of the full Directive, as amended by these changes, is included at Appendix 1. There will be more projects subject to mandatory EIA (Annex I) and discretionary EIA (Annex II). Alternatives also become mandatory, and there is emphasis on consultation and participation. The likely implication is more EIA activity in the EU Member States over the next decade. Member States will also have to face up to some challenging issues when dealing with topics such as alternatives, risk assessment and cumulative impacts.

2.7 An overview of EC systems

Given the flexible wording of the original EC Directive, differences in its implementation by the Member States were bound to emerge. The five-year review identified many of these inconsistencies, and that several Member States had failed to translate the Directive into regulations. As a result, a number of Member States have recently strengthened their regulations to achieve a fuller implementation of the Directive. The amendments of 1997 are likely to reduce many of the remaining differences. In addition to the substantial extensions and modifications to the lists of projects in Annexes I and II the amended Directive (CEC 1997) also introduces a number of procedural changes, including a formal screening procedure for selecting projects for assessment, and EIS content changes, including an obligation on developers to include an outline of the main alternatives studied, and an indication of the main reasons for their choices, taking into account environmental effects. The Directive also enables a developer, if it so wishes, to ask a competent authority for formal advice on the scope of the information that should be included in a particular environmental statement. Member States, if *they* so wish, can require competent authorities to give an opinion on the scope of any proposed environmental statement, whether the developer has requested one or not. The new Directive also strengthens consultation and publicity, obliging competent authorities to take into account the results of consultations with the public, and the reasons and considerations on which the decision on a project proposal has been based.

This section outlines the main current differences between the EIA systems established by the Member States (including the most recent: Austria, Finland and Sweden) in response to the Directive, to the best of the authors' knowledge. Appendix 2 describes the Member States' EIA systems in greater depth (EIA Centre).

- Member States implement Directive 85/337 differently. For some the regulations implementing it come under the broad remit of nature conservation (e.g. France, Greece, the Netherlands, Portugal); for some they come under the planning system (e.g. Denmark, Ireland, Sweden, the UK); in others specific EIA regulations were enacted (e.g. Belgium, Italy). In Belgium, and to an extent in Germany and Spain, the responsibility for EIA has been devolved to the regional level, whereas in most countries the national government retains broad responsibility for it. Some countries (e.g. France, the Netherlands, the UK) have implemented the Directive on time (relatively), some (e.g. Belgium, Portugal) late.
- In all Member States, EIA is mandatory for Annex I projects. However, until the amendments to the Directive were agreed countries differed in their interpretation of which Annex II projects require EIA: some considered that only a few Annex II projects require EIA (e.g. Greece, Italy, Portugal), some that the competent authority decides if an EIA is needed, on a case-by-case basis (e.g. Ireland, the UK), some that lists are compiled that specify Annex II projects requiring EIA (e.g. France, the Netherlands). This affects the number of EIAs prepared. In France, for instance, thresholds for projects requiring EIA are so low that thousands of EIAS are prepared annually. In Denmark, by contrast, until recently only a few dozen EIAs were prepared annually.
- In most Member States, EIAs are carried out and paid for by the developers or consultants commissioned by them. However, in Flanders (Belgium) EIAs are carried out by experts approved by the authority responsible for environmental

49

matters, and in Spain the competent authority carries out an EIA based on studies carried out by the developer.

- Until the amendments, which made it a more formal stage of the EIA process, scoping was carried out as a discrete and mandatory step in some countries or their regions (e.g. Austria, Wallonia in Belgium, the Netherlands), but not in others (e.g. Spain, the UK). The consideration of alternatives to a proposed project was mandatory in only a few countries or regions of countries (e.g. Wallonia). The Netherlands was unique; it also required an analysis of the most environmentally acceptable alternative in each case. Again the amendments to the Directive will change this; developers are now obliged to include an outline of the main alternatives studied.

- The Member States vary considerably in the level of public consultation they require in the EIA process. The Directive requires an EIS to be made available after it is handed to the competent authority. However, some Member States or regions of them go well beyond this. In Denmark, the Netherlands and Wallonia, the public is consulted during the scoping process. In the Netherlands and Flanders, a public hearing must be held after an EIS is handed in. In Spain, the public must be consulted before an EIS is submitted. In Austria, the public can participate at several stages of an EIA, and citizens' groups and the Ombudsman for the Environment have special status. Amendments to the Directive may encourage the spread of good practice.

- In a few countries or national regions, EIA commissions have been established. In the Netherlands, the commission assists in the scoping process, reviews the adequacy of an EIS, and receives monitoring information from the competent authority. In Flanders, it reviews the qualifications of the people carrying out an EIA, determines its scope and reviews an EIS for compliance with legal requirements. Italy also has an EIA commission.

- The decision to proceed with a project is, in the simplest case, the responsibility of the competent authority (e.g. in Flanders, Germany, the UK). However, in some cases the minister responsible for the environment must first decide whether a project is environmentally compatible (e.g. in Denmark, Italy, Portugal).

- Only the Netherlands at present requires the systematic monitoring of a project's actual impacts by the competent authority.

As a result of these differences, some countries (e.g. Austria, the Netherlands, Belgium) seem to have particularly effective and comprehensive EIA systems, whereas others (e.g. Italy, Luxembourg, Portugal) have considerably weaker systems. However, as the amendments become more widely implemented, this may well change.

Finally, Member States differ in the extent to which they have voluntarily broadened out the application of EIA to policies, plans and programmes (PPPs). The Netherlands, Denmark and Finland all require EIA for some PPPs, and other countries have established non-mandatory guidelines for such EIAs. This will be discussed further in Chapter 13.

2.8 Summary

This chapter has reviewed the development of EIA worldwide, from its unexpectedly successful beginnings in the USA to recent developments in the EU. In practice, EIA ranges from the production of very simple *ad hoc* reports to the production of extremely bulky and complex documents, from wide-ranging to non-existent consultation with the public, from detailed quantitative predictions to broad statements about likely future trends. All these systems, however, have the broad aim of improving decision-making by raising decision-makers' awareness of a proposed action's environmental consequences. Over the past twenty-five years, EIA has become an important tool in project planning, and its applications are likely to expand further (see Ch.13). The next chapter focuses on EIA in the UK context.

References

Abracosa, R. & L. Ortolano 1987. Environmental impact assessment in the Philippines: 1977–1985. *Environmental Impact Assessment Review* **7**, 293–310.

Anderson, F. R., D. R. Mandelker, A. D. Tarlock 1984. *Environmental protection: law and policy*. Boston: Little, Brown.

Bear, D. 1990. EIA in the USA after twenty years of NEPA. *EIA newsletter* **4**, EIA Centre, University of Manchester.

Briffett, C. 1995. *The effectiveness of environmental impact assessment in Southeast Asia* (MSc dissertation). Oxford: Oxford Brookes University.

Catlow, J. & C. G. Thirwall 1976. *Environmental impact analysis* (DOE Research Report 11). London: HMSO.

CEC (Commission of the European Communities) 1973. First action programme on the environment. *Official Journal* C112, 20 December.

CEC 1980. Draft directive concerning the assessment of the environmental effects of certain public and private projects. COM(80), 313 final. *Official Journal* C169, 9 July.

CEC 1982. Proposal to amend the proposal for a Council directive concerning the environmental effects of certain public and private projects. COM(82), 158 final. *Official Journal* C110, 1 May.

CEC 1985. On the assessment of the effects of certain public and private projects on the environment. *Official Journal* L175, 5 July.

CEC 1992. *Towards sustainability*. Brussels: CEC.

CEC 1993. *Report from the Commission of the Implementation of Directive 85/337/EEC on the assessment of the effects of certain public and private projects on the environment*. COM (93), 28 Final. Brussels: CEC.

CEC 1994. *Environmental impact assessment review checklist*. Brussels: EC Directorate-General XI.

CEC 1995. *Environmental impact assessment: guidance on screening DG XI*. Brussels: CEC.

CEC 1997 Council Directive 97/11/EC of 3 March 1997 amending Directive 85/337/EEC on the assessment of certain public and private projects on the environment. *Official Journal*. L73/5, 3 March.

CEQ (Council on Environmental Quality) 1978. National Environmental Policy Act. Implementation of procedural provisions: final regulations. *Federal Register* 43(230), 55977–56007, 29 November.

CEQ 1993. *Environmental quality: 23nd annual report*. Washington DC: US Government Printing Office.

Clark, B. D., K. Chapman, R. Bisset, P. Wathern 1976. *Assessment of major industrial applications: a manual* (DOE Research Report 13). London: HMSO.

Clark, B. D., K. Chapman, R. Bisset, P. Wathern, M. Barrett 1981. *A manual for the assessment of major industrial proposals*. London: HMSO.

Clark, B. D. & R. G. H. Turnbull 1984. Proposals for environmental impact assessment procedures in the UK. In *Planning and ecology*, R. D. Roberts & T. M. Roberts (eds), 135–44. London: Chapman & Hall.

CPRE 1991. *The environmental assessment directive: five years on*. London: Council for the Protection of Rural England.

Dobry, G. 1975. *Review of the development control system: final report*. London: HMSO.

DOE 1977. Press Notice 68. London: Department of the Environment.

DOE 1978. Press Notice 488. London: Department of the Environment.

EIA Centre. *EIA Newsletter* (annual). EIA Centre, University of Manchester.

European Bank for Reconstruction and Development 1992. *Environmental procedures*. London: European Bank for Reconstruction and Development.

Hall, E. 1994. *The Environment versus people? A study of the treatment of social effects in EIA* (MSc dissertation). Oxford: Oxford Brookes University.

HMSO 1971. *Report of the Roskill Commission on the Third London Airport*. London: HMSO.

House of Lords 1981a. *Environmental assessment of projects*. Select Committee on the European Communities, 11th Report, Session 1980–81. London: HMSO.

House of Lords 1981b. *Parliamentary Debates (Hansard) Official Report, Session 1980–81*, 30 April, 1311–47. London: HMSO.

Lee, N. & C. M. Wood 1984. Environmental impact assessment procedures within the European Economic Community. In *Planning and ecology*, R. D. Roberts & T. M. Roberts (eds), 128–34. London: Chapman & Hall.

Legore, S. 1984. Experience with environmental impact assessment in the USA. In *Planning and ecology*, R. D. Roberts & T. M. Roberts (eds), 103–12. London: Chapman & Hall.

Moreira, I. V. 1988. EIA in Latin America. In *Environmental impact assessment: theory and practice*, P. Wathern (ed.), 239–53. London: Unwin Hyman.

NEPA (National Environmental Policy Act) 1970. 42 USC 4321–4347, 1 January, as amended.

O'Riordan, T. & W. R. D. Sewell (eds) 1981. *Project appraisal and policy review*. Chichester: Wiley.

Orloff, N. 1980. *The National Environmental Policy Act: cases and materials*. Washington DC: Bureau of National Affairs.

Overseas Development Administration 1996. *Manual of environmental appraisal: revised and updated*. London: ODA.

Petts, J. & P. Hills 1982. *Environmental assessment in the UK*. Institute of Planning Studies, University of Nottingham.

Roe, D., B. Dalal-Clayton, R. Hughes 1995. *A directory of impact assessment guidelines.* London: International Institute for Environment and Development.

Sadler, B. 1996. *Environmental assessment in a changing world: evaluating practice to improve performance*, International Study on the Effectiveness of Environmental Assessment. Ottawa: Canadian Environmental Assessment Agency.

Scottish Development Department 1974. *Appraisal of the impact of oil-related development*, DP/TAN/16. Edinburgh: SDA.

State of California, Governor's Office of Planning and Research 1992. CEQA: *California Environmental Quality Act – statutes and guidelines.* Sacramento: State of California.

Stevens Committee 1976. *Planning control over mineral working.* London: HMSO.

Swaffield, S. 1981. Environmental assessment: by administration or design? *Planning Outlook* **24**(3), 102–9.

Tomlinson, P. 1986. Environmental assessment in the UK: implementation of the EEC Directive. *Town Planning Review* **57**(4), 458–86.

Turner, T. 1988. The legal eagles. *Amicus Journal* (winter), 25–37.

United Nations Environment Programme (UNEP) 1997. *Environmental impact assessment training resource manual.* Stevenage: SMI Distribution.

Wathern P. (ed.) 1988. *Environmental impact assessment: theory and practice.* London: Unwin Hyman.

Welles, H. 1995. EIA capacity-strengthening in Asia: the USAID/WRI model. *Environmental Professional* **17**, 103–16.

White House 1994. *Memorandum from President Clinton for all heads of all departments and agencies on an executive order on federal actions to address environmental injustice in minority populations and low income populations.* Washington DC: White House.

Williams, R. H. 1988. The environmental assessment directive of the European Community. In *The rôle of environmental impact assessment in the planning process*, M. Clark & J. Herington (eds), 74–87. London: Mansell.

Wood, C. 1981. The impact of the European Commission's directive on environmental planning in the United Kingdom. *Planning Outlook* **24**(3), 92–8.

Wood, C. 1988. The genesis and implementation of environmental impact assessment in Europe. In *The rôle of environmental impact assessment in the planning process*, M. Clark & J. Herington (eds), 88–102. London: Mansell.

World Bank 1992. *Environmental assessment sourcebook.* Washington DC: World Bank.

World Bank 1995. *Environmental assessment: challenges and good practice*, Washington DC: World Bank.

Notes

1 E.g. *Ely v. Velds*, 451 F.2d 1130, 4th Cir. 1971; *Carolina Action v. Simon*, 522 F.2d 295, 4th Cir. 1975.

2 *Calvert Cliff's Coordinating Committee, Inc. v. United States Atomic Energy Commission* 449 F.2d 1109, DC Cir. 1971.

3 *Natural Resources Defense Council, Inc. v. Morton*, 458 F.2d 827, DC Cir. 1972.

4 Arkansas, California, Connecticut, Florida, Hawaii, Indiana, Maryland, Massachusetts, Minnesota, Montana, New York, North Carolina, South Dakota, Virginia, Washington and Wisconsin, plus the District of Columbia.

5 Arizona, Delaware, Georgia, Louisiana, Michigan, New Jersey, North Dakota, Oregon, Pennsylvania, Rhode Island and Utah.

CHAPTER 3
UK agency and legislative context

3.1 Introduction

This chapter discusses the legislative framework within which EIA is carried out in the UK. It begins with an outline of the principal actors involved in EIA and in the associated planning and development process. It follows with an overview of relevant regulations and the types of project to which they apply, then of the EIA procedures required by the Town and Country Planning (Assessment of Environmental Effects) Regulations 1988 and the 1998 amendments. These can be considered the "generic" EIA regulations, which apply to most projects and provide a model for the other EIA regulations. The latter are then summarized. Readers should refer to Chapter 8 for a discussion of the main effects and limitations of the application of these regulations.

3.2 The principal actors

An overview

Any proposed major development has an underlying configuration of interests, strategies and perspectives. But whatever the development, be it a motorway, a power station, a reservoir or a forest, it is possible to divide those involved in the planning and development process broadly into four main groups. These are:

- the developers;
- those directly or indirectly affected by or having an interest in the development;
- the government and regulatory agencies;
- various intermediaries (consultants, advocates, advisers) with an interest in the interaction between the developer, the affected parties and the regulators (Fig. 3.1).

An introduction to the range of "actors" involved is an important first step in understanding the UK legislative framework for EIA.

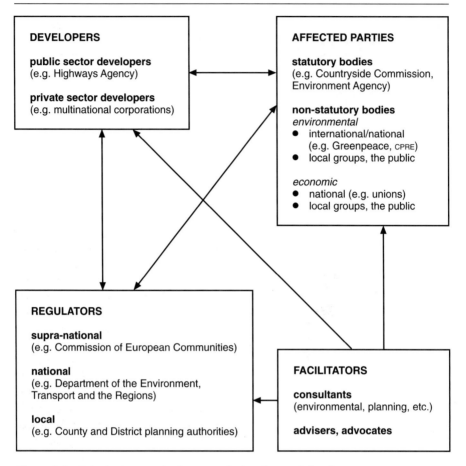

Figure 3.1 Principal actors in the EIA and planning and development processes.

Developers

In the UK, EIA applies to projects in both the public and private sectors, although there are notable exemptions, including Ministry of Defence developments and those of the Crown Commission. Public-sector developments are sponsored by central government departments (such as the former Department of Transport), by local authorities and by statutory bodies, such as the Environment Agency and the Highways Agency. Some were also sponsored by nationalized industries (such as the former British Rail and the nuclear industry), but the rapid privatization programme of the 1980s and 1990s has transferred many former nationalized industries to the private sector. Some, such as the major energy companies (National Power, PowerGen, British Gas) and the regional water authorities, have major and continuing programmes of projects, where it may be possible to develop and refine EIA procedures, learning from experience. Many other private-sector companies, often of multinational form, may also produce a stream of projects. However, for many developers, a major project may be a one-off or "once in a lifetime" activity. For them, the EIA process, and the associated planning and development process, may be much less familiar, requiring quick learning and, it is to be hoped, the provision of some good advice.

Affected parties

Those parties directly or indirectly affected by such developments are many. In Figure 3.1 they have been broadly categorized, according to their role or degree of power (e.g. statutory, advisory), level of operation (e.g. international, national, local) or emphasis (e.g. environmental, economic). The growth in environmental groups, such as Greenpeace, Friends of the Earth, the Council for the Protection of Rural England and the Royal Society for the Protection of Birds, is of particular note and is partly associated with the growing public interest in environmental issues. For instance, membership of the RSPB grew from 100,000 in 1970 to over a million in 1997. Membership of Sustrans, a charity which promotes car-free cycle routes, rose from 4,000 in 1993 to 20,000 in 1996. Such groups, although often limited in resources, may have considerable "moral weight". The accommodation of their interests by a developer is often viewed as an important step in the "legitimization" of a project. Like the developers, some environmental groups, especially at the national level, may have a long-term, continuing role. Some local amenity groups also may have a continuing role and an accumulation of valuable knowledge about the local environment. Others, usually at the local level, may have a short life, being associated with one particular project. In this latter category can be placed local pressure groups, which can spring up quickly to oppose developments. Such groups have sometimes been referred to as NIMBYS ("not in my back yard"), and their aims often include the maintenance of property values and existing lifestyles, and the diversion of any necessary development elsewhere.

Statutory consultees are an important group in the EIA process. The planning authority must consult such bodies before making a decision on a major project requiring an EIA. Statutory consultees in England and Wales include the Countryside Commission, English Nature, the Environment Agency (for certain developments) and the principal local council for the area in which the project is proposed. Other consultees often involved include the local highway authority and the county archaeologist. As noted above, non-statutory bodies, such as the RSPB and the general public, may provide additional valuable information on environmental issues.

Regulators

The government, at various levels, will normally have a significant role in regulating and managing the relationship between the groups previously outlined. As discussed in Chapter 2, the European Commission has adopted a Directive on EIA procedures (CEC 1985 and amendments). The UK government has subsequently implemented these through an array of regulations and guidance (see Section 3.3). The principal department involved is the Department of the Environment, Transport and the Regions (DETR, formerly DOE) through its London headquarters and regional offices. Notwithstanding the government scepticism noted in Chapter 2, William Waldegrave, UK Minister of State for the Environment commented in 1987 that ". . . one of the most important tasks facing Government is to inspire a development process which takes into account not only the nature of any environmental risk but also the perceptions of the risk by the public who must suffer its consequences" (ESRC 1987).

Of particular importance in the EIA process is the local authority, and especially the relevant local planning authority (LPA). This may involve district, county and unitary authorities. Such authorities act as filters through which schemes proposed by developers usually have to pass. In addition, the local planning authority often opens the door for other agencies to become involved in the development process.

Facilitators

A final group, but one of particular significance in the EIA process, includes the various consultants, advocates and advisers who participate in the EIA and the planning and development processes. Such agents are often employed by developers; occasionally they may be employed by local groups, environmental groups and others to help to mount opposition to a proposal. They may also be employed by regulatory bodies to help them in their examination process.

A recent UK survey (Weston 1995) showed that environmental and planning consultancies carry out most of the EIA work, consultancies specializing in such issues as archaeology or noise contributing less. There has been a massive growth in the number of environmental consultancies in the UK (see Fig. 3.2). The numbers have almost trebled since the mid-1980s, and it has been estimated that clients in 1993–4 were spending approximately £400 million on their services, with growth of about 10 per cent per year (ENDS 1995). Major factors underpinning the consultancy growth have been the advent of the UK Environmental Protection Act (EPA) in 1990, EIA regulations, the growing UK business interest in environmental management systems (e.g. BS7750), and the proposed EC regulations on eco-auditing and strategic environmental assessment.

Figure 3.3 provides a summary of the main work areas for environmental consultancies. Although the requirements of the EPA (with its "duty of care" regulations, which came into force in April 1992) and the Water Resources Act of 1991 have concentrated the minds of developers and clients on water pollution and contaminated land in particular, there is no denying the significance of the EIA boom for consultants. Further characteristics of recent consultancy activity are discussed in Chapter 8.

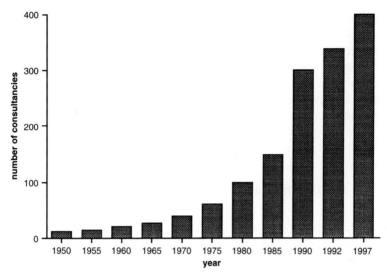

Figure 3.2 Increase in the number of environmental consultancies in the UK (1950–1997). (Based on: ENDS 1993, 1997)

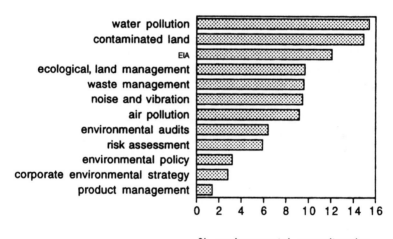

Figure 3.3 Main work areas for environmental consultancies (1987/88–91/92). (Based on: ENDS 1992)

Agency interaction

The various agencies outlined here represent a complex array of interests and aims, any combination of which may come into play for a particular development. This array has several dimensions, and within each there may be a range of often

59

conflicting views. For example, there may be conflict between local and national views, between the interests of profit maximization and those of environmental conservation, between short-term and long-term perspectives and between corporate bodies and individuals. The agencies are also linked in various ways. Some links are statutory, others advisory. Some are contractual, others regulatory. The EIA regulations and guidance provide a set of procedures linking the various actors discussed, and these are now outlined.

3.3   EIA regulations: an overview

In the UK, EC Directive 85/337 is implemented through over forty different secondary regulations under section 2(2) of the European Communities Act 1972: these are listed in Table 3.1. The large number of regulations is symptomatic of how EIA has been implemented in the UK. Different regulations apply to projects covered by the planning system, projects covered by other authorization systems and projects not covered by any authorization system but still requiring EIA. Different regulations apply to England and Wales, to Scotland and to Northern Ireland. Some of the regulations listed in Table 3.1 are the original regulations implementing Directive 85/337; others, especially those enacted since 1994, close loopholes to ensure that all of the Directive's requirements are met. The regulations are supplemented by an array of EIA guidance from government and other bodies (see Table 3.2). In addition, the Planning and Compensation Act 1991 allows the government to require EIA for other projects that fall outside the Directive.

In contrast to the US system of EIA, that of Directive 85/337 applies to both public and private sector development. The developer carries out the EIA, and the resulting EIS must be handed in with the application for authorization. In England and Wales, most of the developments listed in Annexes I and II of Directive 85/337 fall under the remit of the planning system, and are thus covered by the Town and Country Planning (Assessment of Environmental Effects) Regulations 1988 (the T&CP Regulations). Various additions and amendments to these regulations have been enacted over the years to plug loopholes and extend the remit of the regulations, for instance:

- to expand and clarify the original list of projects for which EIA is required (e.g. to include motorway service areas and wind farms);
- to require EIA for projects that would otherwise be permitted (e.g. land reclamation, waste water treatment works, projects in Simplified Planning Zones);
- to require EIA for projects resulting from a successful appeal against a planning enforcement notice;
- to allow the Secretary of State (SOS) for the Environment to direct that a particular development should be subject to EIA even if it is not listed in the regulations.

Table 3.1   UK EIA regulations and dates of implementation.

UK regulations for projects subject to the Town and Country Planning system

England and Wales

Town and Country Planning (Assessment of Environmental Effects) Regulations 1988 (SI 1199)[1]
Town and Country Planning (Assessment of Environmental Effects) (Amendment) Regulations 1990 (SI 367)[1]
Town and Country Planning (Assessment of Environmental Effects) (Amendment) Regulations 1992 (SI 1494)[1]
Town and Country Planning (Simplified Planning Zones) Regulations 1992 (SI 2414)
Town and Country Planning (Assessment of Environmental Effects) (Amendment) Regulations 1994 (SI 677)[1]
Town and Country Planning General Development (Amendment) Order 1994 (SI 678)
Town and Country Planning (Environmental Assessment and Permitted Development) Regulations 1995 (SI 417)[1]
Town and Country Planning (General Permitted Development) Order 1995 (SI 418)[1]
Town and Country Planning (General Development Procedure) Order 1995 (SI 419)[1]
Town and Country Planning (Environmental Assessment and Unauthorised Development) Regulations 1995 (SI 2258)[1]

Scotland

Part II of the Environmental Assessment (Scotland) Regulations 1988 (SI 1221)
Town and Country Planning (General Development Procedure) (Scotland) Order 1992 (SI 224)
Environmental Assessment (Scotland) Amendment Regulations 1994 (SI 20212)

Northern Ireland

Planning (Assessment of Environmental Effects) Regulations (Northern Ireland) 1989 (SR 20)
Planning (Assessment of Environmental Effects) (Amendment) Regulations Northern Ireland) 1994 (SR 395)
Planning (Simplified Planning Zones) (Excluded Development) Order (Northern Ireland) 1994 (SR 426)
Planning (General Development) (Amendment) Order (Northern Ireland) 1995 (SR 356)
Planning (Environmental Assessment and Permitted Development) Regulations (Northern Ireland) 1995 (SR 357)

Gibraltar

Town Planning (Applications) (Amendment) Regulations 1993

UK EIA regulations for projects subject to alternative consent systems

Afforestation

Environmental Assessment (Afforestation) Regulations 1988 (SI 1207)
Environmental Assessment (Afforestation) Regulations (Northern Ireland) 1989 SR 226)

Land drainage improvements

Land Drainage Improvement Works (Assessment of Environmental Effects) Regulations 1988 (SI 1217)
Part V (Drainage Works) of the Environmental Assessment (Scotland) Regulations 1988 (SI 1221)
Drainage (Environmental Assessment) Regulations (Northern Ireland) 1991 (SR 376)
Land Drainage Improvement Works (Assessment of Environmental Effects) (Amendment) Regulations 1995 (SI 2195)

Table 3.1 (cont'd)

Marine salmon farming

Environmental Assessment (Salmon Farming in Marine Waters) Regulations 1988 (SI 1218)

Trunk roads and motorways

Highways (Assessment of Environmental Effects) Regulations 1988 (SI 1241)
Part VI (Amendments of the Roads (Scotland) Act 1984) of the Environmental Assessment (Scotland)
    Regulations 1988 (SI 1221)
Roads (Northern Ireland) Order 1993 (SR 3160(NI 15))
Highways (Assessment of Environmental Effects) Regulations 1994 (SI 1002)
Roads (Assessment of Environmental Effects) Regulations (Northern Ireland) 1994 (SR 316)

Railways, tramways, inland waterways and works interfering with navigation rights

Transport and Works (Application and Objections Procedure) Rules 1992 (SI 2902)
Transport and Works (Assessment of Environmental Effects) Regulations 1995 (SI 1541)

Ports and harbours

Harbour Works (Assessment of Environmental Effects) Regulations 1988 (SI 1336)
Harbour Works (Assessment of Environmental Effects) (No 2) Regulations 1989 (SI 424)
Harbour Works (Assessment of Environmental Effects) Regulations (Northern Ireland) 1990
    (SR 181)
Harbour Works (Assessment of Environmental Effects) Regulations 1992 (SI 1421)
Harbour Works (Assessment of Environmental Effects) (Amendment) Regulations 1996 (SI 1946)

Power stations, overhead power lines and long-distance oil and gas pipeline

Part III (Electricity Applications) of the Environmental Assessment (Scotland) Regulations 1988
    (SI 1221)
Electricity and Pipe-line Works (Assessment of Environmental Effects) Regulations 1990 (SI 442)
Electricity and Pipe-line Works (Assessment of Environmental Effects) (Amendment) Regulations
    1996 (SI 422)
Electricity and Pipe-line Works (Assessment of Environmental Effects) (Amendment) Regulations
    1997 (629)

Projects approved by private Act of Parliament

Standing Order 27a. 20 May 1991
General Order 27a. 20 May 1992. Inserted by the Private Legislation Procedure (Scotland)
    General Order 1992 (SI 1992/1206)

---

[1] These regulations will be replaced by the consolidated Town and Country Planning (Assessment of
Environmental Effects) Regulations 1998.

Table 3.2 UK Government EIA guidance.

DOE Circular 15/88 (Welsh Office 23/88) "Environmental Assessment" – 12 July 1988

SDD Circular 13/88 "Environmental Assessment: Implementation of EC Directive: The Environmental Assessment (Scotland) Regulations 1988" – 12 July 1988

Scottish Office Circular 26/91 "Environmental Assessment and Private Legislation Procedures"

Forestry Commission booklet "Environmental Assessment of Afforestation Projects" – 4 August 1988. Replaced by Forestry Commission booklet on EA of new woodlands

Crown Estate Office note "Environmental Assessment of Marine Salmon Farms" – 15 July 1988

DOE Circular 24/88 (Welsh Office 24/88) "Environmental Assessment of Projects in Simplified Planning Zones and Enterprise Zones" – 25 November 1988

SDD Circular 26/88 "Environmental Assessment of Projects in Simplified Planning Zones and Enterprise Zones" – 25 November 1988

DOE memorandum on new towns "Environmental Assessment" – 30 March 1989

DOT Department Standard HD 18/88 "Environmental Assessment under EC Directive 85/337" – July 1989. Replaced by DOT manual on environmental assessment.

DOE advisory booklet "Environmental Assessment – A Guide to the Procedures" – 6 November 1989

DOE free leaflet "Environmental Assessment" – October 1989

Welsh Office free leaflet "Environmental Assessment/Asesu'r Amgylchedd"

Scottish Office free leaflet "Environmental Assessment – a Guide" – June 1990

DOE (Northern Ireland) Development Control Advice Note No. 10 "Environmental Impact Assessment" – 1989

Overseas Development Administration "Manual of Environmental Appraisal". Revised April 1992

DOE Circular 15/92 (Welsh Office 32/92) "Publicity for Planning Applications" – 3 June 1992

DOE Circular 19/92 (Welsh Office 39/92) "The Town and Country Planning General Regulations 1992/The Town and Country Planning (Development Plans and Consultation) Directions 1992 – 13 July 1992

DOT guide "Transport and Works Act 1992: A Guide to Procedures for Obtaining Orders Relating to Transport Systems, Inland Waterways and Works Interfering with Rights of Navigation" – 1992

DOE Planning Policy Guidance Note 5 "Simplified Planning Zones" (paras 7–9 of Annex A and Appendices 1 & 2) – November 1992

DTI booklet "Guidance on Environmental Assessment of Cross-Country Pipelines" – 1992

Forestry Commission booklet "Environmental Assessment of New Woodlands" – April 1993

DOT/SO/WO/DOE (NI) "Design Manual for Roads and Bridges Vol II: Environmental Assessment" – June 1993

SO Environment Department Circular 26/94 "The Environmental Assessment (Scotland) Amendment Regulations 1994" – 1994

DOE Circular 3/95 (Welsh Office 12/95) "Permitted Development and Environmental Assessment" – 1995

DOE/WO booklet "Your Permitted Development Rights and Environmental Assessment" – March 1995

DOE Circular 11/95 "The Use of Conditions in Planning Permissions" para. 77

DOE Circular 13/95 (Welsh Office 39/95) "The Town and Country Planning (Environmental Assessment and Unauthorised Development) Regulations 1995"

Environment Agency "A Scoping Handbook for Projects" – 1996

"Monitoring Environmental Assessment and Planning" – DOE 1991

"Evaluation of Environmental Information for Planning Projects: A Good Practice Guide" – DOE 1994a

"Good Practice on the Evaluation of Environmental Information for Planning Projects: Research Report" – DOE 1994b

"Preparation of Environmental Statements for Planning Projects that Require Environmental Assessment" – DOE 1995

"Changes in the Quality of Environmental Statements for Planning Projects" – DOE 1996

Other types of projects listed in the Directive require separate legislation, since they are not governed by the planning system. Of the various *transport* projects, local highway developments and airports are dealt with under the T&CP Regulations by the local planning (highways) authority, but motorways and trunk roads proposed and regulated by the Department of Transport (DOT) (now DETR) fall under the Highways (AEE) Regulations 1988 and 1994. Applications for harbours are regulated by the DOT under the various Harbour Works (AEE) Regulations. New railways and tramways require EIA under the Transport and Works (Applications and Objections) Procedure 1992 and (AEE) Regulations 1995.

Energy projects producing less than 50 MW are regulated by the local authority under the T&CP Regulations. Those of 50 MW or over, most electricity power lines, and pipelines (in Scotland as well as in England and Wales) are controlled by the Department of Trade and Industry under the Electricity and Pipeline Works (AEE) Regulations 1990.

New *land drainage* works, including flood defence and coastal defence works, require planning permission and are thus covered by the T&CP Regulations. Improvements to drainage works carried out by the Environment Agency and other drainage bodies require EIA through the Land Drainage Improvement Works (AEE) Regulations and amendments, which are regulated by the Ministry of Agriculture, Fisheries, and Food (MAFF).

Forestry projects for which grants are given by the Forestry Authority require EIA under the Environmental Assessment (Afforestation) Regulations 1988.

Salmon farms within 2 km of the coast of England, Wales or Scotland require a lease from the Crown Estates Commission, but not planning permission. For these developments, EIA is required under the Environmental Assessment (Salmon Farming in Marine Waters) Regulations 1988.

Most other developments in Scotland are covered by the Environmental Assessment (Scotland) Regulations 1988, including developments related to town and country planning, electricity, roads and bridges, development by planning authorities and land drainage. The British regulations apply to harbours, pipelines and afforestation projects. Northern Ireland has separate legislation in parallel with that of England and Wales.

As will be discussed in Chapter 8, about 70 per cent of all the EIAS prepared in the UK fall under the T&CP Regulations, about 10 per cent fall under each of the EA (Scotland) Regulations 1988 and the Highways (AEE) Regulations; almost all the rest involve land drainage, electricity and pipeline works, afforestation projects in England and Wales and planning-related developments in Northern Ireland.

The enactment of this wide range of EIA regulations has made many of the early concerns regarding procedural loopholes (e.g. CPRE 1991, Fortlage 1990) obsolete. However several issues still remain. First is the ambiguity inherent in the term "project". An example of this is the EIA procedures for electricity generation and transmission, in which a power station and the transmission lines to and from it are seen as separate projects for the purposes of EIA, despite the fact that they are inextricably linked (Sheate 1995). See Chapter 10 for further discussion of this issue. Another example is the division of road construction into several separate projects for planning and EIA purposes even though none of them would be independently viable. This is discussed further in Chapter 13. Finally, the amendments to the Directive are likely to have other effects on UK regulations; indeed draft Town and Country Planning (Assessment of Environmental Effects) Regulations 1998 propose some minor changes and consolidation of regulations (replacing six of the regulations for England and Wales outlined in Table 3.1).

3.4 The Town and Country Planning (Assessment of Environmental Effects) Regulations 1988 (ESI 1199)

The T&CP Regulations implement Directive 85/337 for those projects that require planning permission in England and Wales. They are the central form in which Directive 85/337 is implemented in the UK; the other UK EIA regulations were established to cover projects that are not covered by the T&CP regulations. As a result, the T&CP Regulations are the main focus of discussions on EIA procedures and effectiveness. This section presents the procedures of the T&CP Regulations. Figure 3.4 summarizes these procedures; the letters in the figure correspond to the letters in bold preceding the explanatory paragraphs below. Section 3.5 considers other main EIA regulations as variations of the T&CP Regulations and Section 3.6 reports on changes following from the amended EC Directive.

The T&CP Regulations were issued on 15 July 1988, 12 days after Directive 85/337 was to have been implemented. Guidance on the Regulations, aimed primarily at local planning authorities, is given in DOE Circular 15/88 (Welsh Office Circular 23/88). A guidebook entitled *Environmental assessment: a guide to the procedures* (DOE 1989), aimed primarily at developers and their advisers, was released in November 1989. Only the regulations are mandatory: the guidance interprets and advises, but cannot be enforced. However, the reader is strongly advised to read the guidebook. Further DOE guidance on good practice in carrying out and reviewing EIAS was published in 1994 and 1995 (DOE 1994a, 1994b, 1995), and is also strongly recommended reading.

Which projects require EIA?

The T&CP Regulations require EIAS to be carried out for two broad categories of project, given in Schedules 1 and 2. These broadly corresponded[1] to Annexes I and II of Directive 85/337 before it was amended, excluding those projects that do not require planning permission. The amended Schedules are likely to correspond very closely to Annexes 1 and 2 in the amended Directive, as detailed in Appendix 1. Table 3.3 lists Schedule 1 and Schedule 2 projects. For Schedule 1 projects, EIA is

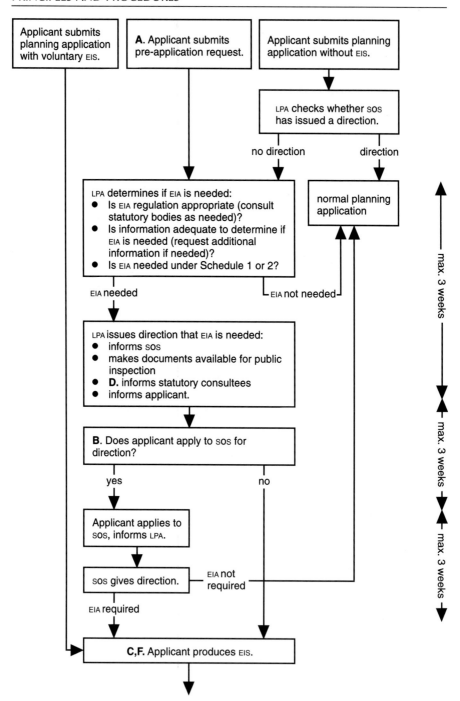

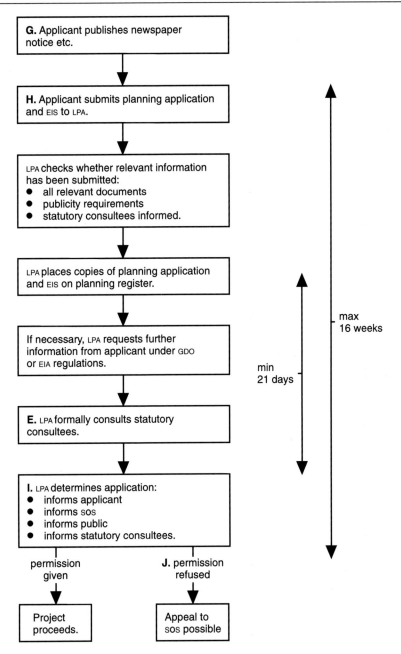

Figure 3.4 Summary of T&CP Regulations EIA procedure. (Based on: DOE 1989)

Table 3.3 Projects requiring EIA under the T&CP Regulations (1988)[1].

Schedule 1

The following types of development ("Schedule 1 projects") require environmental assessment in every case:

(1) The carrying out of building or other operations, or the change of use of buildings or other land (where a material change) to provide any of the following:

1. A crude-oil refinery (excluding an undertaking manufacturing only lubricants from crude oil) or an installation for the gasification and liquefaction of 500 tonnes or more of coal or bituminous shale per day.

2. A thermal power station or other combustion installation with a heat output of 300 MW or more, other than a nuclear power station or other nuclear reactor.

3. An installation designed solely for the permanent storage or final disposal of radioactive waste.

4. An integrated works for the initial melting of cast-iron and steel.

5. An installation for the extraction of asbestos or for the processing and transformation of asbestos or products containing asbestos:
    (a) where the installation produces asbestos–cement products, with an annual production of more than 20,000 tonnes of finished products; or
    (b) where the installation produces friction material, with an annual production of more than 50 tonnes of finished products; or
    (c) in other cases, where the installation will utilize more than 200 tonnes of asbestos per year.

6. An integrated chemical installation, that is to say, an industrial installation or group of installations where two or more linked chemical or physical processes are employed for the manufacture of olefins from petroleum products, or of sulphuric acid, nitric acid, hydrofluoric acid, chlorine or fluorine.

7. A special road; a line for long-distance railway traffic; or an aerodrome with a basic runway length of 2100 m or more.

8. A trading port, an inland waterway which permits the passage of vessels of over 1,350 tonnes or a port for inland waterway traffic capable of handling such vessels.

9. A waste-disposal installation for the incineration or chemical treatment of special waste.

(2) The carrying out of operations whereby land is filled with special waste, or the change of use of land (where a material change) to use for the deposit of such waste.

Schedule 2

The following types of development ("Schedule 2 projects") require environmental assessment if they are likely to have significant effects of the environment by virtue of factors such as their nature, size or location:

1. Agriculture
    (a) water-management for agriculture
    (b) poultry-rearing
    (c) pig-rearing
    (d) a salmon hatchery

Table 3.3   (cont'd)

(e)   an installation for the rearing of salmon
(f)   the reclamation of land from the sea

2.   Extractive industry
   (a)   extracting peat
   (b)   deep drilling, including in particular:

   - geothermal drilling
   - drilling for the storage of nuclear waste material
   - drilling for water supplies
     but excluding drilling to investigate the stability of the soil

   (c)   extracting minerals (other than metalliferous and energy-producing minerals) such as marble, sand, gravel, shale, salt, phosphates and potash
   (d)   extracting coal or lignite by underground or open-cast mining
   (e)   extracting petroleum
   (f)   extracting natural gas
   (g)   extracting ores
   (h)   extracting bituminous shale
   (i)   extracting minerals (other than metalliferous and energy-producing minerals) by open-cast mining
   (j)   a surface industrial installation for the extraction of coal, petroleum, natural gas or ores or bituminous shale
   (k)   a coke oven (dry distillation of coal)
   (l)   an installation for the manufacture of cement

3.   Energy industry
   (a)   a non-nuclear thermal power station, not being an installation falling within Schedule 1, or an installation for the production of electricity, steam and hot water
   (b)   an industrial installation for carrying gas, steam or hot water; or the transmission of electrical energy by overhead cables
   (c)   the surface storage of natural gas
   (d)   the underground storage of combustible gases
   (e)   the surface storage of fossil fuels
   (f)   the industrial briquetting of coal or lignite
   (g)   an installation for the production or enrichment of nuclear fuels
   (h)   an installation for the reprocessing of irradiated nuclear fuels
   (i)   an installation for the collection or processing of radioactive waste, not being an installation falling within Schedule 1
   (j)   an installation for hydroelectric energy production

4.   Processing of metals
   (a)   an ironworks or steelworks including a foundry, forge, drawing plant or rolling mill (not being a works falling within Schedule 1)
   (b)   an installation for the production (including smelting, refining, drawing and rolling) of non-ferrous metals, other than precious metals
   (c)   the pressing, drawing or stamping of large castings
   (d)   the surface treatment and coating of metals
   (e)   boiler-making or manufacturing reservoirs, tanks and other sheet-metal containers
   (f)   manufacturing or assembling motor vehicles or manufacturing motor-vehicle engines
   (g)   a shipyard
   (h)   an installation for the construction or repair of aircraft
   (i)   the manufacture of railway equipment

Table 3.3   (cont'd)

    (j)   swaging by explosives
    (k)   an installation for the roasting or sintering of metallic ores

5. Glass making
the manufacture of glass

6. Chemical industry
    (a)   the treatment of intermediate products and production of chemicals, other than development falling within Schedule 1
    (b)   the production of pesticides or pharmaceutical products, paints or varnishes, elastomers or peroxides
    (c)   the storage of petroleum or petrochemical or chemical products

7. Food industry
    (a)   the manufacture of vegetable or animal oils or fats
    (b)   the packing or canning of animal or vegetable products
    (c)   the manufacture of dairy products
    (d)   brewing or malting
    (e)   confectionery or syrup manufacture
    (f)   an installation for the slaughter of animals
    (g)   an industrial starch manufacturing installation
    (h)   a fish-meal or fish-oil factory
    (i)   a sugar factory

8. Textile, leather, wood and paper industries
    (a)   a wool scouring, degreasing and bleaching factory
    (b)   the manufacture of fibre board, particle board or plywood
    (c)   the manufacture of pulp, paper or board
    (d)   a fibre-dyeing factory
    (e)   a cellulose-processing and production installation
    (f)   a tannery or a leather dressing factory

9. Rubber industry
the manufacture and treatment of elastomer-based products

10. Infrastructure projects
    (a)   an industrial estate development project
    (b)   an urban development project
    (c)   a ski-lift or cable-car
    (d)   the construction of a road, or a harbour, including a fishing harbour, or an aerodrome, not being development falling within Schedule 1
    (e)   canalization or flood-relief works
    (f)   a dam or other installation designed to hold water or store it on a long-term basis
    (g)   a tramway, elevated or underground railway, suspended line or similar line, exclusively or mainly for passenger transport
    (h)   an oil or gas pipeline installation
    (i)   a long-distance aqueduct
    (j)   a yacht marina

11. Other projects
    (a)   a holiday village or hotel complex
    (b)   a permanent racing or test track for cars or motor cycles
    (c)   an installation for the disposal of controlled waste or waste from mines and quarries, not being an installation falling within Schedule 1

Table 3.3   (cont'd)

(d)	a waste water treatment plant
(e)	a site for depositing sludge
(f)	the storage of scrap iron
(g)	a test bench for engines, turbines or reactors
(h)	the manufacture of artificial mineral fibres
(i)	the manufacture, packing, loading or placing in cartridges of gunpowder or other explosives
(j)	a knackers' yard

12. The modification of a development which has been carried out, where that development is within a description mentioned in Schedule 1.

13. Development within a description mentioned in Schedule 1, where it is exclusively or mainly for the development and testing of new methods or products and will not be permitted for longer than one year.

[1] See Section 3.6 and Appendix 1 (Annexes I and II) for changes under amended Regulations (1998).

required in every case. A Schedule 2 project requires EIA if it is deemed "likely to give rise to significant environmental effects". The "significance" of a project's environmental effects is determined on the basis of three criteria:

- whether the project is of more than local importance, principally in terms of physical scale;
- whether the project is intended for a particularly sensitive location, for example, a national park or a [site of special scientific interest] . . .
- whether the project is thought likely to give rise to particularly complex or adverse effects, for example, in terms of the discharge of pollutants. (DOE 1989)

The guidebook (DOE 1989) includes indicative criteria and thresholds for a range of Schedule 2 projects that "are intended to indicate the types of cases in which, in the Secretary of State's view, environmental assessment may be required under the regulations". For instance, pig-rearing installations for more than 400 sows, industrial estate developments of more than 20 ha and new roads of over 10 km not located in a designated area may require EIA, according to the Circular. Further guidance on criteria and thresholds will be produced with the implementation of the amendments to the Directive.

A. A developer may decide that a project requires EIA under the T&CP Regulations, or may want to carry out an EIA even if it is not required. If the developer is uncertain, the LPA can be asked to determine if an EIA is needed. To do this the developer must provide the LPA with a plan showing the development site, a description of the proposed development and an indication of its possible environmental impacts. The LPA must then make a decision within three weeks. The LPA can ask for more information from the developer, but this does not extend the three-week decision-making period.

If the LPA decides that no EIA is needed, the application is processed as a normal planning application. If instead the LPA decides that an EIA is needed, it must explain why, and make both the developer's information and the decision publicly available. If the LPA receives a planning application without an EIS when it feels that it is needed, the LPA must notify the developer within three weeks, explaining why an EIS is needed. The developer then has three weeks in which to notify the LPA of the intention either to prepare an EIS or to appeal to the SOS; if the developer does not do so, the planning application is refused.

B. If the LPA decides that an EIA is needed but the developer disagrees, the developer can refer the matter to the SOS for a ruling.[2] The SOS must give a decision within three weeks. If the SOS decides that an EIA is needed, an explanation is needed; it is published in the *Journal of Planning and Environment Law*. No explanation is needed if no EIA is required. The SOS may make a decision if a developer has not requested an opinion, and may rule, usually as a result of information made available by other bodies, that an EIA is needed where the LPA has decided that it is not needed.

The contents of the EIA

Schedule 3 of the T&CP Regulations, which is shown in Table 3.4, lists the information that should be included in an EIA. Schedule 3 interprets the requirements of Directive 85/337's Annex III (Annex IV, post-amendments) according to the criteria set out in Article 5 of the Directive, namely:

> Member States shall adopt the necessary measures to ensure that the developer supplies in an appropriate form the information specified in Annex III (Annex IV, post-amendments) inasmuch as:
>
> > The Member States consider that the information is relevant to a given state of the consent procedure and to the specific characteristics of a particular project or type of project and of the environmental features likely to be affected;
> >
> > The Member States consider that a developer may reasonably be required to compile this information having regard *inter alia* to current knowledge and methods of assessment.

In Schedule 3, the information required in Annex III has been interpreted to fall into two categories: "specified information", which must be included in an EIA, and "further information", which *may* be included "by way of explanation or amplification of any specified information". This distinction is important: as will be seen in Chapter 8, the EISs prepared to date have generally not included "further information", although this includes such important matters as the alternatives that were considered and the expected wastes or emissions from the development. In addi-

Table 3.4   Content of EIS required by the T&CP Regulations (1988)[1].

The following are the statutory provisions with respect to the content of environmental statements, as set out in Schedule 3 to the Town and Country Planning (Assessment of Environmental Effects) Regulations 1988.

1.   An environmental statement comprises a document or series of documents providing for the purpose of assessing the likely impact upon the environment of the development proposed to be carried out, the information specified in paragraph 2 (referred to in this Schedule as "the specified information").

2.   The specified information is:
   (a)   a description of the development proposed, comprising information about the site and the design and size or scale of the development;
   (b)   the data necessary to identify and assess the main effects which that development is likely to have on the environment;
   (c)   a description of the likely significant effects, direct and indirect, on the environment of the development, explained by reference to its possible impact on: human beings, soil, fauna, flora, water, air, climate, the landscape, the interaction between any of the foregoing, material assets, and the cultural heritage;
   (d)   where significant adverse effects are identified with respect to any of the foregoing, a description of the measures envisaged in order to avoid, reduce or remedy those effects; and
   (e)   a summary in non-technical language of the information specified above.

3.   An environmental statement may include, by way of explanation or amplification of any specified information, further information on any of the following matters:
   (a)   the physical characteristics of the proposed development, and the land-use requirements during the construction and operational phases;
   (b)   the main characteristics of the production processes proposed, including the nature and quantity of the materials to be used;
   (c)   the estimated type and quantity of expected residues and emissions (including pollutants of water, air or soil, noise, vibration, light, heat and radiation) resulting from the proposed development when in operation;
   (d)   (in outline) the main alternatives (if any) studied by the applicant, appellant or authority and an indication of the main reasons for choosing the development proposed, taking into account the environmental effects;
   (e)   the likely significant direct and indirect effects on the environment of the development proposed which may result from:

   • the use of natural resources;
   • the emission of pollutants, the creation of nuisances, and the elimination of waste;

   (f)   the forecasting methods used to assess any effects on the environment about which information is given under subparagraph (e); and
   (g)   any difficulties, such as technical deficiencies or lack of know-how, encountered in compiling any specified information.

   In paragraph (e), "effects" includes secondary, cumulative, short-, medium- and long-term, permanent, temporary, positive and negative effects.

4.   Where further information is included in an environmental statement pursuant to paragraph 3, a non-technical summary of that information shall also be provided.

[1] See Section 3.6 and Appendix 1 (Annex IV) for changes under amended Regulations (1998).

tion, in Appendix 4 of the guidebook (DOE 1989), the DOE has given a longer checklist of matters which may be considered for inclusion in an EIA: this list is for guidance only, but it helps to ensure that all the possible significant effects of the development are considered.

C. There has been no mandatory requirement in the UK for a formal "scoping" stage at which the LPA, the developer and other interested parties agree on what will be included in the EIA. Indeed, there is no requirement for any kind of consultation between the developer and other bodies before the submission of the formal EIS and planning application. However, the DOE guidance stresses the benefits of early consultation and early agreement on the scope of the EIA. It also notes that the preparation of the EIS is the responsibility of the applicant, although the LPA may put forward its views about what it should include.

Statutory and other consultees

Under the T&CP Regulations, a number of statutory consultees are involved in the EIA process, as noted in Section 3.2. These bodies are involved at two stages of an EIA.

D. First, when a LPA determines that an EIA is required, it must inform the statutory consultees of this. The consultees in turn must make available to the developer, if so requested and at a reasonable charge, any relevant environmental information in their possession. This does not include any confidential information or information that the consultees do not already have in their possession.

E. Secondly, once the EIS has been submitted, the LPA or developer must send a free copy to each of the statutory consultees. The consultees may make representations about the EIS to the LPA for at least two weeks after they receive the EIS. The LPA must take account of these representations when deciding whether to grant planning permission. The developer may also contact other consultees and the general public while preparing the EIS. The DOE guidance explains that these bodies may have particular expertise in the subject or may highlight important environmental issues that could affect the project. The developer is under no obligation to contact any of these groups, but again the DOE guidance stresses the benefits of early and thorough consultation.

Carrying out the EIA; preparing the EIS

F. The DOE gives no formal guidance about what techniques and methodologies should be used in EIA, noting only that they will vary depending on the proposed development, the receiving environment and the information available.

Submitting the EIS and planning application; public consultation

G. When the EIS has been completed, the developer must publish a notice in a local newspaper and post notices at the site. These notices must fulfil the requirements of §26 of the Town and Country Planning Act 1971, state that a copy of the EIS is available for public inspection, give a local address where copies may be obtained and state the cost of the EIS, if any. The public can make written representations to the LPA for at least 20 days after the publication of the notice, but within 21 days of the LPA's receipt of the planning application.

H. After the EIS has been publicly available for at least 21 days, the developer submits to the LPA the planning application, copies[3] of the EIS, and certification that the required public notices have been published and posted. The LPA must then send copies of the EIS to the statutory consultees, inviting written comments within a specified time (at least two weeks from receipt of the EIS), forward another copy to the SOS and place the EIS on the planning register. It must also decide whether any additional information about the project is needed before a decision can be made, and, if so, obtain it from the developer. The clock does not stop in this case: a decision must still be taken within the appropriate time.

Planning decision

I. Before making a decision about the planning application, the LPA must collect written representations from the public within three weeks of the receipt of the planning application, and from the statutory consultees at least two weeks from their receipt of the EIS. It must wait at least three weeks after receiving the planning application before making a decision. In contrast to normal planning applications, which must be decided within eight weeks, those accompanied by an EIS must be decided within 16 weeks. If the LPA has not made a decision after 16 weeks, the applicant can appeal to the SOS for a decision. The LPA cannot consider a planning application invalid because the accompanying EIS is felt to be inadequate: it can only ask for further information within the 16-week period.

In making its decision, the LPA must consider the EIS and any comments from the public and statutory consultees, as well as other material considerations. The environmental information is only part of the information that the LPA considers, along with other material considerations. The decision is essentially still a political one, but it comes with the assurance that the project's environmental implications are understood. The LPA may grant or refuse permission, with or without conditions.

J. If a LPA refuses planning permission, the developer may appeal to the SOS, as for a normal planning application. The SOS may request further information before making a decision.

3.5 Other EIA regulations

This section summarizes the procedures of the other EIA regulations under which a large number of EISs have been prepared to date. We discuss the regulations in descending order of frequency of application to date (see Fig. 8.4):

Environmental Assessment (Scotland) Regulations 1988
Highways (AEE) Regulations 1988
Land Drainage Improvement Works (AEE) Regulations 1988
Electricity and Pipe-line Works (AEE) Regulations 1990
Environmental Assessment (Afforestation) Regulations 1988.

Environmental Assessment (Scotland) Regulations 1988 (SI 1221)

The EA (Scotland) Regulations apply to projects covered by the T&CP (Scotland) Act 1972 c. 52, and the T&CP (Development by Planning Authorities) (Scotland) Regulations 1981. They have five main sections, which apply to different types of development. Part II, Planning, resembles the English T&CP regulations, but allows four weeks for decisions by the LPA or SOS instead of three, and allows statutory consultees formally to withdraw from the consultation process. Part III, Electricity Applications, resembles the English electricity regulations (see below). Part IV, New Towns, allows development corporations to act as planning authorities for EIA purposes. Part V, Drainage Works, requires EIA for drainage works that LPAs or statutory bodies consider to be potentially environmentally harmful. Part VI, Trunk Road Projects, requires the SOS to decide whether or not a road proposal is subject to Directive 85/337: if so, the SOS must prepare an EIS, make it available for public consultation and allow time to receive representations before coming to a decision. The EA (Scotland) Regulations have schedules very similar to those of the English T&CP Regulations, except that nuclear power stations are included in Schedule 1.

Highways (Assessment of Environmental Effects) Regulations 1988 (SI 1241)

The Highways (AEE) Regulations apply to motorways and trunk roads proposed by the DOT. The regulations amend the Highways Act 1980 by inserting a new Section 105A, which requires the SOS for Transport to publish an EIS for the proposed route when draft orders for certain new highways, or major improvements to existing highways, are published. The SOS determines whether the proposed project comes under Annex I or Annex II of Directive 85/337, and whether an EIA is needed.

The regulations require an EIS to contain:

- a description of the published scheme and its site;
- a description of measures proposed to mitigate adverse environmental effects;

- sufficient data to identify and assess the main effects that the scheme is likely to have on the environment;
- a non-technical summary.

Before 1993, the requirements of the Highways (AEE) Regulations were further elaborated in DOT standard AD 18/88 (DOT 1989) and the *Manual of Environmental Appraisal* (DOT 1983). In response to strong criticism,[4] particularly by the Standing Advisory Committee on Trunk Road Assessment (1992), these were superseded in 1993 by the *Design manual for roads and bridges,* vol II: *Environmental assessment* (DOT 1993). The manual proposes a three-stage EIA process and gives extensive, detailed advice on how these EIAs should be carried out. It is discussed further in Chapter 10.

Land Drainage Improvement Works (Assessment of Environmental Effects) Regulations 1988 (SI 1217)

The Land Drainage Improvement Works (AEE) Regulations apply to almost all watercourses in England and Wales except public-health sewers. If a drainage body (including a local authority acting as a drainage body) determines that its proposed improvement actions are likely to have a significant environmental effect, it must publish a description of the proposed actions in two local newspapers and indicate whether it intends to prepare an EIS. If it does not intend to prepare one, the public can make representations within 28 days concerning any possible environmental impacts of the proposal; if no representations are made, the drainage body can proceed without an EIS. If representations are made, but the drainage body still wants to proceed without an EIS, MAFF gives a decision on the issue at ministerial level.

The contents required of the EIS under these regulations are virtually identical to those under the T&CP Regulations. When the EIS is complete, the drainage body must publish a notice in two local newspapers, send copies to English Nature, the Countryside Commission and any other relevant bodies and make copies of the EIS available at a reasonable charge. Representations must be made within 28 days and are considered by the drainage body in making its decision. If all objections are then withdrawn, the works can proceed; otherwise the minister gives a decision. Overall, these regulations are considerably weaker than the T&CP Regulations, because of their weighting in favour of consent, unless objections are raised, and their minimal requirements for consultation with environmental organizations.

Electricity and Pipe-line Works (Assessment of Environmental Effects) Regulations 1990 (SI 442)

The Electricity and Pipe-line Works (AEE) Regulations require an EIA to be carried out for the construction or extension of all nuclear power stations and of other generating stations of 300 MW or more in England and Wales under §36 of the Electricity Act 1989. They also require an EIA for:

- the construction or extension of a non-nuclear generating station of less than 300 MW in England and Wales under §36 of the Electricity Act;
- the installation of an overhead transmission line in England or Wales under §37 of the Electricity Act;
- the construction or diversion of a pipeline in Great Britain under the Pipe-lines Act 1962, where the SOS for the Department of Trade and Industry determines that the development would be "likely to have significant effects on the environment by virtue of factors such as its nature, size or location".

The regulations allow a developer to make a written request to the SOS to decide whether an EIA is needed. The SOS must consult with the LPA before making a decision. When a developer gives notice that an EIS is being prepared, the SOS must notify the LPA or the principal council for the relevant area, the Countryside Commission, English Nature, and the Environment Agency in the case of a power station, so that they can provide relevant information to the applicant.

The contents required of the EIA are almost identical to those listed in the T&CP Regulations. When the EIS and the application are handed in, the developer must publish a notice in one or more local papers for two successive weeks, giving details of the proposed project and/or a local address where copies of the EIS can be obtained. The SOS must advise the statutory consultees that the EIS has been completed, determine whether they want copies of the EIS, and inform them that they may make representations. However, the regulations have no clear procedures for consultation with environmental organizations or the public after the EIS is published. Chapter 10 provides further discussion.

Environmental Assessment (Afforestation) Regulations 1988 (SI 1207)

These regulations, which are further explained in a booklet by the Forestry Authority entitled *Environmental assessment of afforestation projects*, apply to applications for grants or loans for afforestation projects, where the Forestry Authority thinks that they are likely to have significant environmental impacts. Afforestation projects that do not require a grant and projects carried out by the Forestry Authority itself do not require EIAS.

When the Forestry Authority receives an application for a forestry planting grant, it informs the applicant if an EIA is needed. An EIA is likely to be needed for any new planting in a national nature reserve or an SSSI, or if the planting is expected to have a major impact because of its size, nature or location. In case of disagreement the applicant may appeal to the minister of the MAFF, SOS for Scotland or SOS for Wales. The contents required of the EIA are almost identical to those required by the T&CP Regulations. The lack of EIA requirements for forestry projects that do not require a grant, the fact that the Forestry Authority reviews EIAS despite its primary role as a promoter of forestry and afforestation and the lack of any requirement that the Forestry Authority should carry out EIAS on its own projects have all been criticized (e.g. by the CPRE (CPRE 1991)).

3.6 UK implementation of Directive 97/11/EC

The UK government has, to date, produced two consultation papers on the implementation of Directive 97/11/EC (DETR, 1997a, 1997b). These papers indicate a generally positive response to the implementation of the amendments:

> The Government believes that the Directive represents a significant improvement on the original, particularly in clarifying a number of ambiguities. The new measures should improve the consistency with which the environment is taken into account in major development decisions throughout the Community. At the same time, the Directive offers Member States sufficient flexibility to achieve this without adding unnecessarily to bureaucracy or burdens on developers. . . . Wherever possible, new measures will be incorporated into existing procedures. (DETR 1997a)

The implementation of the new and amended project classes can, in most cases, be easily incorporated within existing consent systems. Revised screening procedures are proposed to include both new "exclusive" thresholds and enhanced indicative thresholds (see Ch. 4 for further discussion). Both sets of thresholds would take account of the new criteria in Annex III of the Directive. The government also proposes, for the Town and Country Planning System, to require LPAS to place a record of decisions on the need for EIA on the planning register. On scoping, while the government will continue to encourage developers to consult informally with the relevant competent authorities, it will also legislate to allow a developer to ask the competent authority for a formal opinion, within a specified time frame, on the information that should be included in a particular EIS. The applicant will also be required to provide an outline of the main alternatives studied, and LPAS will be required to publicize reasons when consent is given for a project which has been subject to EIA. The amended legislation will also incorporate the requirements of the UNECE Convention on EIA in a Transboundary Context (the Espoo Convention) which was negotiated after 85/337 EEC was adopted.

The draft Town and Country Planning (Assessment of Environment Effects) Regulations (DETR 1998) will consolidate parts of the previous wide array of legislation. They also represent a clearer coming together of the EC Directive and the Regulations. In particular, Schedules 1, 3 and 4 of the amended Regulations replicate very closely Annexes I, III and IV of the Directive. Schedule 2 has only very minor modifications from Annex II; primarily in 2.10, where (b) includes sports stadia, leisure centres and multiplex cinemas. Also, there is a separate category (n) for motorway service areas, and a few other categories are split or relocated. Schedule 2.12 also includes an additional category (f) for golf courses and associated developments. The UK Government intends to review the amended procedures after two years of operation.

3.7 Summary

Directive 85/337 has been implemented in the UK through over forty regulations that link those involved – developers, affected parties, regulators and facilitators – in a variety of ways. The Town and Country Planning (Assessment of Environmental Effects) Regulations are central. Other regulations cover projects that do not fall under the English and Welsh planning system, such as motorways and trunk roads, power stations, electricity transmission lines, pipelines, land drainage works, afforestation projects, development projects in Scotland and Northern Ireland, and amendments to some of these regulations. The Planning and Compensation Act 1991 allows other projects not listed in Directive 85/337 also to be subject to EIA.

The 1988 T&CP Regulations included three schedules, which broadly corresponded to the Annexes of EC Directive 85/337. Schedule 3, which listed the contents required of an EIS, distinguishes between mandatory "specified information" and discretionary "further information"; this distinction weakened the UK's EIA system. LPAs have discretion to determine which Schedule 2 projects require EIA and to recommend the contents of the EIS, but the developer is ultimately responsible for preparing the EIS. Statutory consultees must be sent copies of the EIS, and the public must be allowed to purchase copies; both groups can make representations to the LPA about the EIS. The LPA must consider the EIS and any representations when deciding whether to grant or refuse planning permission. The developer can appeal to the SOS in cases of disagreement with the LPA. The other EIA regulations are broadly similar to those of the T&CP Regulations, although there are some differences, mainly about screening and public consultation.

The amendments to the Directive (97/11/EC) will be implemented from March 1999, and UK legislation will replicate even more closely the (now) four Annexes of the EC Directive, bringing a wider range of projects into the UK EIA system. The consideration of alternatives, scoping, consultation and public participation will, *inter alia*, have a higher profile in the evolving UK EIA system.

References

CEC (Commission of the European Communities) 1985. On the assessment of the effects of certain public and private projects on the environment. *Official Journal* L175, 5 July.

CPRE (Council for the Protection of Rural England) 1991. *The Environmental Assessment Directive: five years on.* London: CPRE.

DETR (Department of the Environment, Transport and the Regions) 1997a. *Consultation paper: implementation of EC Directive (97/11/EC) on environmental assessment.* London: DETR.

DETR 1997b. *Consultation paper: implementation of EC Directive (97/11/EC) – determining the need for environmental assessment.* London: DETR.

DETR 1998. *Draft Town and Country Planning (Assessment of Environmental Effects) Regulations 1998.* London: DETR.

DOE (Department of the Environment) 1989. *Environmental assessment: a guide to the procedures.* London: HMSO.

DOE 1991. *Monitoring environmental assessment and planning.* London: HMSO.

DOE 1994a. *Evaluation of environmental information for planning projects: a good practice guide.* London: HMSO.

DOE 1994b. *Good practice on the evaluation of environmental information for planning projects: research report.* London: HMSO.

DOE 1995. *Preparation of environmental statements for planning projects that require environmental assessment: a good practice guide.* London: HMSO.

DOE 1996. *Changes in the quality of environmental statements for planning projects.* London: HMSO.

DOT (Department of Transport) 1983. *Manual of environmental appraisal.* London: HMSO.

DOT 1989. *Departmental standard HD 18/88: environmental assessment under the EC Directive 85/337.* London: Department of Transport.

DOT 1993. *Design manual for roads and bridges,* vol. 11: *Environmental assessment.* London: HMSO.

ENDS 1992. *Directory of environmental consultants 1992/93.* London: Environmental Data Services.

ENDS 1993. *Directory of environmental consultants 1993/94.* London: Environmental Data Services.

ENDS 1995. *Directory of environmental consultants* 4th edn. London: Environmental Data Services.

ENDS 1997. *Directory of environmental consultants 1997/98.* London: Environmental Data Services.

ESRC 1987. *Newsletter: Environmental Issues.* London: Economic and Social Research Council, February.

Fortlage, C. 1990. *Environmental assessment: a practical guide.* Aldershot: Gower.

SACTRA (Standing Advisory Committee on Trunk Road Assessment) 1992. *Assessing the environmental impact of road schemes.* London: HMSO.

Sheate, W. R. 1995. Electricity generation and transmission: a case study of problematic EIA implementation in the UK. *Environmental Policy and Practice* 5(1), 17–25.

Weston, J. 1995. Consultants in the EIA process. *Environmental Policy and Practice* 5(3), 131–4.

Notes

1   There are some discrepancies. For instance, power stations of 300 MW or more are included in Schedule 1, although they actually fall under the Electricity and Pipe-line

81

Works (AEE) Regulations, and all "special roads" are included, although the regulations should apply to special roads under local authority jurisdiction.

2   The decision is actually made by the relevant regional office of the DOE; there are ten such offices. As will be discussed in Chapter 8, this has led to some discrepancies where two or more offices have made different decisions on very similar projects.

3   This includes enough copies for all the statutory consultees to whom the developer has not already sent copies, one copy for the LPA and several (dependent on the 1993 amendments to the T&CP Regulations) for the secretary of state.

4   The criticism was well deserved. The circular's assertion that ". . . individual highway schemes do not have a significant effect on climatic factors and, in most cases, are unlikely to have significant effects on soil or water" is particularly interesting in view of the cumulative impact of private transport on air quality.

Part 2

Process

This illustration by Neil Bennett is reproduced from Bowers, J. (1990), *Economics of the environment: the conservationist's response to the Pearce Report*, British Association of Nature Conservationists, 69 Regent Street, Wellington, Telford, Shropshire, TF1 1PE.

CHAPTER 4
Starting up; early stages

4.1 Introduction

This is the first of four chapters that discuss how an EIA is carried out. The focus throughout is on both the procedures required by UK legislation and the ideal of best practice. Although Chapters 4–7 seek to provide a logical step-by-step approach through the EIA process, there is no one exclusive approach. Process is set within an institutional context, and the context will vary from country to country (see Ch. 11). As already noted, even in one country, the UK, there may be a variety of regulations for different projects (see Chs 9 and 10). The various steps in the process can be taken in different sequences. Some may be completely missing in certain cases. Also, it is hoped, the process will not just be linear but build in cycles, with feedback from later stages to the earlier ones.

Chapter 4 covers the early stages of the EIA process. These include setting up a management process for the EIA activity, clarifying whether an EIA is required at all ("screening") and an outline of the extent of the EIA ("scoping"), which may involve consultation between several of the key actors outlined in Chapter 3. Early stages of EIA should also include an exploration of possible alternative approaches for a project, although this has not been a mandatory requirement in UK legislation (until 1999) and is missing from many studies. Baseline studies, setting out the parameters of the development action (including associated policy positions) and the present and future state of the environment involved, are also included in Chapter 4. However, the main section in the chapter is devoted to impact identification. This is important in the early stages of the process, but, reflecting the cyclical, interactive nature of the process, some of the impact identification methods discussed here may also be employed in the later stages. Conversely, some of the prediction, evaluation, communication and mitigation approaches discussed in Chapter 5 can be used in the early stages, as can the participation approaches outlined in Chapter 6. The discussion in this chapter starts, however, with a brief introduction to the management of the EIA process.

4.2 Managing the EIA process

The EIA process is a management-intensive process. EIAs often deal with major (and sometimes poorly defined) projects, with many wide-ranging and often controversial impacts. As we noted in Chapter 3, they can involve many participants with very different perspectives on the relative merits and impacts of projects. It is important that the EIA process is well managed. This section notes some of the elements involved in such management.

The EIA process invariably involves an *interdisciplinary* team approach. Early US legislation strongly advocated such an approach:

> Environmental impact statements shall be prepared using an interdisciplinary approach which will ensure the integrated use of the natural and social sciences and the environmental design arts. The disciplines of the preparers shall be appropriate to the scope and issues identified in the scoping approach. (CEQ 1978, para. 1502.6)

Such an interdisciplinary approach not only reflects the normal scope of EIA studies, from the biophysical to the socio-economic, as we note elsewhere in this book, but also brings to the process the advantages of multiple viewpoints and perspectives on the complex issues involved (Canter 1991).

The *team* producing the Environmental Impact Statement may be one, or a combination, of proponent in-house, lead external consultant, external sub-consultants and individual specialists. The size of the team may vary from two (one person, although sometimes used, does not constitute a team), to twelve, and even larger for some projects; the average is three or four. Fortlage (1990) identified 17 relevant specialist types, including town planner, ecologist, chemist, archaeologist and lawyer. A team should cover the main issues involved. A small team of three could, as exemplified by Canter, cover the areas of physical/chemical, biological/ecology and cultural/socio-economic, with a membership that might include, for example, an environmental engineer, an ecologist and a planner, at least one member having training or experience in environmental impact assessment and management *per se*. However, the finalization of a team's membership may be possible only after an initial scoping exercise has been undertaken.

Many EIA teams make a clear distinction between a "core/focal" management team and associated specialists, often reflecting the fact that no one organization can cover all the inputs needed in the production of an ES for a major project. Some commentators (see Weaver et al. 1996) promote the virtues of this approach. On a study for a major open-cast mining project in South Africa, Weaver et al. had a core project team of five people: a project manager, two senior authors, an editorial consultant and a word processor. This team managed the inputs into the EIA process, co-ordinating over sixty scientific and non-scientific contributors, and organized various public participation and liaison programmes.

The team project manager obviously has a pivotal role. In addition to personnel and team management skills, the manager should have a broad appreciation of the project type under consideration, a knowledge of the relevant processes and impacts subject to EIA, the ability to identify important issues and preferably a substantial area of expertise. Petts and Eduljee (1994) identify the following core roles for a project manager:

- selecting an appropriate project team;
- managing specialist inputs;
- liaising with the people involved in the process;
- managing change in the internal and external environment of the project;
- co-ordinating the contributions of the team in the various documentary outputs.

The management team has to *co-ordinate resources* – information, people and equipment – to achieve an EIA study of quality, on time and within its budget. Budgets vary, as we have noted elsewhere (see Chs 1 and 8), but may involve major expenditure for large projects. The time available may also vary; the average is 4–6 months, but it could be much longer for complex projects. National and international quality assurance procedures (for example British Standard 5750/ISO 9000) may also apply for the activities of many companies.

In interdisciplinary team work, *complementarity, comparability and co-ordination* are particularly important. Weaver et al. (1996) stress the importance of complementarity for the technical skills needed to compete the task, and of personal skills for those in the core management team. Fortlage and others rightly stress that where there are various groups of consultants, it is important that findings and data are co-ordinated (e.g. that they should work to agreed map scales and to agreed chapter formats) and can be fed into a central source. "This is one of the weakest aspects of most assessment teams; all consultants must be aware, and stay aware of others' work in order to avoid lacunae, anomalies and contradictions which will be the delight of opposing counsel and the media" (Fortlage 1990).

Of course basic management skills – including team management and time management – must not be overlooked. Cleland & Kerzner (1986) suggested the following factors were important in the successful management of an interdisciplinary team:

(a) a clear, concise statement of the mission or purpose of the team;

(b) a summary of the goals or milestones that the team is expected to accomplish in planning and conducting the environmental impact study;

(c) a meaningful identification of the major tasks required to accomplish the team's purposes, with each task broken down by individual;

(d) a summary delineation of the strategy of the team relative to policies, programs, procedures, plans, budgets, and other resource allocation methods required in the conduct of the environmental impact study;

(e) a statement of the team's organizational design, with information included on the roles and authority and responsibility of all members of the team, including the team leader; and

(f) a clear delineation of the human and non-human resource support services available for usage by the interdisciplinary team.

Good practice also recommends *a clear documentation of the interdisciplinary team approach* in the EIS. This would indicate the specific roles of team members, and their titles, qualifications and experience. The nature of liaison with other parties in the process, including public and other meetings, should also ideally be noted.

4.3   Project screening – is an EIA needed?

The number of projects that could be subject to EIA is potentially very large. Yet many projects have no substantial or significant environmental impact. A screening mechanism seeks to focus on those projects with potentially significant adverse environmental impacts or whose impacts are not fully known. Those with few or no impacts are screened out and allowed to proceed to the normal planning permission and administrative processes without any additional assessment or additional loss of time and expense.

Screening can be partly determined by the EIA regulations operating in a country at the time of an assessment. Chapter 3 indicated that in the EC, including the UK, there are some projects (Annex/Schedule 1) that will always be screened out for full assessment, by virtue of their scale and potential environmental impacts (for example a crude-oil refinery, a sizeable thermal power station, a special road). There are many other projects (Annex/Schedule 2) for which the screening decision is less clear. Here two examples of a particular project may be screened in different ways (one "in", one "out" for full assessment) by virtue of a combination of criteria, including project scale, the sensitivity of the proposed location and the expectation of adverse environmental impacts. Chapter 9 provides examples of variations in the interpretation of need for EISs for new settlements in the UK. In such cases it is important to have working guidelines, indicative criteria and thresholds on conditions considered likely to give rise to significant environmental impacts (see Section 3.4).

In California, the list of projects that must always have the full review is determined by project type, development and location. For example, type includes, *inter alia*, a proposed local general plan; development includes, *inter alia*, a residential development of more than five hundred units, a hotel or motel of more than five hundred rooms, a commercial office building of more than 250,000 square feet of floor space; location includes, *inter alia*, the Lake Tahoe Basin, the California Coastal Zone, an area within a quarter of a mile of a wild and scenic area (State of California 1992). This constitutes an "inclusion list" approach. In addition there may be an "exclusion list", as used in California and Canada, identifying these categories of project for which an EIA is not required because experience has shown that the adverse effects are not significant.

Table 4.1 Thresholds vs. case-by-case approach to screening: advantages and disadvantages.

Advantages	Disadvantages
Thresholds	
Simple to use	Place arbitrary, inflexible rules on a variable environment (unless tiered)
Quick to use; more certainty	Less room for common sense or good judgement
Consistent between locations	May be or become inconsistent with relevant neighbours
Consistent between decisions within locations	Difficult to set and, once set, difficult to change
Consistent between project types	Lead to a proliferation of projects lying just below the thresholds
Case by Case	
Allows common sense and good judgement	Likely to be complex and ambiguous
Flexible – can incorporate variety in project and environment	Likely to be slow and costly
Can evolve (and improve) easily	Open to abuse by decision-makers because of political or financial interests
	Open to poor judgement of decision-makers
	Likely to be swayed by precedent and therefore lose flexibility

Some EIA procedures include an initial outline EIA study to check on likely environmental impacts and on their significance. Under the California Environmental Quality Act a "negative declaration" can be produced by the project proponent, thereby claiming that the project has minimal significant effects and does not require a full EIA. The declaration must be substantiated by an initial study, which is usually a simple checklist against which environmental impacts must be ticked as yes, maybe or no. If the responses are primarily no, and most of the yes and maybe responses can be mitigated, then the project may be screened out from a full EIA. In Canada and Australia, the screening procedures are also well developed (see Ch. 11).

In general there are two main approaches to screening. The *use of thresholds* involves placing projects in categories and setting thresholds for each project type. These may relate, for example, to project characteristics (e.g. 20 ha and over), to anticipated project impacts (e.g. the production of 50,000 tonnes or more of waste per annum to be taken from a site) and to project location (e.g. a designated landscape area). A *case by case* approach involves the appraisal of the characteristics of projects, as they are submitted for screening, against a checklist of guidelines and criteria. Some of the advantages and disadvantages of these two approaches are summarized in Table 4.1. In practice, there are many hybrid approaches with, for

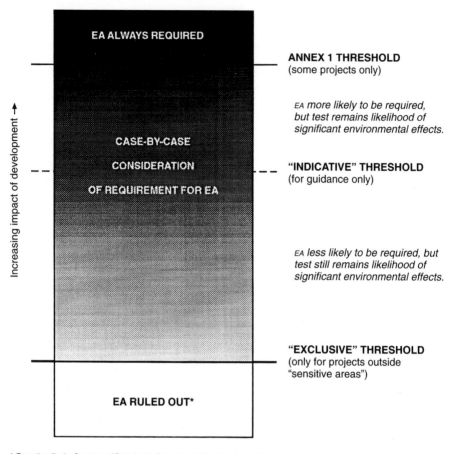

* Exceptionally, the Secretary of State (or the Department of the Environment in
Northern Ireland) may direct that a project requires EA even though it is below the
"exclusive" threshold and outside a "sensitive area".

Figure 4.1   An illustrative guide to the thresholds system. (*Source:* DETR 1998)

example, indicative thresholds used in combination with a flexible case by case
approach. Figure 4.1 provides an illustrative guide to a threshold system proposed
by the UK government for the implementation of the amended Directive 97/11/EC
(DETR 1997).

4.4   Scoping – which impacts and issues to consider?

The scope of an EIA is the impacts and issues it addresses. The process of scoping
is that of deciding, from all of a project's possible impacts and from all the alternatives that could be addressed, which are the significant ones. An initial scoping of

possible impacts may identify those impacts thought to be potentially significant, those thought to be not significant and those whose significance is unclear. Further study should examine impacts in the various categories. Those confirmed by such a study to be not significant are eliminated; those in the uncertain category are added to the initial category of other potentially significant impacts. This refining of focus onto the most significant impacts continues throughout the EIA process.

Scoping is generally carried out in discussions between the developer, the competent authority, other relevant agencies and, ideally, the public. It is often the first stage of negotiations and consultation between a developer and other interested parties. It is an important step in EIA because it enables the limited resources of the team preparing an EIA to be allocated to best effect, and prevents misunderstanding between the parties concerned about the information required in an EIS. Scoping can also identify issues that should later be monitored. Although it is an important step in the EIA process, it has not been a legally mandated step in the UK. The Department of the Environment (now DETR) recommends that a developer should consult with the competent authority and statutory consultees before preparing an EIS, but in practice this happens in only about half of all cases (Fuller 1992). This lack of early discussion is one of the principal limitations to effective EIA to date.

Scoping should begin with the identification of individuals, communities, local authorities and statutory consultees likely to be affected by the project; good practice would be to bring them together in a working group and/or meetings with the developer. One or more of the impact identification techniques discussed in Section 4.8 can be used to structure a discussion and suggest important issues to consider. Other issues could include particularly valued environmental attributes, those impacts considered of particular concern to the affected parties and social, economic, political and environmental issues related to the locality. Reference should be made to relevant structure plans, local plans, subject plans and government policies and guidelines, which we discuss in Section 4.7. Various alternatives should be considered, as discussed in Section 4.5. The result of this process of information collection and negotiation should be the identification of the chief issues and impacts, an explanation of why other issues are not considered significant, and, for each key impact, a defined temporal and spatial boundary within which it will be measured. Some developers, such as the Highways Agency for England, produce a scoping report as a matter of good practice. This indicates the proposed coverage of the EIA and the uncertainties that have been identified and can act as a basis for further studies and for public participation.

As we shall discuss in Chapter 11, other countries (e.g. Canada and the Netherlands) have a formal scoping stage, in which the developer agrees with the competent authority or an independent EIA commission, sometimes after public consultation, on the subjects the EIA will cover. Increasingly, sources of guidance on impacts normally associated with particular types of projects are being developed by various government and other regulatory agencies (see, for example, Environment Agency (1996) *Scoping guidance notes* (UK), and Government of New South Wales (1996), *EIS guidelines*).

91

As part of its five-year review of Directive 85/337, the EC proposed the inclusion of a mandatory scoping stage. As we noted in Chapters 2 and 3, there will be a provision for mandatory scoping in the amended Directive if Member States so wish. The importance of scoping and consultation early in the EIA process was highlighted in a recent research report for the UK Department of the Environment (DOE 1996). While the research did show some positive change to the use of scoping and consultation early in the process, the activity was still unsatisfactory in almost half the post-1991 cases investigated. The research identified such early consultation and scoping as very important for the quality of the EIS, for all participants in the EIA process. Indeed it can be argued that one of the most valuable roles of the EIA process is to encourage such consultation.

4.5   The consideration of alternatives

If a project is not screened out and is believed to have potentially significant impacts on the environment, then an EIA is undertaken for the project *and*, ideally, for feasible alternatives. During the course of project planning, many decisions are made concerning the type and scale of the project proposed, its location, and the processes involved. Most of the possible alternatives that arise will be rejected by the developer on economic, technical or regulatory grounds. The role of EIA is to ensure that environmental criteria are also considered at these early stages. The US Council on Environmental Quality (CEQ 1978) calls the discussion of alternatives "the heart of the environmental impact statement": how an EIA addresses alternatives will determine its relation to the subsequent decision-making process. A discussion of alternatives ensures that the developer has considered both other approaches to the project and the means of preventing environmental damage. A consideration of alternatives also encourages analysts to focus on the *differences* between real choices. It can allow people who were not directly involved in the decision-making process to evaluate various aspects of a proposed project and how decisions were arrived at. It also provides a framework for the competent authority's decision, rather than merely a justification for a particular action. Finally, if unforeseen difficulties arise during the construction or operation of a project, a re-examination of these alternatives may help to provide rapid and cost-effective solutions.

UK regulatory requirements

The original EC Directive 85/337 stated that alternative proposals should be considered in an EIA, subject to the requirements of Article 5 (if the information is relevant and if the developer may reasonably be required to compile this information). Annex III required "where appropriate, an outline of the main alternatives studied by the developer and an indication of the main reasons for this choice, taking into account the environmental effects". In the UK, this requirement has been interpreted as discretionary. The Town and Country Planning (Assessment of Environmental Effects) Regulations 1988 noted that:

> An environmental statement may include, by way of explanation or amplification of any specified information, further information on . . . (in outline) the main alternatives (if any) studied by the applicant, appellant or authority and an indication of the main reasons for choosing the development proposed, taking into account the environmental effects.

With minor changes of wording, this clause is repeated in the other UK EIA regulations. To date in the UK, about two-thirds of EISs have not considered alternatives at all.[1] The one-third of EISs that have considered alternatives have mostly been for linear developments (e.g. roads, rail, transmission lines) that consider different routes between two given points. A few others, particularly for energy projects, mention alternative sites that have been considered for development projects; some EISs for extraction projects also include alternatives of scale and operating conditions. The Department of Transport's *Manual of environmental appraisal* (DOT 1983) required that alternatives should be considered: without this requirement, which was in existence before Directive 85/337 became operational, very few EISs would consider alternatives at all. However, the amended Directive (CEC 1997) "requires environmental statements to include an outline of the main alternatives studied by the developer, and an indication of the main reasons for the developer's choice, taking into account the environmental effects". In the UK, since the mid-1990s, government good practice guidance (DOE 1994) has encouraged developers to show what alternatives (of site or process selection), if any, have been considered. It is recognized that the early and proper consideration of alternatives can be very useful when presenting a robust application for development consent. Under the amended Directive, where such alternatives have been considered, developers will have to include information on them in the environmental statement.

Types of alternative

A thorough consideration of alternatives would begin early in the planning process, before the type and scale of development and its location have been agreed on. A number of broad types of alternative can be considered: the "no action" option, alternative locations, alternative scales of the project, alternative processes or equipment, alternative site layouts, alternative operating conditions, alternative ways of dealing with environmental impacts. We shall discuss the last of these in Section 5.4.

The *"no action" option* refers to environmental conditions if a project were not to go ahead. Consideration of this option is required in some countries, e.g. the USA.[2] In essence, consideration of the "no action" option is equivalent to a discussion of the need for the project: do the benefits of the project outweigh its costs? This option is rarely discussed in UK EISS.

The consideration of alternative *locations* is an essential component of the project planning process. In some cases, a project's location is constrained in varying degrees: for instance, gravel extraction can take place only in areas with sufficient gravel deposits, and wind-farms require locations with sufficient wind speed. In other cases, the best location can be chosen to maximize, for example, economic, planning and environmental considerations. For industrial projects, for instance, economic criteria such as land values, the availability of infrastructure, the distance from sources and markets, and the labour supply, are likely to be important (Fortlage 1990). For road projects, engineering criteria strongly influence the alignment. In all these cases, however, siting in "environmentally robust" areas, or away from designated or environmentally sensitive areas, should be considered.

The consideration of different *scales* of development is also integral to project planning. In some cases, a project's scale will be flexible. For instance, the scale of a waste-disposal site can be changed, depending, for example, on the demand for landfill space, the availability of other sites and the presence of nearby residences or environmentally sensitive sites. The number of turbines on a wind-farm could vary widely. In other cases, the developer will need to decide whether an entire unit should be built or not. For instance, the reactor building of a PWR nuclear power station is a large discrete structure that cannot easily be scaled down. Pipelines or bridges, to be functional, cannot be broken down into smaller sections.

Alternative *processes and equipment* involve the possibility of achieving the same objective by a different method. For instance, 1500 MW of electricity can be generated by one combined-cycle gas turbine power station, by a tidal barrage, by several waste-burning power stations or, in the extreme, by thousands of wind turbines. Gravel can be directly extracted or recycled. Waste may be incinerated or put in a landfill.

Once the location, scale and processes of a development have been decided upon, different *site layouts* can still have different impacts. For instance, noisy plant can be sited near or away from residences. Power-station cooling towers can be few and tall (using less land) or many and short (causing less visual impact). Buildings can be sited either prominently or to minimize their visual impact. Similarly, *operating conditions* can be changed to minimize impacts. For instance, a level of noise at night is usually more annoying than the same level during the day, so night-time work could be avoided. Establishing designated routes for project-related traffic can help to minimize disturbance to local residents. Construction can take place at times of the year that minimize environmental impacts, for example on migratory and nesting birds.

The presentation and comparison of alternatives

The costs of alternatives vary for different groups of people and for different environmental components. Discussions with local residents, statutory consultees and special interest groups may rapidly eliminate some alternatives from consideration and suggest others. However, it is unlikely that one alternative will emerge as being most acceptable to all the parties concerned. The EIS should distil information about a reasonable number of realistic alternatives into a format that will facilitate public discussion and, finally, decision-making. Methods for comparing and presenting alternatives span the range from simple, non-quantitative descriptions, through increasing levels of quantification, to a complete translation of all impacts into their monetary values.

To date in the UK, those EISs for non-road proposals that *have* discussed alternatives, have merely described them, their main impacts and the reasons for their rejection. Many of the impact identification methods discussed later in this chapter are relevant to this stage of decision-making. Overlay maps compare the impacts of various locations in a non-quantitative manner. Checklists or less complex matrices can also be applied to various alternatives and compared; this may be the most effective way to present the impacts of alternatives visually. Some of the other techniques used for impact identification – the threshold of concern checklist, weighted matrix, and EES – allow alternatives to be implicitly compared. Broadly they do this by assigning quantitative importance weightings to environmental components, rating each alternative (quantitatively) according to its impact on each environmental component, multiplying the ratings by their weightings to obtain a weighted impact, and aggregating these weighted impacts to obtain a total score for each alternative. These scores do not correspond to real-life monetary value, but can be compared with each other to identify preferable alternatives. With the exception of the threshold-of-concern checklist, they do not lend themselves to the clear presentation of the alternatives in question, and none of them clearly states who will be affected by the different alternatives.

The UK Department of Transport tries to tread an uneasy middle path between the various techniques. Its *Manual of environmental appraisal* (DOT 1983) and its replacement, the *Design manual for roads and bridges*, vol. II (DOT 1993) use a framework to appraise the impacts of road proposals against a "no action" option for various affected groups (see Section 10.2 for further details).

4.6 Understanding the project/development action

Understanding the dimensions of the project

Schedule 3 of the Town and Country Planning (Assessment of Environmental Effects) Regulations 1988 (and as amended) requires "a description of the development proposed, comprising information about the site and the design and scale or size of the development" and "the data necessary to identify and assess the main effects which that development is likely to have on the environment". *Environmental assessment: a guide to the procedures* (DOE 1989) provides a brief listing of information that may be used to describe the project. At first glance, this description of a proposed development would appear to be one of the more straightforward steps in the EIA process. However, projects have many dimensions, and relevant information may be limited. As a consequence, this first step may pose some challenges. Crucial dimensions to be clarified include the purpose of the project, its life-cycle, physical presence, process(es), policy context and associated policies.

An outline of *the purpose and rationale* of a project provides a useful introduction to the project description. This may, for example, set the particular project in a wider context – the missing section of a major motorway, a power station in a programme of developments, a new settlement in an area of major population growth. A discussion of purpose may include the rationale for the particular type of project, for the choice of the project's location and for the timing of the development. It may also provide background information on planning and design activities to date.

As we noted in Section 1.4, all projects have a *life-cycle of activities*, and a project description should clarify the various stages in the life-cycle, and their relative duration, of the project under consideration. A minimum description would usually involve the identification of construction and operational stages and associated activities. Further refinement might include planning and design, project commissioning, expansion, close-down and site rehabilitation stages. The size of the development at various stages in its life-cycle should also be specified. This can include reference to inputs, outputs, physical size and the number of people to be employed.

The *location and physical presence* of a project should also be clarified at an early stage. This should include its general location on a base map in relation to other activities and to administrative areas. A more detailed site layout of the proposed development, again on a large-scale base map, should illustrate the land area and the main disposition of the elements of the project (e.g. storage areas, main processing plant, waste-collection areas, transport connections to the site). Where the site layout may change substantially between different stages in the life-cycle, it is valuable to have a sequence of anticipated layouts. Location and physical presence should also identify elements of a project that, although integral, may be detached from the main site (e.g. the construction of a barrage in one area may involve opening up a major quarry development in another area). A description of the physical presence of a project is invariably improved by a three-dimensional visual image, which may include a photo-montage of what the site layout may look like at, for example, full operation. A clear presentation of location and physical presence is important for an assessment of change in land-uses, any physical disruption to other infrastructures, severance of activities (e.g. agricultural holdings, villages) and visual intrusion and landscape changes.

Understanding a project also involves an understanding of the *processes* integral to it. The nature of processes varies between industrial, service and infrastructure projects, but many can be described as a flow of inputs through a process and their transformation into outputs. The nature, origins and destinations of the inputs and outputs and the timescale over which they are expected should be identified. This systematic identification should be undertaken for both physical and socio-economic characteristics, although the interaction should be clearly recognized, with many of the socio-economic characteristics following from the physical.

Physical characteristics may include:

- the land take and physical transformation of a site (e.g. clearing, grading), which may vary between different stages of a project's life-cycle;
- the total operation of the process involved (usually illustrated with a process-flow diagram);
- the types and quantities of resources used (e.g. water abstraction, minerals, energy);
- transport requirements (of inputs and outputs);
- the generation of wastes, including estimates of types, quantity and strength of aqueous wastes, gaseous and particulate emissions, solid wastes, noise and vibration, heat and light, radiation, etc;
- the potential for accidents, hazards and emergencies;
- processes for the containment, treatment and disposal of wastes and for the containment and handling of accidents; monitoring and surveillance systems.

Socio-economic characteristics may include:

- the labour requirements of a project – including size, duration, sources, particular skills categories and training;
- the provision or otherwise of housing, transport, health and other services for the workforce;
- the direct services required from local businesses or other commercial organizations;
- the flow of expenditure from the project into the wider community (from the employees and subcontracting);
- the flow of social activities (service demands, community participation, community conflict).

Figure 4.2 shows the interaction between the physical (ecological in this case) and socio-economic processes that may be associated with an industrial plant.

Projects should also be seen in their *planning policy context*. In the UK, the main local policy context is outlined and detailed in structure and local plans. The description of location must pay regard to land-use designations and development constraints that may be implicit in some of the designations. Of particular importance is a project's location in relation to various environmental zones (e.g. areas of outstanding natural beauty (AONBs), sites of special scientific interest (SSSIs), green belts and local and national nature reserves). Attention should also be given to national planning guidance, provided in the UK by an important set of Department of Environment, Transport and Regions Planning Policy Guidance Notes (PPGs).

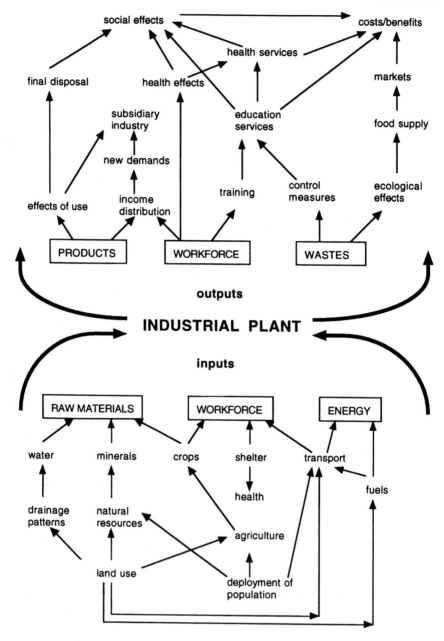

Figure 4.2 Interaction between an industrial plant and its socio-ecological environment. (*Source:* Marstrand 1976)

The projects may also have *associated policies*, not obvious from site layouts and process-flow diagrams, that are nevertheless significant for subsequent impacts. For example, shift-working will have implications for transport and noise that may be very significant for nearby residents. The use of a construction site hostel, camp or village can significantly internalize impacts on the local housing market and on the local community. The provision of on- or off-site training can greatly affect the mixture of local and non-local labour and the balance of socio-economic effects.

Types and sources of data

Various *types of data* are used. The life-cycle of a project can be illustrated on a linear bar chart. Particular stages may be identified in more detail where the impacts are considered of particular significance; this is often the case for the construction stage of major projects. Location and physical presence are best illustrated on a map base, with varying scales to move from the broad location to the specific site layout. This may be supplemented by aerial photographs, photo-montages and visual mock-ups according to the resources and issues involved (see Figs 4.3–4.6).

A process diagram for the different activities associated with a project should accompany the location and site-layout maps. This may be presented in the form of a simplified pictorial diagram or in a block flow chart. The latter can be presented simply to show the main interconnections between the elements of a project (see Fig. 4.4 for socio-economic processes) or in sufficient detail to provide a comprehensive picture. Figure 4.5 shows a materials flow chart for a petroleum refinery; it outlines all the raw materials, additives, end-products, byproducts and atmospheric, liquid and solid wastes. A comprehensive flow chart of a production process should include the types, quantities and locations of resource inputs, intermediate and final product outputs and wastes generated by the total process.

The various information and illustrations should clearly identify the main variations between a project's stages. Figure 4.6 illustrates a labour-requirements diagram that identifies the widely differing requirements, in absolute numbers and in skill categories, of the construction and operational stages. In addition, more sophisticated flow diagrams could indicate the type, frequency (normal, batch, intermittent or emergency) and duration (minutes or hours per day or week) of each operation. Seasonal and material variations, including time periods of peak pollution loads, can also be documented.

The form and *sources of data* vary according to the degree of detail required and the stage in the assessment process. Site-layout diagrams and process-flow charts may be only in outline, provisional form at the initial design stage. Subsequent investigations and the identification of sources of potential significant impacts may lead to changes in layout and process.

The initial brief from the developer provides the starting point. Ideally, the developer may have detailed knowledge of the proposed project's characteristics, likely

99

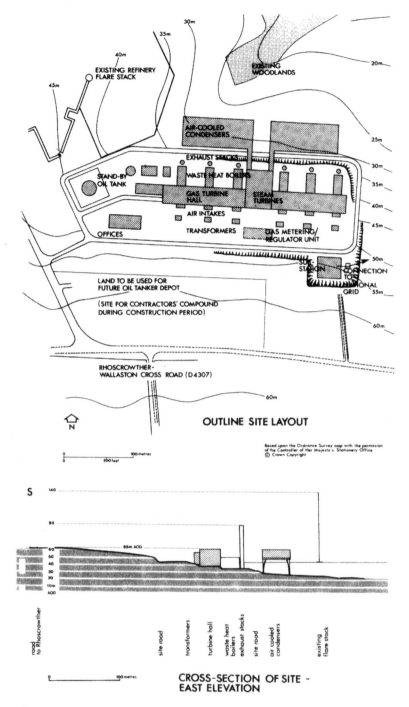

Figure 4.3 Example of a project site layout. (*Source:* Rendel Planning 1990, *Angle Bay Energy Project environmental statement*)

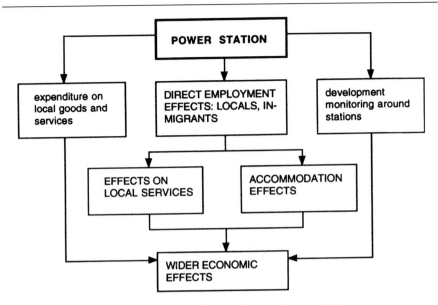

Figure 4.4 Socio-economic process diagram for a major project.

layout and production processes, drawing on previous experience. With the rapid development of EIA and the production of EISs, an analyst can also supplement such information with reference to comparative studies as sources for project profiles, although the availability of such statements in the UK is still far from satisfactory, and their predictions are untested (see Chs 7 and 8). Use can also be made of other published data (e.g. published emission and effluent factors for the components of a project, published data on accident rates; TNO 1983, VROM 1985). Site visits can be made to comparable projects, and advice can be gained from consultants with experience of the type of project under consideration. As the project design and assessment process develops, so the developer will have to provide more detailed information on the characteristics specific to the project.

Even in the ideal situation, there will be considerable interaction between the analyst and developer to refine the project's characteristics. Unfortunately, the situation may often be far from ideal; Mills (1992) and Frost (1994) provide interesting examples of major changes from the project description in EISs to the actual implemented action. The initial brief may leave a lot to be desired, and the analyst will have to draw on the other sources noted above to clarify the details of the project. The analyst may also draw on EIA literature (books and journals), guidelines, manuals and statistical sources. Further useful sources include United Nations Environment Programme (1981), Lee (1987), Wood & Lee (1987) and CEC (1993).

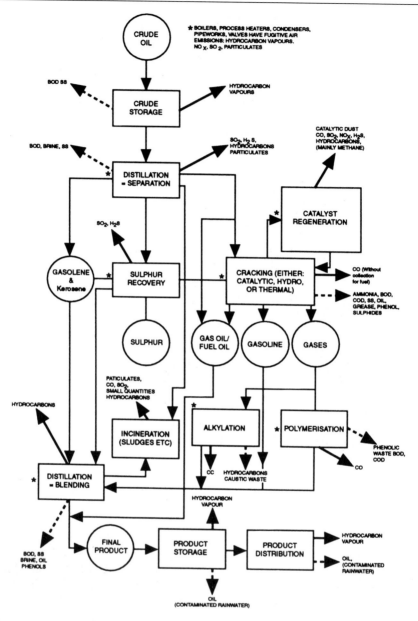

Figure 4.5 Materials flow-chart for a petroleum refinery. (*Source:* UNEP 1981)

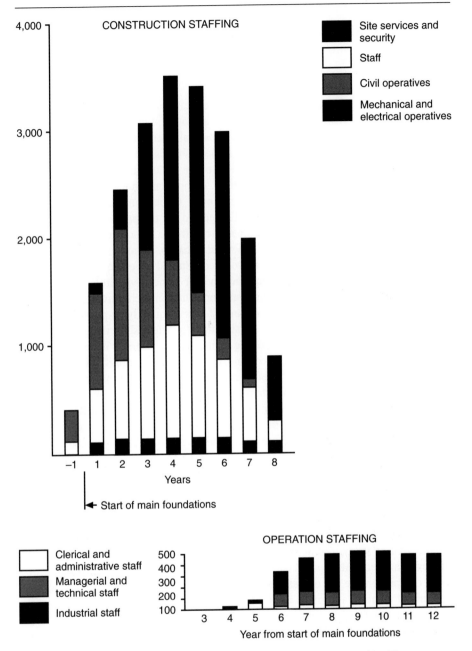

Figure 4.6 Labour requirements for a project over several stages of its life.

4.7   Establishing the environmental baseline

General considerations

The establishment of an environmental baseline includes both the present and likely future state of the environment, assuming that a proposed project is not undertaken, taking into account changes resulting from natural events and from other human activities. For example, the population of a species of fish in a lake may already be declining before the proposed introduction of an industrial project on the lake shore. Figure 1.6 illustrated the various time, component and scale dimensions of the environment, and all these dimensions need to be considered in the establishment of the environmental baseline. The period for the prediction of the future state of the environment should be comparable with the life of the proposed development; this may mean predicting for several decades. Components include both the biophysical and socio-economic environment. Spatial coverage may focus on the local, but refer to the wider region and beyond for some environmental elements.

Initial baseline studies may be wide-ranging, but comprehensive overviews can be wasteful of resources. The studies should focus as quickly as possible on those aspects of the environment that may be significantly affected by the project, either directly or indirectly. The rationale for the choice of focus should be explained with reference to the nature of the project and to the initial scoping and consultation exercises. Although the studies would normally take the various environmental elements separately, it is also important to understand the interaction between them and the functional relationships involved; for instance, flora will be affected by air and water quality, and fauna will be affected by flora. This will facilitate prediction. As with most aspects of the EIA process, establishing the baseline is not a "one-off" activity. Studies will move from broad-brush to more detailed and focused approaches. The identification of new potential impacts may open up new elements of the environment for investigation; the identification of effective measures for mitigating impacts may curtail certain areas of investigation.

Baseline studies can be presented in the EIS in a variety of ways. These often involve either a brief overview of the biophysical and socio-economic environments for the area of study, following the project description, with the detailed focused studies in subsequent impact chapters (e.g. air quality, geology, employment), or a more comprehensive set of detailed studies at an early stage providing a point of reference for future and often briefer impact chapters.

Environmental components or elements can be described simply in broad categories, as outlined in Table 1.3. *Environmental assessment: a guide to the procedures* (DOE 1989) also provides a relatively short list (see Table 4.2), including an important distinction between physical features and policy framework. In contrast, Leopold has 88 components in his interactive matrix (see Fig. 4.12 below), and each of these could be subdivided further. Several UN publications provide a more balanced listing of both the biophysical and socio-economic elements (see United

Table 4.2 Information describing the site and its environment.

Physical features

1. Population – proximity and numbers.
2. Flora and fauna (including both habitats and species) – in particular, protected species and their habitats.
3. Soil; agricultural quality, geology and geomorphology.
4. Water; aquifers, water courses and shoreline, including the type, quantity, composition and strength of any existing discharges.
5. Air; climatic factors, air quality, etc.
6. Architectural and historic heritage, archaeological sites and features and other material assets.
7. Landscape and topography.
8. Recreational uses.
9. Any other relevant environmental features.

The policy framework

10. Where applicable, the information considered under this section should include all relevant statutory designations such as national nature reserves, sites of special scientific interest, national parks, areas of outstanding natural beauty, heritage coasts, regional parks, country parks, national forest parks and designated areas, local nature reserves, areas affected by tree preservation orders, water protection zones, nitrate sensitive areas, conservation areas, listed buildings, scheduled ancient monuments, and designated areas of archaeological importance. It should also include references to structure, unitary and local plan policies applying to the site and surrounding area which are relevant to the proposed development.
11. Reference should also be made to international designations, e.g. those under the EC "Wild Birds" Directive, the World Heritage Convention, the UNEP Man and Biosphere Programme and the Ramsar Convention.

(*Source:* DOE 1989)

Nations Environment Programme 1981). Table 4.3 provides an example of a framework for analyzing each baseline sub-element.

Data sources and issues

Data on environmental conditions vary in availability and in quality. Important data sources for most locations in the UK are the statutory development plans (local plans, structure plans, unitary development plans). These usually provide a range of very useful data on the physical, social and economic environment; they are reasonably up to date. In a few locations, such data are supplemented by environmental audits or state-of-the-environment reports. The focus of these new studies is normally on the physical environment. The Environmental Audit for Oxfordshire, for example, includes detailed studies on land-use, landscape, open space, forestry, wildlife, agriculture, noise, air quality, water pollution, waste management, transport, energy and environmental management (Aspinwall & Company Ltd 1991). Chapter 12 provides further discussion of environmental audits. Local data can be

Table 4.3 Framework for analysing baseline subelements: example of use.

Sub-element	Objectives	Required information/ specialist(s)	Methodology	Findings/measurements
Water quality	Protection of human health and aquatic life	Existing water quality; possible sources of pollution: run-off, leakage from waste treatment system, surface seepage of pollutants, intrusion of saline or polluted water; capacity of treatment system Water quality analyst; aquatic biologist; water pollution control engineer; sanitary and civil engineers	Laboratory analyses or field measurement of water quality; pollution indices	Potential for degradation of water quality; safety of potable water
Surface waters	Protection of: plant and animal life; water supply for domestic and industrial needs; natural water purification systems; groundwater recharge and discharge; recreation and aesthetic values	Location of surface waters – streams, rivers, ponds, lakes, etc.; surface water volume, flow rates, frequency and duration of seasonal variations; 7-day, 10-year low flow; water uses; ecological characteristics; recreation and aesthetic uses Hydrologist; ecologist	Measurement of proximity of site to surface waters; field measurement of volume, rate and direction of water movement; categories of water usage; ecological assessment – see ecology element	Potential modification of volume, rate and direction of water movement; impact on ecological character; degree and type of water usage

(*Source:* United Nations Environment Programme (Industry and Environment Office) 1981)

supplemented in the UK with published data from a wide range of national government sources – including the Census of population, *Regional Trends, Digest of Environmental Protection and Water Standards, Transport statistics* – and increasingly from EU sources. Some countries have guidelines and manuals for EIA that list principal sources of information for different environmental elements, and these can be very useful (see Environment Agency (1996) and Government of New South Wales (1996) noted earlier).

However, much useful information is unpublished or "semi-published" and internal to various organizations. In the UK, under the EIA regulations, statutory consultees (e.g. the Countryside Commission, English Nature, the Environment Agency) are obliged

> to provide the developer (on request) with any information in their possession which is likely to be relevant to the preparation of the environmental statement ... The obligation on statutory consultees relates only to information already in their possession: they are not required to undertake research on behalf of the developer. (DOE 1989)

There are of course many other useful non-statutory consultees, at local and other levels, who may be able to provide valuable information. Local history, conservation and naturalist societies may have a wealth of information on, for example, local flora and fauna, rights of way and archaeological sites. National bodies, such as the Royal Society for the Protection of Birds (for bird populations and habitats) and the Forestry Authority (for tree surveys), may have particular knowledge and expertise to offer. Consultation with local amenity groups at an early stage in the EIA process can help not only with data but also with the identification of those key environmental issues for which data should be collected.

Every use should be made of data from existing sources, but there will invariably be gaps in the required environmental baseline data for the project under consideration. Environmental monitoring and surveys may be necessary, although the UK DOE notes: "While a careful study of the proposed location will generally be needed (including environmental survey information), original scientific research will not normally be necessary" (DOE 1989). Surveys and monitoring raise a number of issues. They are inevitably constrained by budgets and time, and must be selective. However, such selectivity must ensure that the length of time over which monitoring and surveys are undertaken is appropriate to the task in hand. For example, for certain environmental features (e.g. many types of flora and fauna) a survey period of 12 months or more may be needed to take account of seasonal variations. Sampling procedures will often be used for surveys; the extent and implications of the sampling error involved should be clearly established.

The quality and reliability of environmental data vary a great deal, and this can influence the use of such data in the assessment of impacts. Fortlage (1990) clarifies this in the following useful classification:

- "hard" data from reliable sources which can be verified and which are not subject to short-term change, such as geological records and physical surveys of topography and infrastructure;

107

- "intermediate" data which are reliable but not capable of absolute proof, such as water quality, land values, vegetation condition, and traffic counts, which have variable values;
- "soft" data which are a matter of opinion or social values, such as opinion surveys, visual enjoyment of landscape, and numbers of people using amenities, where the responses depend on human attitudes and the climate of public feeling.

A valuable innovation in the provision of environmental data is the increasing provision through the Internet and the development of computerized data banks. The use of geographical information systems, for particular sets of data and for particular locations, is increasing. Geographical information systems (GIS) are computer-based databases that include spatial references for the different variables stored, so that maps of such variables can be displayed, combined and analysed with speed and ease (see Rodriguez-Bachiller in Morris and Therivel 1995). The GIS market is developing apace, but initial setting-up costs are usually expensive, depending on the accessibility of relevant data. Of particular relevance are digital data because they are data already existing in a computer-readable format (i.e. data which can be "recycled" very quickly and very cheaply. *Directory of digital data sources in the UK* (O'Carroll et al. 1994) is particularly oriented to the needs of EIA; it seeks to support the efficient development of GIS by signposting existing digital map bases and data with which GIS can be integrated, and to reduce the redundant regeneration of data sets already in existence. The analyst should however be wary of the seductive attraction of quantitative data at the expense of qualitative data; each type has a valuable role in establishing baseline conditions. Finally, it should be remembered that all data sources suffer from some uncertainty, and this needs to be explicitly recognized in the prediction of environmental effects (see Ch. 5).

4.8 Impact identification

Aims and methods

Impact identification brings together project characteristics and baseline environmental characteristics with the aim of ensuring that all potentially significant environmental impacts (adverse or favourable) are identified and taken into account in the EIA process. A wide range of methods have been developed. Sorensen & Moss (1973) note that the present diversity "should be considered as a healthy condition in a newly formed and growing discipline".

When choosing a method, the analyst needs to consider more specific aims, some of which conflict:

- to ensure compliance with regulations;
- to provide a comprehensive coverage of a full range of impacts, including social, economic and physical;

- to distinguish between positive and negative, large and small, long-term and short-term, reversible and irreversible impacts;
- to identify secondary, indirect and cumulative impacts as well as direct impacts;
- to distinguish between significant and insignificant impacts;
- to allow a comparison of alternative development proposals;
- to consider impacts within the constraints of an area's carrying capacity;
- to incorporate qualitative as well as quantitative information;
- to be easy and economical to use;
- to be unbiased and to give consistent results;
- to be of use in summarizing and presenting impacts in the EIS.

Many of the methods were developed in response to the NEPA and have since been expanded and refined. The simplest involve the use of lists of impacts to ensure that none has been forgotten. The most complex include the use of interactive computer programmes, networks showing energy flows and schemes to allocate significance weightings to various impacts. Many of the more complex methods were developed for (usually US) government agencies that deal with large numbers of fairly similar project types (e.g. the US Bureau of Land Reclamation and the US Forest Service).

In the UK, the use of impact identification techniques is less well developed. Simple checklists or, at best, simple matrices are used to identify and summarize impacts. This may be attributable to the high degree of flexibility and discretion in the UK's implementation of Directive 85/337, to a general unwillingness in the UK to make the EIA process over-complex or to disillusionment with the more complex approaches that are available.

The aim of this section is to present a range of these methods, from the simplest checklists needed for compliance with regulations to complex approaches that developers, consultants and academics who aim to further "best practice" may wish to investigate further. The methods are divided into the following categories:

- checklists
- matrices
- quantitative methods
- networks
- overlay maps.

The discussion of the methods here relates primarily to impact identification, but most of the approaches are also of considerable (and sometimes more) use in other stages of the EIA process – in impact prediction, evaluation, communication, mitigation, presentation, monitoring and auditing. As such, there is considerable interaction between Chapters 4, 5, 6 and 7, paralleling the interaction in practice between these various stages.

For further information on the range of methods available we refer the reader to Morris & Therivel (1995), Bregman & Mackenthun (1992), Bisset (1983, 1989), Wathern (1984), Sorensen & Moss (1973), Munn (1979), Rau & Wooten (1980) and Jain et al. (1977).

Table 4.4 Part of a descriptive checklist.

Data required	Information sources, predictive techniques
Nuisance	
Change in occurrence of odour, smoke, haze, etc., and number of people affected	Expected industrial processes and traffic volumes, citizen surveys
Water quality	
For each body of water, changes in water uses, and number of people affected	Current water quality, current and expected effluent
Noise	
Change in noise levels, frequency of occurrence, and number of people bothered	Current noise levels, changes in traffic or other noise sources, changes in noise mitigation measures, noise propagation models, citizen surveys

(Adapted from Schaenman 1976)

Checklists

Most checklists are based on a list of special biophysical, social and economic factors that may be affected by a development. The *simple checklist* can help only to identify impacts and ensure that impacts are not overlooked. Checklists do not usually include direct cause–effect links to project activities. Nevertheless, they have the advantage of being easy to use. Appendix 3 of the Town and Country Planning (Assessment of Environmental Effects) Regulations 1988 (see Table 5.1) is an example of a simple checklist.

Descriptive checklists (e.g. Schaenman 1976) give guidance on how to assess impacts. They can include data requirements, information sources, and predictive techniques. An example of part of a descriptive checklist is shown in Table 4.4.

Questionnaire checklists are based on a set of questions to be answered. Some of the questions may concern indirect impacts and possible mitigation measures. They may also provide a scale for classifying estimated impacts, from highly adverse to highly beneficial. Figure 4.7 shows part of a questionnaire checklist.

Threshold of concern checklists consist of a list of environmental components and, for each component, a threshold at which those assessing a proposal should become concerned with an impact. The implications of alternative proposals can be seen by examining the number of times that an alternative exceeds the threshold of concern. For example, Figure 4.8 shows part of a checklist developed by the US Forest Service; it compares three alternative development proposals on the basis of various components. For the component of economic efficiency, a benefit : cost ratio of 1:1 is the threshold of concern; for spotted owls, 35 pairs is the threshold. In the example, alternative X causes two thresholds of concern to be exceeded, alternative Y one, alternative Z four; this would indicate that alternative Y is the least

110

Disease vectors

(a) Are there known disease problems in the project area transmitted through vector species such as mosquitoes, flies, snails, etc.?　　yes　　no　　not known

(b) Are these vector species associated with:

- aquatic habitats?　　yes　　no　　not known
- forest habitats?　　yes　　no　　not known
- agricultural habitats?　　yes　　no　　not known

...

(f) Will the project provide opportunities for vector control through improved standards of living?　　yes　　no　　not known

Estimated impact on disease vectors?

high adverse ◄ - - - - - - - - - - - - - insignificant - - - - - - - - - - - - - ► high benefit

Figure 4.7 Part of a questionnaire checklist. Adapted from US Agency for International Development 1981.

Environmental component	Criterion	TOC	Alt X		Alt Y		Alt Z	
			Imp	Imp > TOC?	Imp	Imp > TOC?	Imp	Imp > TOC?
Air quality	emission standards	1	2C	yes	1C	no	2C	yes
Economics	benefit: cost ratio	1:1	3:1	no	4:1	no	2:1	no
Endangered species	no. pairs of spotted owls	35	50D	no	35D	no	20D	yes
Water quality	water quality standards	1	1C	no	2C	yes	2C	yes
Recreation	no. camping sites	5000	2800C	yes	5000C	no	3500C	yes

Figure 4.8 Part of a threshold of concern (TOC) checklist. (Adapted from Sassaman 1981)

detrimental. Impacts are also rated according to their duration: A for 1 year or less, B for 1–10 years, C for 10–50 years, and D for irreversible impacts. Of the impacts listed, a reduction in the number of spotted owls would be irreversible, and the other impacts would last 10–50 years (Sassaman 1981).

111

Environmental component	Project action				
	Construction		Operation		
	Utilities	Residential and commercial buildings	Residential buildings	Commercial buildings	Parks and open spaces
Soil and geology	X	X			
Flora	X	X			X
Fauna	X	X			X
Air quality				X	
Water quality	X	X	X		
Population density			X	X	
Employment		X		X	
Traffic	X	X	X	X	
Housing			X		
Community structure		X	X		X

Figure 4.9 Part of a simple matrix.

Matrices

Matrices are the most commonly used method of impact identification in EIA. Simple matrices are merely two-dimensional charts showing environmental components on one axis and development actions on the other. They are, essentially, expansions of checklists that acknowledge the fact that various components of a development project (e.g. construction, operation, decommissioning; buildings, access road) have different impacts. An action likely to have an impact on an environmental component is identified by placing a cross in the appropriate cell. The main advantage is the incorporation of cause–effect relationships. Figure 4.9 shows an example of a *simple matrix*. Three-dimensional matrices have also been developed in which the third dimension refers to economic and social institutions: such an approach identifies the institutions from which data are needed for the EIA process, and highlights areas in which knowledge is lacking.

The *time-dependent matrix* (e.g. Parker & Howard 1977) includes a number sequence to represent the timescale of the impacts (e.g. one figure per year). Figure 4.10 shows an example, where magnitude is represented by numbers from 0 (none) to 4 (high), over the course of seven years.

Magnitude matrices go beyond the mere identification of impacts by describing them according to their magnitude, importance and/or time frame (e.g. short-, medium, or long-term). Figure 4.11 is an example of a magnitude matrix.

The best known type of quantified matrix is the *Leopold matrix*, which was developed for the US Geological Survey by Leopold et al. (1971). It is based on a horizontal list of 100 project actions and a vertical list of 88 environmental components. Figure 4.12 shows a section of this matrix and lists all its elements. Of the

Environmental component	Project action				
	Construction (3 years)		Operation (25 years, evens out after 4 years)		
	Utilities	Residential and commercial buildings	Residential buildings	Commercial buildings	Parks and open spaces
Soil and geology	211	321	0000	0000	0001
Flora	221	422	1223	1111	1123
Fauna	221	311	1100	1100	1122
Air quality	000	000	0123	0034	0011
Water quality	010	022	1223	0111	0000
Population density	011	112	2344	0222	0011
Employment	120	342	1111	1334	1111
Traffic	220	332	2333	2333	1111
Housing	010	121	2344	0000	0000
Community structure	010	232	2344	1111	1233

Figure 4.10   Part of a time-dependent matrix.

Environmental component	Project action				
	Construction		Operation		
	Utilities	Residential and commercial buildings	Residential buildings	Commercial buildings	Parks and open spaces
Soil and geology	•	•			
Flora	•	●			○
Fauna	•	•			o
Air quality				•	
Water quality	○	•	•		
Population density			o	o	
Employment		○		○	
Traffic	•	•	•	●	
Housing			○		
Community structure		•	○		o

• = small negative impact   o = small positive impact
● = large negative impact   ○ = large positive impact

Figure 4.11   Part of a magnitude matrix.

113

(a)

1. Identify all actions (located across the top of the matrix) that are part of the proposed project

2. Under each of the proposed actions, place a slash at the intersection with each item on the side of the matrix if an impact is possible

3. Having completed the matrix, in the upper left hand corner of each box with a slash, place a number from 1 to 10 which indicates the MAGNITUDE of the possible impact; 10 represents the greatest magnitude of impact and 1, the least (no zeroes). Before each number place + (if the impact would be beneficial). In the lower right hand corner of the box place a number from 1 to 10 which indicates the IMPORTANCE of the possible impact (e.g. regional vs. local); 10 represents the greatest importance and 1 the least (no zeroes)

4. The text which accompanies the matrix should be a discussion of the significant impacts, those columns and rows with large numbers of boxes marked and individual boxes with large numbers

	a	b	c	d	e
a		2\1			8\3
b	7\8		3\1		9\7

Sample matrix

A. Modification of regime

a. Exotic flora or fauna introduction
b. Biological controls
c. Modification of habitat
d. Alteration of ground cover
e. Alteration of ground water hydrology
f. Alteration of drainage
g. River control and flow modification
h. Canalization
i. Irrigation
j. Weather modification
k. Burning
l. Surface or paving
m. Noise and vibration

B. Land transformation and construction

a. Urbanization
b. Industrial sites and buildings
c. Airports
d. Highways and bridges
e. Roads and trails
f. Railroads
g. Cables and lifts
h. Transmission lines, pipelines, corridors
i. Barriers including fencing
j. Channel dredging and straightening
k. Channel revetments
l. Canals
m. Dams and impoundments
n. Piers, seawall, marinas and sea terminals
o. Offshore structures
p. Recreational structures
q. Blasting and drilling
r. Cut and fill
s. Tunnels and underground structures

C. Resource extraction

a. Blasting and drilling
b. Surface excavation
c. Subsurface excavation and retorting
d. Well drilling and fluid removal
e. Dredging
f. Clear cutting and other lumbering
g. Commercial fishing and hunting

Proposed actions

CHEMICAL CHARACTERISTICS

1. Earth
a. Mineral resources
b. Construction material
c. Soils
d. Land form
e. Force fields and background radiation
f. Unique features

2. Water
a. Surface
b. Ocean
c. Underground
d. Quality
e. Temperature
f. Recharge
g. Snow, ice and permafrost

(b)

Part 1. Project actions

A. *Modification of regime*
a) exotic flora or fauna introduction
b) Biological controls
c) Modification of habitat
d) Alteration of ground cover
e) Alteration of groundwater hydrology
f) Alteration of drainage
g) River control and flow modification
h) Canalization
i) Irrigation
j) Weather modification
k) Burning
l) Surface or paving
m) Noise and vibration

B. *Land transformation and construction*
a) Urbanization
b) Industrial sites and buildings
c) Airports
d) Highways and bridges
e) Roads and trails
f) Railroads
g) Cables and lifts
h) Transmission lines, pipelines and corridors
i) Barriers, including fencing
j) Channel dredging and straightening
k) Channel revetments
l) Canals
m) Dams and impoundments
n) Piers, seawalls, marinas, and sea terminals
o) Offshore structures
p) Recreational structures
q) Blasting and drilling
r) Cut and fill
s) Tunnels and underground structures

C. *Resource extraction*
a) Blasting and drilling
b) Surface excavation
c) Subsurface excavation and retorting
d) Well drilling and fluid removal
e) Dredging
f) Clear cutting and other lumbering
g) Commercial fishing and hunting

D. *Processing*
a) Farming
b) Ranching and grazing
c) Feed lots
d) Dairying
e) Energy generation
f) Mineral processing
g) Metallurgical industry
h) Chemical industry
i) Textile industry
j) Automobile and aircraft
k) Oil refining
l) Food
m) Lumbering
n) Pulp and paper
o) Product storage

E. *Land alteration*
a) Erosion control and terracing
b) Mine sealing and waste control
c) Strip-mining rehabilitation
d) Landscaping
e) Harbour dredging
f) Marsh fill and drainage

F. *Resource renewal*
a) Reforestation
b) Wildlife stocking and management
c) Groundwater recharge
d) Fertilization application
e) Waste recycling

G. *Changes in traffic*
a) Railway
b) Automobile
c) Trucking
d) Shipping
e) Aircraft
f) River and canal traffic
g) Pleasure boating
h) Trails
i) Cables and lifts
j) Communication
k) Pipeline

H. *Waste emplacement and treatment*
a) Ocean dumping
b) Landfill
c) Emplacement of tailings, spoil and overburden
d) Underground storage
e) Junk disposal
f) Oil well flooding
g) Deep well emplacement
h) Cooling water discharge
i) Municipal waste discharge, including spray irrigation
j) Liquid effluent discharge
k) Stabilization and oxidation ponds
l) Septic tanks, commercial and domestic
m) Stack and exhaust emission
n) Spent lubricants

I. *Chemical treatment*
a) Fertilization
b) Chemical de-icing of highways, etc.
c) Chemical stabilization of soil
d) Weed control
e) Insect control (pesticides)

J. *Accidents*
a) Explosions
b) Spills and leaks
c) Operational failure

Others

Part 2. Natural and human environmental elements

A. *Physical and chemical characteristics*
1. Earth
a) Mineral resources
b) Construction material
c) Soils
d) Landform
e) Force fields and background radiation
f) Unique physical features
2. Water
a) Surface
b) Ocean
c) Underground
d) Quality
e) Temperature
f) Recharge
g) Snow, ice and permafrost
3. Atmosphere
a) Quality (gases, particulates)
b) Climate (micro, macro)
c) Temperature
4. Processes
a) Floods
b) Erosion
c) Deposition (sedimentation, precipitation)
d) Solution
e) Sorption (ion exchange, complexing)
f) Compaction and setting
g) Stability (slides, slumps)
h) Stress-strain (earthquakes)
i) Air movements

B. *Biological conditions*
1. Flora
a) Trees
b) Shrubs
c) Grass
d) Crops
e) Microflora
f) Aquatic plants
g) Endangered species
h) Barriers
i) Corridors
2. Fauna
a) Birds
b) Land animals, including reptiles
c) Fish and shellfish
d) Benthic organisms
e) Insects
f) Microfauna
g) Endangered species
h) Barriers
i) Corridors

C. *Cultural factors*
1. Land-use
a) Wilderness and open spaces
b) Wetlands
c) Forestry
d) Grazing
e) Agriculture
f) Residential
g) Commercial
h) Industrial
i) Mining and quarrying
2. Recreation
a) Hunting
b) Fishing
c) Boating
d) Swimming
e) Camping and hiking
f) Picnicking
g) Resorts
3. Aesthetics and human interest
a) Scenic views and vistas
b) Wilderness qualities
c) Open space qualities
d) Landscape design
e) Unique physical features
f) Parks and reserves
g) Monuments
h) Rare and unique species or ecosystems
i) Historical or archaeological sites and objects
j) Presence of misfits
4. Cultural status
a) Cultural patterns, lifestyle
b) Health and safety
c) Employment
d) Population density
5. Man-made facilities and activities
a) Structures
b) Transportation network (movement access)
c) Utility networks
d) Waste disposal
e) Barriers
f) Corridors

D. *Ecological relationships, such as*
a) Salinization of water resources
b) Eutrophication
c) Disease – insect vectors
d) Food chains
e) Salinization of surficial material
f) Brush encroachment
g) Other
Others

Figure 4.12 (a) Part of Leopold Matrix; (b) Leopold Matrix elements.

8,800 possible interactions between project action and environmental component, Leopold et al. estimate that an individual project is likely to result in 25–50. In each appropriate cell, two numbers are recorded. The number in the top left-hand corner represents the impact's magnitude, from +10 (very positive) to −10 (very negative). That in the bottom right-hand corner represents the impact's significance, from 10 (very significant) to 1 (insignificant); there is no negative significance. This distinction between magnitude and significance is important: an impact could be large but insignificant, or small but significant. For instance, in ecological terms, paving over a large field of intensively used farmland may be quite insignificant compared with the destruction of even a small area of a SSSI.

The Leopold matrix is easily understood, can be applied to a wide range of developments, and is reasonably comprehensive for first-order, direct impacts. However, it has disadvantages. The fact that it was designed for use on many different types of project makes it unwieldy for use on any one project. It cannot reveal indirect effects of developments: like checklists and most other matrices, it does not relate environmental components to one another, so the complex interactions between ecosystem components that lead to indirect impacts are not assessed. The inclusion of magnitude/significance scores has additional drawbacks: it gives no indication whether the data on which these values are based are qualitative or quantitative; it does not specify the probability of an impact occurring; it excludes details of the techniques used to predict impacts; and the scoring system is inherently subjective and open to bias. People may also attempt to add the numerical values to produce a composite value for the development's impacts and compare this with that for other developments; this should not be done because the matrix does not assign weightings to different impacts to reflect their relative importance (Clark et al. 1979).

Weighted matrices were developed in an attempt to respond to some of the above problems. Importance weightings are assigned to environmental components, and sometimes to project components. The impact of the project (component) on the environmental component is then assessed and multiplied by the appropriate weighting(s), to obtain a total for the project. Figure 4.13 shows a small weighted

Environmental component	(a)	Alternative sites					
		Site A		Site B		Site C	
		(c)	(axc)	(c)	(axc)	(c)	(axc)
Air quality	21	3	63	5	105	3	63
Water quality	42	6	252	2	84	5	210
Noise	9	5	45	7	63	9	81
Ecosystem	28	5	140	4	112	3	84
Total	100		500		364		438

(a) = relative weighting of environmental component (total 100)
(c) = impact of project at particular site on environmental component (0–10)

Figure 4.13 A weighted matrix: alternative project sites.

	Importance weighting (a)	Treatment plant	Pumping station	Interceptor	Outfall	Total
Air quality	21	10(b) 8(c)	0 –	50 7	40 8	15,750
Water quality	42	100 9	0 –	0 –	0 –	37,800
Noise	9	0 –	100 3	0 –	0 –	2700
Ecosystem	28	10 5	20 4	40 8	30 8	19,320
Total	100					75,570

(a) = relative weighting of environmental component (total 100)
(b) = relative weighting of project component (total 100)
(c) = impact of project on environmental component (0–10)

Figure 4.14 A weighted matrix: weighted project components. (Based on Wenger & Rhyner 1972)

matrix that compares three alternative project sites. Each environmental component is assigned an importance weighting (a), relative to other environmental components: in the example, air quality is weighted 21 per cent of the total environmental components. The magnitude (c) of the impact of each project on each environmental component is then assessed on a scale 0–10, and multiplied by (a) to obtain a weighted impact (a × c): for instance, site A has an impact of 3 out of 10 on air quality, which is multiplied by 21 to give the weighted impact, 63. For each site, the weighted impacts can then be added up to give a project total. The site with the lowest total, in this case site B, is the least environmentally harmful.

Figure 4.14 shows a similar abbreviated weighted matrix for a sewage treatment facility, broken down into its components. An importance weighting (b), out of 100, has been determined for each project component, and the magnitude (c) of the impact of that project component on each environmental component is then assessed, out of 10. The factors (b) and (c) are then multiplied with the importance weighting (a) of the relevant environmental component to give the weighted impact of each project component on each environmental component. All of these can be added to represent the total impact of the project. This can then be compared with those of other projects. In the example, the treatment plant is the only project component to affect water quality (b = 100), and it has a large impact (c = 9) on water quality; the weighted impact of the treatment plant on water quality is thus 900. This is multiplied by the importance weighting of water quality, 42, to get the weighted impact on water quality, 37,800. In the case of air quality, which more than one project component affects, the weighted impacts of the various components are first added up (e.g. 80 for the treatment plant plus 350 for the interceptor plus 320 for the outfall), then multiplied by the importance weighting of air quality,

21, to get the weighted impact on air quality, 15,750. The project's total weighted impact could then be compared with that of other project alternatives. This method has the advantage of allowing various alternatives to be compared numerically. However, the evaluation procedure depends heavily on the weightings and impact scales assigned. The main problems implicit in such weighting approaches are considered further in Chapter 5. Also the method does not consider indirect impacts.

Distributional impact matrices represent another possible development of the matrix approach. Such matrices can broadly identify who might lose and who might gain from the potential impacts of a development. This is useful information, which is rarely included in the matrix approach, and indeed is often missing from many EISs. Impacts can have varying spatial impacts – varying, for example, between urban and rural areas. Spatial variations may be particularly marked for a linear project, such as a Light Rapid Transit (LRT) system. A project can also have different impacts on different groups in society (for example the impacts of a proposed new settlement on old people, retired with their own houses, and young people, perhaps with children, seeking affordable housing and a way into the housing market) (see Fig. 5.7, Ch. 5).

Quantitative methods

Quantitative methods attempt to compare the relative importance of all impacts by weighting, standardizing and aggregating them to produce a composite index. The best known of these methods is the *environmental evaluation system* (EES), devised by the Battelle Columbus Laboratories for the US Bureau of Land Reclamation to assess water resource developments, highways, nuclear power plants and other projects (Dee et al. 1973). It consists of a checklist of 74 environmental, social and economic parameters that may be affected by a proposal; these are shown in Figure 4.15. It assumes that these parameters can be expressed numerically and that they represent an aspect of environmental quality. For instance, the concentration of dissolved oxygen is a parameter that represents an aspect of the quality of an aquatic environment. For each parameter, functions were designed by experts to express environmental quality on a scale 0–1 (degraded – high quality). Two examples are shown in Figure 4.16. For instance, a stream with more than 10 mg/l of dissolved oxygen is felt to have a high level of environmental quality (1.0), whereas one with only 4 mg/l is felt to have an environmental quality of only about 0.35. Impacts are measured in terms of the likely change in environmental quality for each parameter. Two environmental quality scores are determined for each parameter, one for the current state of the environment and one for the state predicted once the project is in operation. If the post-development score is lower than the pre-development score, the impact is negative, and vice versa. To enable impacts to be compared directly, each parameter is given an importance weighting, which is then multiplied by the appropriate environmental quality score. The importance weightings (shown in parentheses in Fig. 4.15) are determined by having a panel of experts distribute 1,000 points among the parameters. For instance, dissolved oxygen is considered quite important, at 31 points out of 1,000. A composite score for the beneficial and

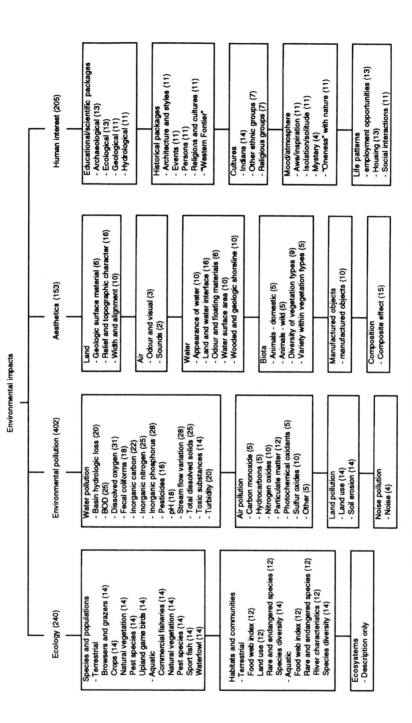

Figure 4.15 Framework for the Battelle Environmental Evaluation system. (*Source:* Dee et al. 1973)

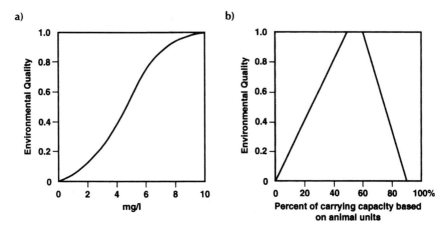

a)

b)

Figure 4.16 Environmental parameter functions for the Environmental Evaluation System: dissolved oxygen and deer: rangeland ratios: (a) dissolved oxygen; (b) browsers and grazers. (*Source:* Dee et al. 1973)

adverse effects of a single project, or for the net impact of alternative projects, can be obtained by adding up the weighted impact scores.

As an example of the full use of the EES, assume that the existing deer : rangeland ratio means that 40 per cent of the annual plant production is consumed (environmental quality score 0.8 in Fig. 4.16). A project likely to halve the deer population would cause the score to drop to 0.4. The post-development score would be lower than the pre-development score, so the impact would be negative. This parameter's importance is 14 points out of 1,000, so the pre- and post-development scores would be multiplied by 14, and could then be compared with other parameters (Dee et al. 1973).

After examining 54 methods of impact identification, the US Army Corps of Engineers, which is responsible for many water-resource projects requiring EIA, decided that methods such as EES had most potential, and used the principles of EES to form its *water resources assessment methodology* (WRAM). The WRAM approach assigns project impacts to four accounts: environmental quality, regional development, national economic development and social wellbeing. Factors in each account are weighted and expressed in common terms by the use of functional curves similar to EES value functions. Aggregate impact scores are then obtained for each account (Solomon et al. 1977).

Another quantitative method was developed to assess alternative highway proposals (Odum et al. 1975). Unlike EES, this method considers impact duration: long-term irreversible impacts are considered to be more important than short-term reversible impacts and are given ten times more weight. A sensitivity analysis showed that errors in impact estimation and weighting could significantly affect the rankings of alternative highway routes. Another method (Stover 1972) considers future impacts to be more important and gives them higher values than short-term impacts: it multiplies the numerical rating of each future impact by its duration in years.

The attraction of these quantitative methods lies in their ability to "substantiate" numerically that a particular course of action is better than others. This may save decision-makers considerable work, and it ensures consistency in assessment and results. However, these methods also have some fundamental weaknesses. They effectively take decisions away from decision-makers (Skutsch & Flowerdew 1976). The methods are difficult for lay-people to understand, and their acceptability depends on the assumptions, especially the weighting schemes, built into them.[3] People carrying out assessments may manipulate results by changing assumptions (Bisset 1978). Quantitative methods also treat the environment as if it consisted of discrete units. Impacts are related only to particular parameters, and much information is lost when impacts are reduced to numbers.

Networks

Network methods explicitly recognize that environmental systems consist of a complex web of relationships, and try to reproduce that web. Impact identification using networks involves following the effects of development through changes in the environmental parameters in the model. The *Sorensen network* was the first network method to be developed; it aimed to help planners reconcile conflicting land-uses in California. Figure 4.17 shows a section of the network dealing with impacts on water quality. Water is one of the six environmental components, the others being climate, geophysical conditions, biota, access conditions and aesthetics.

The Sorensen method begins by identifying potential causes of environmental change associated with a proposed development action, using a matrix format; for instance, forestry potentially results in the clearing of vegetation and the use of herbicides and fertilizers. These environmental changes in turn result in specific environmental impacts; in the example, the clearing of vegetation could result in an increased flow of fresh water, which in turn could imperil cliff structures. The analyst stops following the network when an initial cause of change has been traced through all subsequent impacts and changes in environmental conditions, to its final impacts. Environmental impacts can result either directly from a development action, or indirectly through induced changes in environmental conditions. A change in environmental conditions may result in several different types of impact. Sorensen argues that the method should lead to the identification of remedial measures and monitoring schemes (Sorensen 1971).

The Sorensen network does not establish the magnitude or significance of interrelationships between environmental components, or the extent of change. It requires much time and considerable knowledge of the environment under consideration to construct the network, and it is time-consuming to use manually. Its main advantage is its ability to trace the higher-order impacts of proposed developments. A similar network method, the computerized *IMPACT network*, was designed to assess the impacts of developments on forests and rangelands controlled by the US Forest Service (Thor et al. 1978).

121

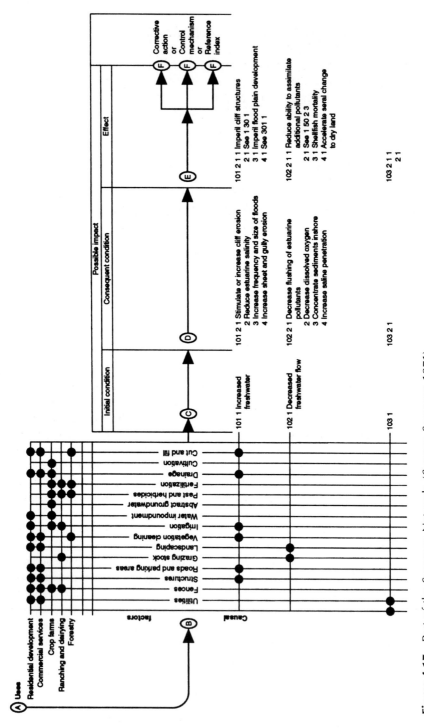

Figure 4.17 Part of the Sorensen Network. (*Source:* Sorensen 1971)

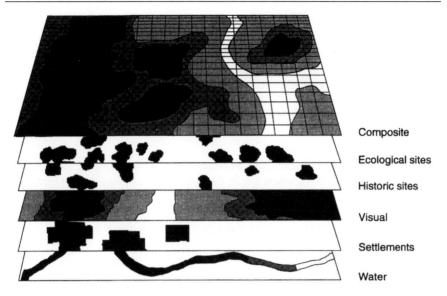

Composite

Ecological sites

Historic sites

Visual

Settlements

Water

Figure 4.18   An example of overlay maps.

Overlay maps

Overlay maps have been used in environmental planning since the 1960s (McHarg 1968), before the NEPA was enacted. A series of transparencies is used to identify, predict, assign relative significance to and communicate impacts, normally at a scale larger than local. A base map is prepared, showing the general area within which the project may be located. Successive transparent overlay maps are then prepared for the environmental components that, in the opinion of experts, are likely to be affected by the project (e.g. agriculture, woodland, noise). The project's degree of impact on the environmental feature is shown by the intensity of shading, darker shading representing a greater impact. The composite impact of the project is found by superimposing the overlay maps and noting the relative intensity of the total shading. Unshaded areas are those where a development project would not have a significant impact. Figure 4.18 shows an example of this technique.

Overlay maps are easy to use and understand and are popular. They are an excellent way of showing the spatial distribution of impacts. They also lead intrinsically to a low-impact decision. The overlay maps method is particularly useful for identifying optimum corridors for developments such as electricity lines and roads, for comparisons between alternatives, and for assessing large regional developments. The development of computer mapping, and in particular geographical information systems, allows more information to be handled. It also allows different importance weightings to be assigned to the impacts: this enables a sensitivity analysis to be carried out, to see whether changing assumptions about impact importance would alter the decision. However, the method is limited in that it does not consider factors such as the likelihood of an impact, secondary impacts, or the

123

difference between reversible and irreversible impacts. It requires the clear classification of often indeterminate boundaries (such as between forest and field), and so is not a true representation of conditions on the ground. It relies on the user to identify likely impacts before it can be used. The manual use of a large number of overlays is also often difficult; it is usually limited to about ten transparencies.

Summary

Table 4.5 summarizes the respective advantages of the main impact identification methods discussed in this section. Given the complexity of many impact identification techniques, it is understandable that most EIAS in the UK use checklists or simple matrices, or some hybrid combination including elements from several of the methods discussed. However, as more EIAS are carried out, as legislation concerning indirect and cumulative impacts (see Ch. 13) is enacted, and especially as large developers begin to establish patterns in preparing EIAS, the use of more sophisticated methods for impact identification may increase. This may of course not wholly benefit the EIA process: EIA methods are not politically neutral, and the

Table 4.5 Comparison of impact identification methods.

	Criterion										
	1	2	3	4	5	6	7	8	9	10	11
Checklists											
Simple/descriptive/question	✔	✔						✔	✔	✔	✔
Threshold	✔	✔	✔		✔	✔	✔	✔		✔	✔
Matrices											
Simple	✔	✔						✔	✔	✔	✔
Magnitude/time-dependent	✔	✔	✔					✔	✔	✔	✔
Leopold	✔	✔	✔		✔			✔	✔		✔
Weighted	✔	✔			✔	✔		✔			✔
Quantitative											
EES/WRAM	✔		✔		✔	✔	✔				
Network											
Sorensen	✔			✔		✔		✔		✔	
Overlay maps		✔	✔		✔	✔		✔	✔	✔	✔

1. Compliance with regulations; 2. comprehensive coverage (social, economic and physical impacts); 3. positive v. negative, reversible v. irreversible impacts, etc.; 4. secondary, indirect, cumulative impacts; 5. significant v. insignificant impacts; 6. compare alternative options; 7. compare against carrying capacity; 8. uses qualitative and quantitative information; 9. easy to use; 10. unbiased, consistent; 11. summarizes impacts for use in EIS.

more sophisticated the method becomes, often the more difficult becomes clear communication and effective participation (see Ch. 6 for more discussion).

4.9 Summary

The early stages of the EIA process are typified by several interacting steps. These include deciding whether an EIA is needed at all (screening), consulting with the various parties involved to produce an initial focus on some of the chief impacts (scoping), and an outline of possible alternative approaches to the project, including alternative locations, scales and processes. Scoping and alternatives have not been mandatory in an EIA in the UK, but both can greatly improve the quality of the process. Early in the process an analyst will also wish to understand the nature of the project concerned, and the environmental baseline conditions in the likely affected area. Projects have several dimensions (e.g. purpose, physical presence, processes and policies) over several stages in their life-cycles; a consideration of the environmental baseline also involves several dimensions. For both projects and the affected environment, obtaining relevant data may present challenges.

Impact identification includes most of the activities already discussed. It also usually involves the use of impact identification methods, ranging from simple checklists and matrices to complex computerized models and networks. The simpler methods are generally easier to use, more consistent and more effective in presenting information in the EIS, but their coverage of impact significance, indirect impacts or alternatives is either very limited or non-existent. The more complex models incorporate these aspects, but at the cost of immediacy. In the UK, if any formal impact identification methods are used, they are normally of a simpler type. The methods discussed here have relevance also to the prediction, assessment, communication and mitigation of environmental impacts, which are discussed in the next chapter.

References

Aspinwall & Company Ltd 1991. *Environmental audit of Oxfordshire*. Oxford: Oxfordshire County Council.

Bisset, R. 1978. Quantification, decision-making and environmental impact assessment in the United Kingdom. *Journal of Environmental Management* 7(1), 43–58.

Bisset, R. 1983. A critical survey of methods for environmental impact assessment. In *An annotated reader in environmental planning and management*, T. O'Riordan & R. K. Turner (eds), 168–85. Oxford: Pergamon.

Bisset, R. 1989. Introduction to EIA methods. Paper presented at the 10th International Seminar on Environmental Impact Assessment, University of Aberdeen, 9–22 July.

Bregman, J. I. & K. M. Mackenthun 1992. *Environmental impact statements*. Chelsea, Michigan: Lewis.

Canter, L. W. 1991. Interdisciplinary teams in environmental impact assessment. *Environmental Impact Assessment Review* **11**, 375–87.

CEC (Commission of the European Communities) 1993. *Environmental manual: user's guide; sectoral environment assessment sourcebook*. Brussels: CEC DG VIII.

CEC (Commission of the European Communities) 1997. Council Directive 97/11/EC amending Directive 85/337/EEC on the assessment of certain public and private projects on the environment. *Official Journal* L73/5, 3 March.

CEQ (Council on Environmental Quality) 1978. National Environmental Policy Act. Implementation of procedural provision: final regulations. *Federal Register* **43**(230), 55977–56007, 29 November.

Clark, B. D., K. Chapman, R. Bisset, P. Wathern 1979. Environmental impact assessment. In *Land-use and landscape planning*, D. Lovejoy ed., 53–87. Glasgow: Leonard Hill.

Cleland, D. I. & H. Kerzner 1986. Engineering team management. New York: Van Nostrand Rheinhold.

Dee, N., J. K. Baker, N. L. Drobny, K. M. Duke, I. Whitman, D. C. Fahringer 1973. An environmental evaluation system for water resources planning. *Water Resources Research* **3**, 523–35.

DETR (Department of the Environment, Transport and the Regions) 1997. *Consultation paper: implementation of EC Directive 97/11/EC: determining the need for EA*. London: DETR.

DOE (Department of the Environment) 1989. *Environmental assessment: a guide to the procedures*. London: HMSO.

DOE 1991. *Policy appraisal and the environment*. London: HMSO.

DOE 1994. *Preparation of environmental statements for planning projects that require environmental assessment: a good practice guide*. London: HMSO.

DOE 1996. *Changes in the quality of environmental statements for planning projects*. London: HMSO.

DOT (Department of Transport) 1983. *Manual of environmental appraisal*. London: HMSO.

DOT (1993), *Design manual for roads and bridges*, vol. 11: *Environmental assessment*. London: HMSO.

Eastman, C. 1997. *The treatment of alternatives in the environmental assessment process* (MSc dissertation) Oxford: Oxford Brookes University.

Environment Agency 1996. *Scoping guidance notes*. London: HMSO.

Fortlage, C. 1990. *Environmental assessment: a practical guide*. Aldershot: Gower.

Frost, R. 1993. Planning beyond environmental statements (MSc dissertation). Oxford: Oxford Brookes University, School of Planning.

Fuller, K. 1992. Working with assessment. In *Environmental assessment and audit, a user's guide 1992–1993* (special supplement to *Planning*), 14–15.

Government of New South Wales 1996. *EIS guidelines*. Sydney: Department of Urban Affairs and Planning.

Jain, R. K., L. V. Urban, G. S. Stacey 1977. *Environmental impact assessment*. New York: Van Nostrand Reinhold.

Jones, C. E., N. Lee, C. Wood 1991. *UK environmental statements 1988–1990: an analysis*. Occasional Paper 29, Department of Town and Country Planning, University of Manchester.

Lee, N. 1987. *Environmental impact assessment: a training guide*. Occasional Paper 18, Department of Town and Country Planning, University of Manchester.

Leopold, L. B., F. E. Clarke, B. B. Hanshaw, J. R. Balsley 1971. *A procedure for evaluating environmental impact*. Washington DC: US Geological Survey Circular 645.

McHarg, I. 1968. *A comprehensive route selection method*. Highway Research Record 246. Washington DC: Highway Research Board.

Marstrand, P. K. 1976. Ecological and social evaluation of industrial development. *Environmental Conservation* **3**(4), 303–8.

Mills, J. 1992. *Monitoring the visual impacts of major projects* (MSc dissertation). Oxford: Oxford Brookes University, School of Planning.

Morris, P. & R. Therivel (eds) 1995. *Methods of environmental impact assessment*, London: UCL Press.

Munn, R. E. 1979. *Environmental impact assessment: principles and procedures*, 2nd edn. New York: Wiley.

O'Carroll, P., A. Rodriguez-Bachiller, J. Glasson 1994. *Directory of digital data sources in the UK*, Working Paper No. 149. Oxford: Oxford Brookes University, Impact Assessment Unit, School of Planning.

Odum, E. P., J. C. Zieman, H. H. Shugart, A. Ike, J. R. Champlin 1975. In *Environmental impact assessment*, M. Blisset (ed.). Austin: University of Texas Press.

Parker, B. C. & R. V. Howard 1977. The first environmental monitoring and assessment in Antarctica: the Dry Valley drilling project. *Biological Conservation* **12**(2), 163–77.

Petts, J. & G. Eduljee 1994. Integration of monitoring, auditing and environmental assessment: waste facility issues. *Project Appraisal* **9**(4), 231–41.

Rau, J. G. & D. C. Wooten 1980. *Environmental impact analysis handbook*. New York: McGraw-Hill.

Rendel Planning 1990. *Angle Bay energy project environmental statement*. London: Rendel Planning.

Sassaman, R. W. 1981. Threshold of concern: a technique for evaluating environmental impacts and amenity values. *Journal of Forestry* **79**, 84–6.

Schaenman, P. W. 1976. *Using an impact measurement system to evaluate land development*. Washington DC: Urban Institute.

Skutsch, M. M. & R. T. N. Flowerdew 1976. Measurement techniques in environmental impact assessment. *Environmental Conservation* **3**(3), 209–17.

Solomon, R. C., B. K. Colbert, W. J. Hanson, S. E. Richardson, L. W. Canter, E. C. Vlachos 1977. *Water resources assessment methodology (WRAM) – impact assessment and alternative evaluation*, Technical Report no. Y-77-1. Vicksburg, Mississippi: Army Corps of Engineers.

Sorensen, J. C. 1971. *A framework for the identification and control of resource degradation and conflict in multiple use of the coastal zone*. Berkeley: Department of Landscape Architecture, University of California.

Sorensen J. C. & M. L. Moss 1973. *Procedures and programmes to assist in the environmental impact statement process*. Berkeley: Institute of Urban and Regional Development, University of California.

State of California, Governor's Office of Planning and Research 1992. *California Environmental Quality Act: statutes and guidelines*. Sacramento: State of California.

Stover, L. V. 1972. *Environmental impact assessment: a procedure*. Pottstown, Pennsylvania: Sanders & Thomas.

Thor, E. C., G. H. Elsner, M. R. Travis, K. M. O'Loughlin 1978. Forest environmental impact analysis – a new approach. *Journal of Forestry* **76**, 723–5.

TNO 1983. *Handbook of emission factors*. The Hague: Netherlands Ministry of Public Housing, Physical Planning and Environmental Affairs.

United Nations Environment Programme 1981. *Guidelines for assessing industrial environmental impact and environmental criteria for the siting of industry*. New York: United Nations.

US Agency for International Development 1981. *Environmental design considerations for rural development projects*. Washington DC: US Agency for International Development.

VROM 1985. *Handling uncertainty in environmental impact assessment* (MER Series, vol. 18). The Hague: Netherlands Ministry of Public Housing, Physical Planning and Environmental Affairs.

Wathern, P. 1984. Ecological modeling in impact analysis. In *Planning and ecology*, R. D. Roberts & T. M. Roberts (eds), 80–93. London: Chapman & Hall.

Weaver, AvB., T. Greyling, B. W. Van Wilyer, F. J. Kruger 1996. Managing the EIA process: logistics and team management of a large environmental impact assessment – proposed dune mine at St. Lucia, South Africa. *Environmental Impact Assessment Review* **16**, 103–13.

Wenger, R. B. & C. R. Rhyner 1972. Evaluation of alternatives for solid waste systems. *Journal of Environmental Systems* **2**(2), 89–108.

Wood, C. & N. Lee (eds) 1987. *Environmental impact assessment: five training case studies*. Occasional Paper 19, Department of Town and Country Planning, University of Manchester.

Notes

1   Jones et al. (1991) note that of 100 EISs prepared between 1988 and 1989 34 discussed alternatives. A more recent study by Eastman (1997) indicates a rise in this proportion to over 50 per cent for a sample of EISs submitted since 1991.

2   In the US, "agencies should: consider the option of doing nothing; consider alternatives outside the remit of the agency; and consider achieving only a part of their objectives in order to reduce impact".

3   For instance, the EES's assumption that individual indicators of water quality (such as dissolved oxygen at 31 points) are more important than employment opportunities and housing put together (at 26 points) would certainly be challenged by large sectors of the public.

4   Another category of techniques, simulation models, was not discussed because they are still relatively undeveloped and have, to date, been applied only to problems involving a few environmental impacts.

CHAPTER 5
Impact prediction, evaluation and mitigation

5.1 Introduction

The focus of this chapter is the central steps of impact prediction, evaluation and mitigation. This is the heart of the EIA process, although as we have already noted the process is not linear. Indeed the whole EIA exercise is about prediction. It is needed at the earliest stages, when a project, including its alternatives, is being planned and designed, and it continues through to mitigation, monitoring and auditing. Yet, despite the centrality of prediction in EIA, there is a tendency for many studies to underemphasize it at the expense of more descriptive studies. Prediction is often not treated as an explicit stage in the process; clearly defined models are often missing from studies. Even when used, models are not detailed, and there is little discussion of limitations. Section 5.2 examines the dimensions of prediction (what to predict), the methods and models used in prediction (how to predict), and the limitations implicit in such exercises (living with uncertainty).

Evaluation follows from prediction and involves an assessment of the relative significance of the impacts. Methods range from the simple to the complex, from the intuitive to the analytical, from qualitative to quantitative, from formal to informal. Cost–benefit analysis, monetary valuation techniques and multi-criteria/multi-attribute methods, with their scoring and weighting systems, provide a number of ways into the evaluation issue. The chapter concludes with a discussion of approaches to the mitigation of significant adverse effects. This may involve measures to avoid, reduce, remedy or compensate for the various impacts associated with projects.

5.2 Prediction

Dimensions of prediction (what to predict)

The object of prediction is to identify the magnitude and other dimensions of identified change in the environment *with* a project or action, in comparison with the situation *without* that project or action. Predictions also provide the basis for the assessment of significance, which we discuss in Section 5.3.

One starting point to identify the dimensions of prediction in the UK is the *legislative requirements* (see Table 3.4, paras 2c and 3c). These basic specifications are amplified in guidance given in *Environmental assessment: a guide to the procedures* (DOE 1989) as outlined in Table 5.1. As already noted, this listing is limited on the assessment of socio-economic impacts. Table 1.3 provides a broader view of the scope of the environment, and of the environmental receptors that may be affected by a project.

Prediction involves the identification of potential change in indicators of such environment receptors. Scoping will have identified the broad categories of impact in relation to the project under consideration. If a particular environmental indicator (e.g. SO_2 levels in the air) revealed an increasing problem in an area, irrespective of the project or action (e.g. a power station), this should be predicted forwards as the baseline for this particular indicator. These indicators need to be disaggregated and specified to provide variables that are measurable and relevant. For example, an economic impact could be progressively specified as

direct employment → local employment → local skilled employment.

In this way, a list of significant impact indicators of policy relevance can be developed.

An important distinction is often made between the prediction of the likely *magnitude* (i.e. size) and the *significance* (i.e. the importance for decision-making) of the impacts. Magnitude does not always equate with significance. For example, a large increase in one pollutant may still result in an outcome within generally accepted standards, whereas a small increase in another may take it above the applicable standards (see Fig. 5.1). In terms of the Sassaman checklist (see Fig. 4.8), the latter is crossing the threshold of concern and the former is not. This also highlights the distinction between *objective* and *subjective* approaches. The prediction of the magnitude of an impact should be an objective exercise, although it is not always easy. The determination of significance is a more subjective exercise, as it normally involves value judgements.

As Table 1.4 showed, prediction should also identify *direct* and *indirect* impacts (simple cause–effect diagrams may be useful here), the *geographical extent* of impacts (e.g. local, regional, national), whether the impacts are *beneficial* or *adverse*, and the *duration* of the impacts. In addition to prediction over the life of a project (including, for example, its construction, operational and other stages), the analyst should also be alert to the "rate of change" of impacts. A slow build-up in an impact may be more acceptable than a rapid change; the development of tourism projects in formerly remote or undeveloped areas provides a topical example of the damaging impacts of rapid change. Projects may be characterized by non-linear processes, by delays between cause and effect, and the intermittent nature of some impacts should be anticipated. The reversibility or otherwise of impacts, their permanency, and their cumulative and synergistic impacts should also be predicted. Cumulative (or additive) impacts are the collective effects of impacts that may be

Table 5.1 Assessment of effects, as outlined in UK regulations.

Assessment of effects

(including direct and indirect, secondary, cumulative, short-, medium- and long-term, permanent and temporary, positive and negative effects of project)

Effects on human beings, buildings and man-made features
1   Change in population arising from the development, and consequential environment effects.
2   Visual effects of the development on the surrounding area and landscape.
3   Levels and effects of emissions from the development during normal operation.
4   Levels and effects of noise from the development.
5   Effects of the development on local roads and transport.
6   Effects of the development on buildings, the architectural and historic heritage, archaeological features, and other human artefacts, e.g. through pollutants, visual intrusion, vibration.

Effects on flora, fauna and geology
7   Loss of, and damage to, habitats and plant and animal species.
8   Loss of, and damage to, geological, palaeotological and physiographic features.
9   Other ecological consequences.

Effects on land
10 Physical effects of the development, e.g. change in local topography, effect of earth-moving on stability, soil erosion, etc.
11 Effects of chemical emissions and deposits on soil of site and surrounding land.
12 Land-use/resource effects:
    (a) quality and quantity of agricultural land to be taken;
    (b) sterilization of mineral resources;
    (c) other alternative uses of the site, including the "do nothing" option;
    (d) effect on surrounding land-uses including agriculture;
    (e) waste disposal.

Effects on water
13 Effects of development on drainage pattern in the area.
14 Changes to other hydrographic characteristics, e.g. ground water level, water courses, flow of underground water.
15 Effects on coastal or estuarine hydrology.
16 Effects of pollutants, waste, etc., on water quality.

Effects on air and climate
17 Level and concentration of chemical emissions and their environmental effects.
18 Particulate matter.
19 Offensive odours.
20 Any other climatic effects.

Other indirect and secondary effects associated with the project
21 Effects from traffic (road, rail, air, water) related to the development.
22 Effects arising from the extraction and consumption of materials, water, energy or other resources by the development.
23 Effects of other development associated with the project, e.g. new roads, sewers, housing power lines, pipelines, telecommunications, etc.
24 Effects of association of the development with other existing or proposed development.
25 Secondary effects resulting from the interaction of separate direct effects listed above.

(*Source:* DOE 1989)

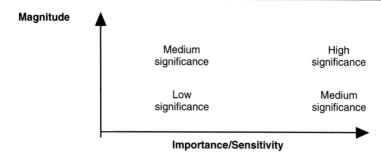

Figure 5.1 Significance expressed as a function of impact magnitude and the importance/sensitivity of the resources or receptors. (Adapted from English Nature 1994, and Institute of Environmental Assessment (IEA) and Landscape Institute 1995)

individually minor but in combination, often over time, major. Such cumulative impacts are difficult to predict, and are often poorly covered or are missing altogether from EIA studies (see Ch. 12).

Another dimension is the unit of measurement, and the distinction between *quantitative* and *qualitative* impacts. Some indicators are more readily quantifiable than others (e.g. a change in the quality of drinking water, in comparison, for example, with changes in community stress associated with a project). Where possible, predictions should present impacts in explicit units, which can provide a basis for evaluation and trade-off. Quantification can allow predicted impacts to be assessed against various local, national and international standards. Predictions should also include estimates of the *probability* that an impact will occur, which raises the important issue of uncertainty.

Methods and models for prediction (how to predict)

There are many possible methods to predict impacts; a study undertaken by Environmental Resources Ltd for the Dutch government in the early 1980s identified 150 different prediction methods used in just 140 EIA studies from the Netherlands and North America (VROM 1984). None provides a magic solution to the prediction problem.

> All predictions are based on conceptual models of how the universe functions; they range in complexity from those that are totally intuitive to those based on explicit assumptions concerning the nature of environmental processes . . . the environment is never as well behaved as assumed in models, and the assessor is to be discouraged from accepting off-the-shelf formulae (Munn 1979).

Predictive methods can be classified in many ways; they are not mutually exclusive. In terms of *scope*, all methods are *partial* in their coverage of impacts, but some seek to be more *holistic* than others. Partial methods may be classified according to type of project (e.g. retail impact assessment), and type of impact (e.g.

wider economic impacts). Some may be *extrapolative*, others may be more *normative*. For extrapolative methods, predictions are made that are consistent with past and present data. Extrapolative methods include, for example: trend analysis (extrapolating present trends, modified to take account of changes caused by the project), scenarios (common-sense forecasts of future state based on a variety of assumptions), analogies (transferring experience from elsewhere to the study in hand) and intuitive forecasting (e.g. the use of the Delphi technique to achieve group consensus on the impacts of a project) (Green et al. 1989). Normative approaches work backwards from desired outcomes to assess whether a project, in its environmental context, is adequate to achieve them. For example, a desired socio-economic outcome from the construction stage of a major project may be 50 per cent local employment. The achievement of this outcome may necessitate modifications to the project and/or to associated employment policies (e.g. on training). Various scenarios may be tested to determine the one most likely to achieve the desired outcomes.

Methods can also be classified according to their *form*, as the following six types of model illustrate.

Mechanistic or mathematical models

Mechanistic or mathematical models describe cause–effect relationships in the form of flow charts or mathematical functions. The latter can range from simple direct input–output relationships to more complex dynamic mathematical models with a wide array of interrelationships. Mathematical models can be spatially aggregated (e.g. a model to predict the survival rate of a cohort population, or an economic multiplier for a particular area), or more locationally based, predicting net changes in detailed locations throughout a study area. Of the latter, retail impact models, which predict the distribution of retail expenditure using gravity model principles, provide a simple example; the comprehensive land-use locational models of Harris, Lowry, Cripps et al., provide more holistic examples (*Journal of American Institute of Planners* 1965). Mathematical models can also be divided into deterministic and stochastic models. Deterministic models, like the gravity model, depend on fixed relationships. In contrast, a stochastic model is probabilistic, and indicates "the degree of probability of the occurrence of a certain event by specifying the statistical probability that a certain number of events will take place in a given area and/or time interval" (Loewenstein 1966).

There are many mathematical models available for particular impacts. Reference to various EISs, especially from the USA, and to the literature (e.g. Bregman & Mackenthun 1992, Hansen & Jorgensen 1991, Rau & Wooten 1980, Suter 1993, US Environmental Protection Agency 1993, Westman 1985) reveals the availability of a rich array. For instance, Kristensen et al. (1990) list 21 mathematical models for phosphorus retention in lakes alone. Figure 5.2 provides a simple flow diagram for the prediction of the local socio-economic impacts of a power station development. Key determinants in the model are the details of the labour requirements for the project, the conditions in the local economy, and the policies of the relevant local

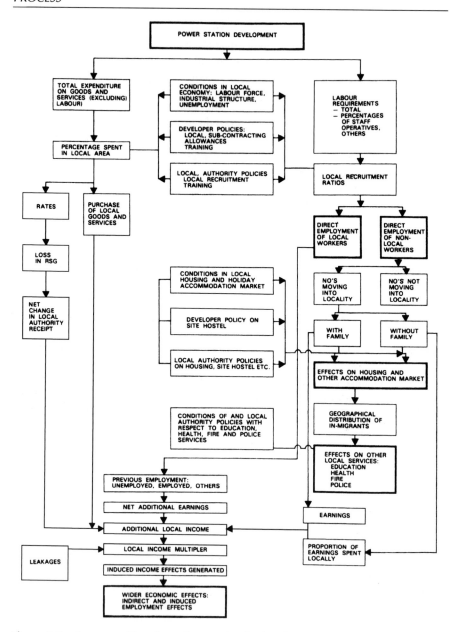

Figure 5.2 A cause–effect flow diagram for the local socio-economic impacts of a power station proposal. (*Source:* Glasson et al. 1987)

$$Y_r = \frac{1}{1-(1-s)(1-t-u)(1-m)} J$$

where

Y_r = change in level of income (Y) in region (r), in £

J = initial income injection (or multiplicand)

t = proportion of additional income paid in direct taxation and National Insurance contributions

s = proportion of income saved (and therefore not spent locally)

u = decline in transfer payments (e.g. unemployment benefits) which result from the rise in local income and employment

m = proportion of additional income spent on imported consumer goods

Figure 5.3 A simple multiplier model for the prediction of local economic impacts.

authority and developer on topics such as training, local recruitment and travel allowances. The local recruitment ratio is a crucial factor in the determination of subsequent impacts.

An example of a deterministic mathematical model, often used in socio-economic impact predictions, is the multiplier (Lewis 1988), an example of which is shown in Figure 5.3. The injection of money into an economy – local, regional or national – will increase income in the economy by some multiple of the original injection. Modification of the basic model allows it to be used to predict income and employment impacts for various groups over the stages of the life of a project (Glasson et al. 1988). The more disaggregated (by industry type) input–output member of the multiplier family provides a particularly sophisticated method for predicting economic impacts, but with major data requirements.

Mass balance models
Mass balance models establish a mass balance equation for a given "compartment", namely a defined physical entity, such as the water in a stream, a volume of soil, or an organism. Inputs to the compartment could be, for instance, water, energy, food or chemicals; outputs could be outflowing water, wastes, or diffusion to another compartment. Changes in the contents of the compartment equal the sum of the inputs minus the sum of the outputs, as illustrated in Figure 5.4. Mass balance models are particularly effective for describing physical changes such as the flow of water in a river basin or the flow of energy through an ecosystem.

Statistical models
Statistical models use statistical techniques such as regression or principal components analysis to describe the relationship between data, to test hypotheses or to extrapolate data. For instance, they can be used in a pollution-monitoring study to describe the concentration of a pollutant as a function of the stream-flow rates and the distance down stream. They can compare conditions at a contaminated site and a control site to determine the significance of any differences in monitoring data.

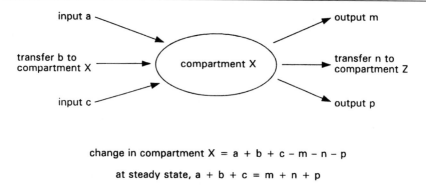

change in compartment X = a + b + c − m − n − p

at steady state, a + b + c = m + n + p

Figure 5.4   An example of a mass balance model.

They can extrapolate a model to conditions outside the data range used to derive the model – e.g. from toxicity at high doses of a pollutant to toxicity at low doses – or from data that are available to data that are unavailable – e.g. from toxicity in rats to toxicity in humans.

Physical, image or architectural models
Physical, image or architectural models are illustrative or scale models that replicate some element of the project–environment interaction. For example a scale
model (or computer graphics) could be used to predict the impacts of a development on the landscape or built environment.

Field and laboratory experimental methods
Field and laboratory experimental methods use existing data inventories, often supplemented by special surveys, to predict impacts on receptors. Field tests are carried out in unconfined conditions, usually at approximately the same scale as the
predicted impact; an example would be the testing of a pesticide in an outdoor
pond. Laboratory tests, such as the testing of a pollutant on seedlings raised in a
hydroponic solution, are usually cheaper to run but may not extrapolate well to
conditions in natural systems.

Analogue models
Analogue models make predictions based on analogous situations. They include
comparing the impacts of a proposed development with a similar existing development; comparing the environmental conditions at one site with those at similar sites
elsewhere; comparing an unknown environmental impact (e.g. of wind turbines on
radio reception) with a known environmental impact (e.g. of other forms of development on radio reception). Analogue models can be developed from site visits,
literature searches, or the monitoring of similar projects. Expert opinion, based, it is
to be hoped, on previous relevant experience, may also be used.

Other methods for prediction

The various impact identification methods discussed in Chapter 4 may also be of value in impact prediction. The Sassaman threshold of concern checklist has already been noted; the Leopold matrix also includes magnitude predictions, although the objectivity of a system where each analyst is allowed to develop a ranking system on a scale of 1 to 10 is somewhat doubtful. Overlays can be used to predict spatial impacts, and the Sorensen network is useful in tracing through indirect impacts.

Choice of prediction methods

When choosing prediction methods, an assessor should be concerned about their appropriateness for the task involved, in the context of the resources available (Lee 1987). Will the methods produce what is wanted (e.g. a range of impacts, for the appropriate geographical area, over various stages), from the resources available (including time, data, range of expertise)? In addition, the criteria of replicability (method is free from analyst bias), consistency (method can be applied to different projects to allow predictions to be compared) and adaptability should also be considered in the choice of methods. In many cases more than one method may be appropriate. For instance the range of methods available for predicting impacts on air quality is apparent from the 165 closely typed pages on the subject in Rau & Wooten (1980). Table 5.2 provides an overview of some of the methods of predicting the initial emissions of pollutants, which, with atmospheric interaction, may degrade air quality, which may then have adverse effects, for example on humans.

In practice, there has been a tendency to use the less formal predictive methods, and especially expert opinion (VROM 1984). Even where more formal methods have been used, they have tended to be simple, for example the use of photo-montages for visual impacts, or of simple dilution and steady-state dispersion models for water quality. However, simple methods need not be inappropriate, especially for early stages in the EIA process, nor need they be applied uncritically or in a simplistic way. Lee (1987) provides the following illustration:

(a) a single expert may be asked for a brief, qualitative opinion; or
(b) the expert may also be asked to justify that opinion (i) by verbal or mathematical description of the relationships he has taken into account and/or (ii) by indicating the empirical evidence which supports that opinion; or
(c) as in (b), except that opinions are also sought from other experts; or
(d) as in (c), except that the experts are also required to reach a common opinion, with supporting reasons, qualifications, etc.; or
(e) as in (d), except that the experts are expected to reach a common opinion using an agreed process of consensus building (e.g. based on "Delphi" techniques (Golden et al. 1979)).

The development of more complex methods can be very time-consuming and expensive, especially since many of these models are limited to specific environmental components and physical processes, and may only be justified when a number

137

Table 5.2 Examples of methods used in predicting air quality impacts.

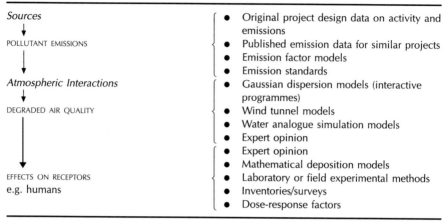

Sources ↓	
POLLUTANT EMISSIONS ↓	• Original project design data on activity and emissions • Published emission data for similar projects • Emission factor models • Emission standards
Atmospheric Interactions ↓	• Gaussian dispersion models (interactive programmes) • Wind tunnel models • Water analogue simulation models • Expert opinion
DEGRADED AIR QUALITY ↓	
EFFECTS ON RECEPTORS e.g. humans	• Expert opinion • Mathematical deposition models • Laboratory or field experimental methods • Inventories/surveys • Dose-response factors

(*Source:* VROM 1984, Rau & Wooten 1980)

of relatively similar projects are proposed. However, notwithstanding the emphasis on the simple informal methods, there is scope for mathematical simulation models in the prediction stage. Munn (1979) identifies a number of criteria for situations in which computer-based simulation or mathematical models would be useful. The following are some of the most relevant:

• the assessment requires the handling of large numbers of simple calculations;
• there are many complex links between the elements of the EIA;
• the affected processes are time-dependent;
• increased definitions of assumptions and elements will be valuable in drawing together the many disciplines involved in the assessment;
• some or all of the relationships of the assessment can only be defined in terms of statistical probabilities.

Living with uncertainty

Environmental impact statements often appear more certain in their predictions than they should. This may reflect a concern not to undermine credibility and/or an unwillingness to attempt to allow for uncertainty. All predictions have an element of uncertainty, but it is only in recent years that such uncertainty has begun to be acknowledged in the EIA process (De Jongh 1988). There are many sources of uncertainty relevant to the EIA process as a whole. In their classic works on strategic choice, Friend & Jessop (1977) and Friend & Hickling (1987) identified three broad classes of uncertainty: uncertainties about the physical, social and economic environment (UE), uncertainties about guiding values (UV) and uncertainties about related decisions (UR) (see Fig. 5.5). All three classes of uncertainty may affect the

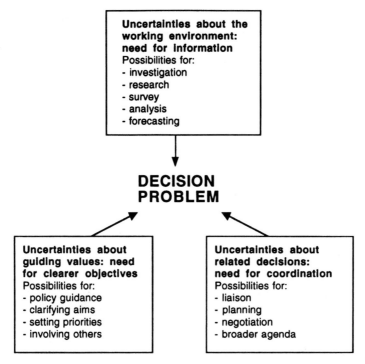

Figure 5.5 The types of uncertainty in decision-making. (*Source:* Friend & Hickling 1987)

accuracy of predictions, but the focus in an EIA study is usually on uncertainty about the environment. This may include the use of inaccurate and/or partial information on the project and on baseline–environmental conditions, unanticipated changes in the project during one or more of the stages of the life-cycle, and oversimplification and errors in the application of methods and models. Socio-economic conditions may be particularly difficult to predict, as underlying societal values may change quite dramatically over the life, say 30–40 years, of a project.

Uncertainty in EIA predictive exercises can be handled in several ways. The assumptions underpinning predictions should be clearly stated (Voogd 1983). Issues of probability and confidence in predictions should be addressed, and ranges may be attached to predictions within which the analyst is *n* per cent confident that the actual outcome will lie. For example, scientific research may conclude that the 95 per cent confidence interval for the noise associated with a new industrial project is 65–70 dBA, which means that only 5 times out of 100 would the dBA be expected to be outside this range. Tomlinson (1989) draws attention to the twin issues of probability and confidence involved in predictions.

These twin factors are generally expressed through the same word. For example, in the prediction "a major oil spill would have major ecological consequences", a high degree of both probability and confidence exists. Situations may arise, however, where a low probability event based upon a low level of confidence is predicted. This is potentially more serious than a higher probability event with high confidence, since low levels of confidence may preclude expenditure on mitigating measures, ignoring issues of significance. Monitoring measures may be an appropriate response in such situations.

It may also be useful to show impacts under "peak" as well as "average" conditions for a particular stage of a project; this may be very relevant in the construction stage of major projects.

Sensitivity analysis may be used to assess the consistency of relationships between variables. If the relationship between input A and output B is such that whatever the changes in A there is little change in B, then no further information may be needed. However, where the effect is much more variable, there may be a need for further information. Of course, the best check on the accuracy of predictions is to check on the outcomes of the implementation of a project after the decision. This is too late for the project under consideration, but could be useful for future projects. Conversely, the monitoring of outcomes of similar projects may provide useful information for the project in hand. Holling (1978), who believes that the "core issue of EIA is how to cope with decision-making under uncertainty", recommends a policy of adaptive environmental impact assessment, with periodic reviews of the EIA through a project's life-cycle. Another procedural approach would be to require an *uncertainty report* as one step in the process; such a report would bring together the various sources of uncertainty associated with a project and the means by which they might be reduced (uncertainties are rarely eliminated).

5.3 Evaluation

Evaluation in the EIA process

Once impacts have been predicted, there is a need to assess their relative significance. Criteria for significance include the magnitude and likelihood of the impact and its spatial and temporal extent, the likely degree of the affected environment's recovery, the value of the affected environment, the level of public concern, and political repercussions. As with prediction, the choice of evaluation method should be related to the task in hand and to the resources available. Evaluation should feed into most stages of the EIA process, but the nature of the methods used may vary, for example according to the number of alternatives under consideration, according to the level of aggregation of information, and according to the number and type of parties involved (e.g. "in house" and/or "external" consultation).

Evaluation methods can be of various types, including simple or complex, formal or informal, quantitative or qualitative, aggregated or disaggregated (see Voogd

Table 5.3 Determinants of environmental significance.

Environmental significance is a judgement made by the Authority (West Australian Environmental Protection Authority) and is based upon the following factors:

(a) character of the receiving environment and the use and value which society has assigned to it;
(b) magnitude, spatial extent and duration of anticipated change;
(c) resilience of the environment to cope with change;
(d) confidence of prediction of change;
(e) existence of policies, programmes, plans and procedures against which the need for applying the environmental impact assessment proposal to a process can be determined;
(f) existence of environmental standards against which a proposal can be assessed; and
(g) degree of public interest in environmental issues likely to be associated with a proposal.

(*Source:* West Australian Environmental Protection Authority 1993)

1983, Maclaren & Whitney 1985). Much, if not most, current evaluation of significance in EIA is simple and often pragmatic, drawing on experience and expert opinion rather than on complex and sophisticated analysis. Table 5.3 provides an example of key factors used in Western Australia, where there is a particularly well-developed EIA system (see Ch. 11 also). To the factors in Table 5.3 could also be added scope for mitigation, sustainability and reversibility. The factor of public interest or perception (g in Table 5.3) is an important consideration, and past and current perceptions of the significance of particular issues and impacts can raise their profile in the evaluation.

The most formal evaluation method is the *comparison of likely impacts against legal requirements and standards* (e.g. air quality standards, building regulations). Table 5.4 illustrates some of the standards which may be used to evaluate the traffic noise impacts of projects in Britain. Table 5.5 provides an example of more general guidance on standards and on environmental priorities and preferences, from the European Commission, for tourism developments. Beyond this, all assessments of significance either implicitly or explicitly apply weights to the various impacts (i.e. some are assessed as more important than others). This involves interpretation and the application of judgement. Such judgement can be rationalized in various ways and a range of methods are available, but all involve values and all are subjective. Parkin (1992) sees judgements as being on a continuum between an analytical mode and an intuitive mode. In practice, many are at the intuitive end of the continuum, but such judgements, made without the benefit of analysis, are likely to be flawed, inconsistent and biased. The "social effects of resource allocation decisions are too extensive to allow the decision to 'emerge' from some opaque procedure free of overt political scrutiny" (Parkin 1992). Analytical methods seek to introduce a rational approach to evaluation.

Two sets of methods are distinguished: those that assume a common utilitarian ethic with a single evaluation criterion (money), and those based on the measurement of personal utilities, including multiple criteria. The *cost-benefit analysis* (CBA) approach, which seeks to express impacts in monetary units, falls into the former

Table 5.4 Examples of standards in relation to impacts of projects on traffic noise in Britain.

- BS 7445 is the standard for description and measurement of environmental noise. It is in three parts: Part 1: Guide to quantities and procedures, Part 2: Guide to acquisition of data, and Part 3: Guide to application of noise limits.
- Noise is measured in decibels (dB) at a given frequency. This is an objective measure of sound pressure. Measurements are made using a calibrated sound meter.
- Human hearing is approximately in the range 0–140 dBA.

  dB   Example of noise

  <40   quiet bedroom
   60   busy office
   72   car at 60 km/h at a distance of 7 metres
   85   HGV at 40 km/h at a distance of 7 m
   90   hazardous to hearing from continuous exposure
  105   jet flying overhead at 250 m
  120   threshold of pain

- Traffic noise is perceived as a nuisance even at low dB levels. Noise comes from tyres on the road, engines, exhausts, brakes and HGV bodies. Poor maintenance of roads and vehicles and poor driving also increase road noise. Higher volumes of traffic and higher proportions of HGVs increase the noise levels. In general, annoyance is proportional to traffic flow for noise levels above 55 dB(A). People are sensitive to a change in noise levels of 1 dB (about 25% change in flow).
- Assessment of traffic noise is assessed in terms of impacts within 300 metres of the road. The EIA will estimate the number of properties and relevant locations (e.g. footpaths and sports fields) in bands of distance from the route: 0–50 m, 50–100 m, 100–200 m, 200–300 m, and then classify each group according to the baseline ambient noise levels (in bands of <50, 50–60, 60–70, >70 dB(A)) and the increase in noise (1–3, 3–5, 5–10, 10–15 and >15 dB(A)).
- Façade noise levels are measured at 1.7 m above ground, 1 m from façade or 10 m from kerb, and are usually predicted using the DTP's Calculation of Road Traffic Noise (CRTN) which measures dB(A) $L_{A10,18hour}$. This is the noise level exceeded 10% of the time between 6:00 and 24:00. Noise levels at the façade are approximately 2 dB higher than 10 m from the building. PPG 13 uses dB(A) $L_{Aeq,16hour}$. This is between 7:00 and 23:00. Most traffic noise meters use dB(A) L_{A10}, and an approximate conversion is:

  $$L_{Aeq,16hour} = L_{A10,18hour} - 2dB.$$

- The DTP recommends an absolute upper limit for noise of 72 dB(A) $L_{eq,18hour}$ (=70 dB(A) $L_{A10,18hour}$) for residential properties. Compensation is payable to properties within 300 metres of a road development for increases greater than 1 dB(A) which result in $L_{A10,18hour}$ above 67.5.
- The DTP considers a change of 30% slight, 60% moderate and 90% substantial. PPG 13 considers 5% to be significant.

  There are four categories of noise in residential areas
      day (16hr)   *night (8hr)*

  A   <55 L_{Aeq} < 42 L_{Aeq} Not determining the application
  B   55–63 L_{Aeq} 42–57 L_{Aeq} Noise control measures are required
  C   63–72 L_{Aeq} 57–66 L_{Aeq} Strong presumption against developer
  D   >72 L_{Aeq} >66 L_{Aeg} Normally refuse the application

  For night-time noise, unless the noise is already in category D, a single event occurring regularly (eg HGV movements) where $L_{Aeq} > 82$ dB puts the noise in category C.

(*Source:* Bourdillon 1996)

Table 5.5 Example of EC guidance on assessing significance of impacts for tourism projects for Asian, Caribbean and Pacific countries.

The significance of certain environmental impacts can be assessed by contrasting the predicted magnitude of impact against a relevant environmental standard or value. For tourism projects in particular, impact significance should also be assessed by taking due regard of those environmental priorities and preferences held by society but for which there are no quantifiable objectives. Particular attention needs to be focused upon the environmental preferences and concerns of those likely to be directly affected by the project.

Environmental Standards

● Water quality standards
◊ potable water supplies (*apply country standards; see also Section 1.3.2,* WHO *(1982) Guidelines for Drinking Water Quality Directives 80/778/*EEC *and 75/440/*EEC*).*
◊ wastewater discharge (*apply country standards for wastewaters and fisheries; see also 76/160/*EEC *and 78/659/*EEC*).*

● National and local planning regulations
◊ legislation concerning change in land-use
◊ regional/local land-use plans (particularly management plans for protected areas and coastal zones).

● National legislation to protect certain areas
◊ national parks
◊ forest reserves
◊ nature reserves
◊ natural, historical or cultural sites of importance.

● International agreements to protect certain areas
◊ World Heritage Convention
◊ Ramsar Convention on wetlands.

● Conservation/preservation of species likely to be sold to tourists or harmed by their activities
◊ national legislation
◊ international conventions
◊ CITES Convention on trade in endangered species.

Environmental Priorities and Preferences

● Participation of affected people in project planning to determine priorities for environmental protection, including:
◊ public health
◊ revered areas, flora and fauna (e.g. cultural/medicinal value, visual landscape)
◊ skills training to undertake local environmental mitigation measures
◊ protection of potable water supply
◊ conservation of wetland/tropical forest services and products, e.g. hunted wildlife, fish stocks
◊ issues of sustainable income generation and employment (including significance of gender – see WID *manual*)

● Government policies for environmental protection (including, where appropriate, incorporation of objectives from Country Environmental Studies/Environmental Action Plans etc.)

● Environmental priorities of tourism boards and trade associations representing tour operators.

(*Source:* CEC 1993)

category. A variety of methods, including *multi-criteria analysis, decision analysis* and *goals achievement,* fall into the latter. The very growth of EIA is partly a response to the limitations of CBA and to the problems of the monetary valuation of environmental impacts. Yet, after two decades of limited concern, there is renewed interest in the monetizing of environmental costs and benefits (DOE 1991). The multi-criteria/multi-attribute methods involve scoring and weighting systems that are also not problem-free. The various approaches are now outlined. In practice, there are many hybrid variations between these two main categories, and these are referred to in both categories.

Cost-benefit analysis and monetary valuation techniques

Cost-benefit analysis itself lies in a range of project and plan appraisal methods that seek to apply monetary values to costs and benefits (Lichfield et al. 1975). At one extreme are *partial* approaches, such as financial-appraisal, cost-minimization and cost-effectiveness methods, which consider only a subsection of the relevant population or only a subsection of the full range of consequences of a plan or project. *Financial appraisal* is limited to a narrow concern, usually of the developer, with the stream of financial costs and returns associated with an investment. *Cost-effectiveness* involves selecting an option that achieves a goal at least cost (for example devising a least-cost approach to produce coastal bathing waters that meet the CEC Blue Flag criteria). The cost-effectiveness approach is more problematic where there are a number of goals and where some actions achieve certain goals more fully than others (Winpenny 1991).

Cost-benefit analysis is more *comprehensive* in scope. It takes a long view of projects (farther as well as nearer future) and a wide view (in the sense of allowing for side-effects). It is based in welfare economics and seeks to include all the relevant costs and benefits to evaluate the net social benefit of a project. It was used extensively in the UK in the 1960s and early 1970s for public-sector projects, the most famous being the Third London Airport (HMSO 1971). The methodology of CBA has several stages: project definition, the identification and enumeration of costs and benefits, the evaluation of costs and benefits, the discounting and presentation of results. Several of the stages are similar to those in EIA. The basic evaluation principle is to measure in monetary terms where possible – as money is the common measure of value and monetary values are best understood by the community and decision-makers – and then reduce all costs and benefits to the same capital or annual basis. Future annual flows of costs and benefits are usually discounted to a net present value (see Table 5.6). A range of interest rates may be used to show the sensitivity of the analysis to changes. If the net social benefit minus cost is positive, then there may be a presumption in favour of a project. However, the final outcome may not always be that clear. The presentation of results should distinguish between tangible and intangible costs and benefits, as relevant, allowing the decision-maker to consider the trade-offs involved in the choice of an option.

144

Table 5.6 Cost-benefit analysis: presentation of results: tangibles and intangibles.

Category	Alternative 1	Alternative 2
Tangibles		
Annual benefits	£ B1	£ b1
	£ B2	£ b2
	£ B3	£ b3
Total annual benefit	£ B1 + B2 + B3	£ b1 + b2 + b3
Annual costs	£ C1	£ c1
	£ C2	£ c2
	£ C3	£ c3
Total annual costs	£ C1 + C2 + C3	£ c1 + c2 + c3
Net discounted present value (NDPV) of benefits and costs over "m" years at X%*	£ D	£ E
Intangibles		
Intangibles are likely to include costs and benefits	I1	i1
	I2	i2
	I3	i3
	I4	i4
Intangibles summation (undiscounted)	I1 + I2 + I3 + I4	i1 + i2 + i3 + i4

* e.g. NPDV (Alt 1)

$$D = \sum \left[\frac{B1}{(1+X)^1} + \frac{B1}{(1+X)^2} + \ldots + \frac{B1}{(1+X)^n} + \frac{B2}{(1+X)^1} + \ldots + \frac{B2}{(1+X)^n} + \frac{B3}{(1+X)^1} + \ldots + \frac{B3}{(1+X)^n} \right]$$

$$- \sum \left[\frac{C1}{(1+X)^1} + \frac{C1}{(1+X)^2} + \ldots + \frac{C1}{(1+X)^n} + \frac{C2}{(1+X)^1} + \ldots + \frac{C2}{(1+X)^n} + \frac{C3}{(1+X)^1} + \ldots + \frac{C3}{(1+X)^n} \right]$$

CBA has excited both advocates (e.g. Dasgupta & Pearce 1978, Pearce et al. 1989, Pearce 1989) and opponents (e.g. Bowers 1990). It does have many problems, including identifying, enumerating and monetizing intangibles. Many environmental impacts fall into the intangible category, for example the loss of a rare species, the urbanization of a rural landscape and the saving of a human life. The incompatibility of monetary and non-monetary units makes decision-making problematic (Bateman 1991). Another problem is the choice of discount rate: for example, should a very low rate be used to prevent the rapid erosion of future costs and benefits in the analysis? This choice of rate has profound implications for the evaluation of resources for future generations. There is also the underlying and fundamental problem of the use of the single evaluation criterion of money, and the assumption that £1 is worth the same to any person, whether a tramp or a millionaire, a resident of a rich commuter belt or of a poor and remote rural community. CBA also ignores distributional effects and aggregates costs and benefits to estimate the change in the welfare of society as a whole.

The *planning balance sheet* (PBS) is a variation on the theme of CBA, and it goes beyond CBA in its attempts to identify, enumerate and evaluate the distribution of costs and benefits between the affected parties. It also acknowledges the difficulty of attempts to monetize the more intangible impacts. It was developed by Lichfield

	Plan A				Plan B			
	Benefits		Costs		Benefits		Costs	
	Capital	Annual	Capital	Annual	Capital	Annual	Capital	Annual
Producers								
X	£a	£b	–	£d	–	–	£b	£c
Y	i_1	i_2	–	–	i_3	i_4	–	–
Z	M_1	–	M_2	–	M_3	–	M_4	–
Consumers								
X´	–	£e	–	£f	–	£g	–	£h
Y´	i_5	i_6	–	–	i_7	i_8	–	–
Z´	M_1	–	M_3	–	M_2	–	M_4	–

£ = benefits and costs that can be monetized
M = where only a ranking of monetary values can be estimated
i = intangibles

Figure 5.6 Example of structure of a planning balance sheet.

et al. (1975) to compare alternative town plans. PBS is basically a set of social accounts structured into sets of "producers" and "consumers" engaged in various transactions. The transaction could, for example, be an adverse impact, such as noise from an airport (the producer) on the local community (the consumers), or a beneficial impact, such as the time savings resulting from a new motorway development (the producer) for users of the motorway (the consumers). For each producer and consumer group, costs and benefits are quantified per transaction, in monetary terms or otherwise, and weighted according to the numbers involved. The findings are presented in tabular form, leaving the decision-maker to consider the trade-offs, but this time with some guidance on the distributional impacts of the options under consideration (Fig. 5.6). More recently, Lichfield (1996) has sought to integrate EIA and PBS further in an approach he calls community impact evaluation (CIE).

Partly in response to the "intangibles" problem in CBA, there has also been considerable interest in the development of *monetary valuation techniques* to improve the economic measurement of the more intangible environmental impacts (DOE 1991, Winpenny 1991, Barde & Pearce 1991). The techniques can be broadly classified into direct and indirect, and they are concerned with the measurement of preferences about the environment rather than with the intrinsic values of the environment. The direct approaches seek to measure directly the monetary value of environmental gains – for example better air quality or an improved scenic view. Indirect approaches measure preferences for a particular effect via the establishment of a "dose-response"-type relationship. The various techniques found under the direct and indirect categories are summarized in Table 5.7. Such techniques can contribute to the assessment of the total economic value of an action or project, which should not only include user values (preferences people have for using an environmental asset, such as a river for fishing) but also non-user values (where

Table 5.7 Summary of environmental monetary valuation techniques.

Direct household production function (HPF)
HPF methods seek to determine expenditure on commodities that are substitutes or complements for an environmental characteristic to value changes in that characteristic. Subtypes include:
- Avertive expenditures: expenditure on various substitutes for environmental change (e.g. noise insulation as an estimate of the value of peace and quiet).
- Travel cost method: expenditure, in terms of cost and time, incurred in travelling to a particular location (e.g. a recreation site) is taken as an estimate of the value placed on the environmental good at that location (e.g. benefit arising from use of the site).

Direct hedonic price methods (HPM)
HPM methods seek to estimate the implicit price for environmental attributes by examining the real markets in which those attributes are traded. Again there are two main subtypes:
- Hedonic house/land prices: these prices are used to value characteristics such as "clean air" and "peace and quiet", through cross sectional data analysis (e.g. on house price sales in different locations).
- Wage risk premia: the extra payments associated with certain higher risk occupations are used to value changes in morbidity and mortality (and implicitly human life) associated with such occupations.

Direct experimental markets
Survey methods are used to elicit individual values for non-market goods. Experimental markets are created to discover how people would value certain environmental changes. Two kinds of questioning, of a sample of the population, may be used:
- Contingent valuation method (CVM): people are asked what they are willing to pay (WTP) for keeping X (e.g. a good view, an historic building) or preventing Y, or what they are willing to accept (WTA) for losing A, or tolerating B.
- Contingent ranking method (CRM) or stated preference: people are asked to rank their preferences for various environmental goods, which may then be valued by linking the preferences to the real price of something traded in the market (e.g. house prices).

Indirect methods
Indirect methods seek to establish preferences through the estimation of relationships between a "dose" (e.g. reduction in air pollution) and an effect (e.g. health improvement). Approaches include:
- Indirect market price approach: the dose–response approach seeks to measure the effect (e.g. value of loss of fish stock) resulting from an environmental change (e.g. oil pollution of a fish farm), by using the market value of the output involved. The replacement-cost approach uses the cost of replacing or restoring a damaged asset as a measure of the benefit of restoration (e.g. of an old stone bridge eroded by pollution and wear and tear).
- Effect on production approach: where a market exists for the goods and services involved, the environmental impact can be represented by the value of the change in output that it causes. It is widely used in developing countries, and is a continuation of the dose–response approach.

(Adapted from DOE 1991, Winpenny 1991, Pearce & Markandya 1990, Barde & Pearce 1991)

people value an asset but do not use it, although some may wish to do so some day). Of course, such techniques have their problems, for example the potential bias in people's replies in the contingent valuation method (CVM) approach. However, simply through the act of seeking a value for various environmental features, such

Table 5.8 A comparison of different scoring systems.

Method	Alternatives				Basic of score
	A (no action)	B	C	D	
Ratio	65	62	71	75	Absolute $L_{10}dB_A$ measure
Interval	0	−3	+6	+10	Difference in $L_{10}dB_A$ using alternative A as base
Ordinal	B	A	C	D	Ranking according to ascending value of $L_{10}dB_A$
Binary	0	0	1	1	0 = less than $70L_{10}dB_A$ 1 = $70L_{10}dB_A$ or more

Base on Lee (1987).

techniques help to reinforce the understanding that such features are not "free" goods and should not be treated as such.

Scoring and weighting and multi-criteria methods

Multi-criteria and multi-attribute methods seek to overcome some of the deficiencies of CBA; in particular they seek to allow for a pluralist view of society, composed of diverse "stakeholders" with diverse goals and with differing values concerning environmental changes. Most of the methods use – and sometimes misuse – some kind of simple scoring and weighting system; such systems generate considerable debate. Here we discuss some key elements of good practice, and then offer a brief overview of the range of multi-criteria/multi-attribute methods available to the analyst.

Scoring may use quantitative or qualitative scales, according to the availability of information on the impact under consideration. Lee (1987) provides an example (see Table 5.8) of how different levels of impact (in this example noise, whose measurement is in units of $L_{10}dB_A$) can be scored in different systems. These systems seek to standardize the impact scores for purposes of comparison. Where quantitative data are not available, ranking of alternatives may use other approaches, for example using letters (A, B, C, etc.) or words (not significant, significant, very significant).

Weighting seeks to identify the relative importance of the various impact types for which scores of some sort may be available (for example the relative importance of a water pollution impact; the impact on a rare flower). Different impacts may be allocated weights (normally numbers) out of a total budget (e.g. 10 points to be allocated between 3 impacts). But by whom?

Multi-criteria/multi-attribute methods seek to recognize the plurality of views and weights in their methods; the Delphi approach also uses individuals' weights,

Table 5.9 Weighting, scoring and trade-offs.

Impact	Weight (w)	Scheme A Score (a)	(aw)	Scheme B Score (b)	(bw)
Noise	2	5	10	1	2
Loss of flora	5	1	5	4	20
Air pollution	3	2	6	2	6
Total			21		28

from which group weights are then derived. In many studies, however, the weights are those produced by the technical team. Indeed the decision-makers may be unwilling to reveal all their personal preferences, for fear of undermining their negotiating positions. This internalization of the weighting exercise does not destroy the use of weights, but it does emphasize the need for clarification of scoring and weighting systems and, in particular, for the identification of the origin of the weightings used in an EIA. Wherever possible, scoring and weighting should be used to reveal the trade-offs in impacts involved in particular projects or in alternatives. For example, Table 5.9 shows that the main issue is the trade-off between the impact on flora of one scheme and the impact on noise of the other scheme.

Several approaches to the scoring and weighting of impacts have already been introduced in the outline of impact identification methods in Chapter 4. The Leopold matrix includes measures of the significance of impacts (on a scale of 1 to 10) as well as of their magnitude. The matrix approach can also be usefully modified to identify the distribution of impacts among geographical areas and/or among various affected parties (Fig. 5.7). The quantitative EES and WRAM methods generate weights

Group environmental component	Project Action							
	Construction stage actions				Operational stage actions			
	A	B	C	D	a	b	c	d
Group 1 (e.g. indigenous population ≥ 45 years old) various								
• Social								
• Physical								
• Economic components								
Group 2 (e.g. indigenous population < 45 years old) various								
• Social								
• Physical								
• Economic components								

Figure 5.7 Simple matrix identification of distribution of impacts.

149

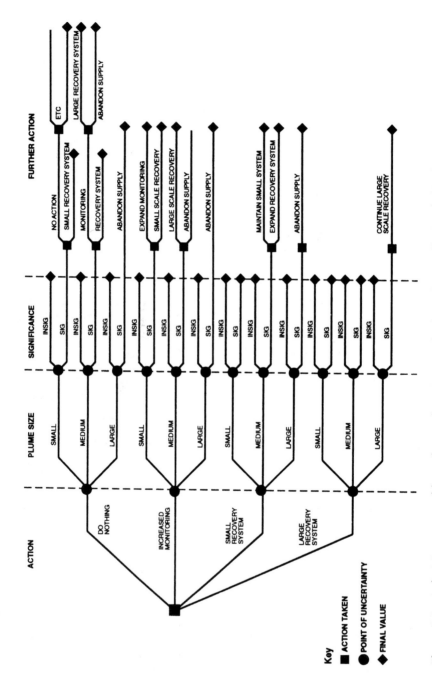

Figure 5.8 A decision tree: problem of groundwater contamination. (*Source:* De Jongh 1988)

Key

■ ACTION TAKEN

● POINT OF UNCERTAINTY

◆ FINAL VALUE

ACTION

PLUME SIZE

SIGNIFICANCE

FURTHER ACTION

DO NOTHING

INCREASED MONITORING

SMALL RECOVERY SYSTEM

LARGE RECOVERY SYSTEM

SMALL

MEDIUM

LARGE

INSIG

SIG

NO ACTION

SMALL RECOVERY SYSTEM

ETC

LARGE RECOVERY SYSTEM

MONITORING

RECOVERY SYSTEM

ABANDON SUPPLY

EXPAND MONITORING

SMALL SCALE RECOVERY

LARGE SCALE RECOVERY

ABANDON SUPPLY

ABANDON SUPPLY

MAINTAIN SMALL SYSTEM

EXPAND RECOVERY SYSTEM

ABANDON SUPPLY

CONTINUE LARGE SCALE RECOVERY

| Goal description: | | | α | | | β | |
| Relative weight: | | | 2 | | | 3 | |
Incidence	Relative weight	Costs	Benefits	Relative weight	Costs	Benefits
Group a	1	A	D	5	E	1
Group b	3	H	J	4	M	2
Group c	1	L	J	3	M	3
Group d	2	–	J	2	V	4
Group e	1	–	K	1	T	5
		Σ	Σ	Σ	Σ	

Figure 5.9 Goals achievement matrix (section of). (Adapted from Hill 1968)

for different environmental parameters, drawn up by panels of experts. Weightings can also be built into overlay maps to identify areas with the most development potential according to various combinations of weightings. Some of the limitations of such approaches have already been noted in Chapter 4.

Other methods in the multi-criteria/multi-attribute category include decision analysis, the goals achievement matrix, multi-attribute utility theory and judgement analysis. *Decision analysis* is the operational form of decision theory, a theory of how individuals make decisions in the face of uncertainty, which owes its modern origins to Von Neuman & Morgenstern (1953). Decision analysis usually involves the construction of a decision tree, an example of which is shown in Figure 5.8. Each branch represents a potential action, with a probability of achievement attached to it.

The *goals achievement matrix* (GAM) was developed as a planning tool by Hill (1968) to overcome the perceived weaknesses of the planning balance sheet approach. GAM makes the goals and objectives of a project/plan explicit, and the evaluation of alternatives is accomplished by measuring the extent to which they achieve the stated goals. The existence of many diverse goals leads to a system of weights. Since all interested parties are not politically equal, the identified groups should also be weighted. The end result is a matrix of weighted objectives and weighted interests/agencies (Fig. 5.9). The use of goals and value weights to evaluate plans in the interests of the community, and not just for economic efficiency, has much to commend it. The approach also provides an opportunity for public participation. Unfortunately, the complexity of the approach has limited its use, and the weights and goals used may often reflect the views of the analyst more than those of the interests and agencies involved.

Multi-attribute utility theory (MAUT) has gained a certain prominence in recent years as an evaluation method that can incorporate the values of the key interests involved (Edwards & Newman 1982, Bisset 1988, Parkin 1992). MAUT involves a number of steps, including the identification of the entities (alternatives, objects) to

151

be evaluated, and the identification and structuring of environmental attributes (e.g. noise level) to be measured. The latter may include a "value tree" with general objectives (values) at the top and specific attributes at the bottom. The ranking of attributes is by the central stakeholder/expert group, whose values are to be maximized. Attributes are scaled and formal value or utility models developed to quantify trade-offs among attribute scales and attributes. For further reference, see Parkin (1992) for an outline of the main steps and an application of a "relatively" simple and well-proven version of MAUT known as the SMART method.

Finally brief reference is made to the *Delphi method*, which can be used to incorporate the views of various stakeholders into the evaluation process. The method is an established means of collecting expert opinion and of gaining consensus among experts on various issues under consideration. It has the advantage of obtaining expert opinion from the individual, with guaranteed anonymity, avoiding the potential distortion caused by peer pressure in group situations. Compared with other evaluation methods it can also be quicker and cheaper.

There have been a number of interesting applications of the Delphi method in EIA (Richey et al. 1985, Green et al. 1989, 1990). Green et al. used the approach to assess the environmental impacts of the redevelopment and reorientation of Bradford's famous Salt Mill. The method involves drawing up a Delphi panel. In the Salt Mill case, the initial panel of 40 included experts with a working knowledge of the project (e.g. planners, tourism officers), councillors, employees, academics, local residents and traders. This was designed to provide a balanced view of interests and expertise. The Delphi exercise usually has a three-stage approach: (1) a general questionnaire asking panel members to identify important impacts (positive and negative); (2) a first-round questionnaire asking panel members to rate the importance of a list of impacts identified from the first stage; (3) a second-round questionnaire, asking panel members to re-evaluate the importance of each impact in the light of the panel's response to the first round. However, the method is not without its limitations. The potential user should be aware that it is difficult to draw up a "balanced" panel in the first place, and to avoid distorting the assessment by the varying drop-out rates of panel members between stages of the exercise, and by an overzealous structuring of the exercise by the organizers.

5.4 Mitigation

Types of mitigation measures

Mitigation is defined in EC Directive 85/337 as "measures envisaged in order to avoid, reduce and, if possible remedy significant adverse effects" (CEC 1985). In similar vein, the US Council on Environmental Quality, in its regulations implementing the National Environmental Policy Act, defines mitigation as including: "not taking certain actions; limiting the proposed action and its implementation;

Table 5.10 Mitigation measures, as outlined in UK *guide to procedures.*

Where significant adverse effects are identified, [describe] the measures to be taken to avoid, reduce or remedy those effects, e.g.:
(a) Site planning
(b) Technical measures, e.g.:
  (i) process selection
  (ii) recycling
  (iii) pollution control and treatment
  (iv) containment (e.g. bunding of storage vessels)
(c) Aesthetic and ecological measures, e.g.:
  (i) mounding
  (ii) design, colour, etc.
  (iii) landscaping
  (iv) tree plantings
  (v) measure to preserve particular habitats or create alternative habitats
  (vi) recording of archaeological sites
  (vii) measures to safeguard historic building or sites
[Assess] the likely effectiveness of mitigating measures.

(*Source:* DOE 1989)

repairing, rehabilitating, or restoring the affected environment; presentation and maintenance actions during the life of the action; and replacing or providing substitute resources or environments" (CEQ 1978). The guidance on mitigation measures provided by the UK government is set out in Table 5.10. It is not possible to specify here all the types of mitigation measures that could be used. Instead the following subsections provide a few examples, relating to biophysical and socio-economic impacts. The reader is also referred to Fortlage (1990) and Morris & Therivel (1995) for useful coverage of mitigation measures. A review of EISs for developments similar to the development under consideration may also suggest useful mitigation measures.

At one extreme, the prediction and evaluation of impacts may reveal an array of impacts with such significant adverse effects that the only effective mitigation measure may be to abandon the proposal altogether. A less draconian, and more normal, situation would be to modify aspects of the development action to avoid various impacts. Examples of methods to *avoid* impacts include:

- the control of solid and liquid wastes by recycling on site or by removing them from the site for environmentally sensitive treatment elsewhere;
- the use of a designated lorry route, and day-time working only, to avoid disturbance to village communities from construction lorry traffic and from night construction work;
- the establishment of buffer zones and the minimal use of toxic substances, to avoid impacts on local ecosystems.

Some adverse effects may be less easily avoided; there may also be less need to avoid them completely. Examples of methods to *reduce* adverse effects include:

- the sensitive design of structures, using simple profiles, local materials and muted colours, to reduce the visual impact of a development, and landscaping to hide it or blend it into the local environment;
- the use of construction site hostels, and coaches for journeys to work, to reduce the impact on the local housing market, and on the roads, of a project employing many workers during its construction stage;
- the use of silting basins or traps, the planting of temporary cover crops and the scheduling of activities during the dry months, to reduce erosion and sedimentation.

During one or more stages of the life of a project, certain environmental components may be temporarily lost or damaged. It may be possible to *repair, rehabilitate or restore* the affected component to varying degrees. For example:

- agricultural land used for the storage of materials during construction may be fully rehabilitated; land used for gravel extraction may be restored to agricultural use, but over a much longer period, and with associated impacts according to the nature of the landfill material used;
- a river or stream diverted by a road project can be unconverted and re-established with similar flow patterns as far as is possible;
- a local community astride a route to a new tourism facility could be relieved of much of the adverse traffic effects by the construction of a bypass (which, of course, introduces a new flow of impacts).

There will invariably be some adverse effects that cannot be reduced. In such cases, it may be necessary to *compensate* people for adverse effects. For example:

- for the loss of public recreational space or a wildlife habitat, the provision of land with recreation facilities, or the creation of a nature reserve, elsewhere;
- for the loss of privacy, quietness and safety in houses next to a new road, the provision of sound insulation and/or the purchase by the developer of badly affected properties.

Mitigation measures can become linked with discussions between a developer and the local planning authority on what is known in the UK as "planning gain". Fortlage (1990) talks of some of the potential complications associated with such discussions, and of the need to distinguish between mitigation measures and planning gain:

> Before any mitigating measures are put forward, the developer and the local planning authority must agree as to which effects are to be regarded as adverse, or sufficiently adverse to warrant the expense of remedial work, otherwise the whole exercise becomes a bargaining game which is likely to be unprofitable to both parties . . .
>
> Planning permission often includes conditions requiring the provision of planning gains by the developer to offset some deterioration of the area caused by the development, but it is essential to distinguish very clearly between those

benefits offered by way of compensation for adverse environmental effects and those which are a formal part of planning consent. The local planning authority may decide to formulate the compensation proposals as a planning condition in order to ensure that they are carried out, so the developer should beware of putting forward proposals that he does not really intend to implement.

Mitigation measures must be planned in an integrated and coherent fashion to ensure that they are effective, that they do not conflict with each other, and that they do not merely shift a problem from one medium to another. A project may also benefit an area, often socio-economically; where such benefits are identified, as a minimum there should be a concern to ensure that they do occur and do not become diluted, and that they may even be enhanced. For example, the potential local employment benefits of a project can be encouraged by the offer of appropriate skills training programmes to local people; various tenure arrangements can be used to make houses in new housing schemes available to local people in need.

The results of a recent research project on the treatment of mitigation within EIA (DETR 1997) still found that UK practice varied considerably. For example, there was too much emphasis on physical measures, rather than on operational or management controls, and a lack of attention to the impacts of construction and to residual impacts after mitigation. A draft good practice guide, resulting from the research project, introduces the concept of the *mitigation hierarchy*, namely to:

- avoid impacts at the source;
- reduce impacts at source;
- abate impacts on site;
- abate impacts at receptor;
- repair impacts;
- compensate in kind;
- compensate by other means; and
- enhance. (Mitchell 1997)

Mitigation in the EIA process

Like many elements in the EIA process, mitigation is not limited to one point in the assessment. Although it may follow logically from the prediction and assessment of the relative significance of impacts, it is in fact inherent in all aspects of the process. An original project design may already have been modified, possibly in the light of mitigation changes made to earlier comparable projects or perhaps as a result of early consultation with the LPA or with the local community. The consideration of alternatives, initial scoping activities, baseline studies and impact identification studies may suggest further mitigation measures. Although more in-depth studies may identify new impacts, mitigation measures may alleviate others. The prediction and evaluation exercise can thus focus on a limited range of potential impacts.

Table 5.11 Example of a section of a summary table for impacts and mitigation measures.

Impact	Mitigation measure(s)	Level of significance after mitigation
1. 400 acres of prime agricultural land would be lost from the County to accommodate the petrochemical plant.	The only *full* mitigating measure for this impact would be to abandon the project.	SU
2. Additional lorry and car traffic on the adjacent hilly section of the motorway will increase traffic volumes by 10–20% above those predicted on the basis of current trends.	A lorry crawler lane on the motorway, funded by the developer, will help to spread the volume, but effects may be *partial* and short lived.	SU
3. The project would block the movement of most terrestrial species from the hilly areas to the east of the site to the wetlands to the west of the site.	A wildlife corridor should be developed and maintained along the entire length of the existing stream which runs through the site. The width of the corridor should be a minimum of 75 ft. The stream bed should be cleaned of silt and enhanced through the construction of occasional pools. The buffer zone should be planted with native riparian vegetation, including sycamore and willow.	LS

Note: SU = Significant unavoidable impact; LS = Less than significant impact

Mitigation measures are normally discussed and documented in each topic section of the EIS (e.g. air quality, visual quality, transport, employment). Those discussions should clarify the extent to which the significance of each adverse impact has been offset by the mitigation measures proposed. A summary chart (see Table 5.11) can provide a clear and very useful overview of the envisaged outcomes, and may be a useful basis for agreement on planning consents. Residual unmitigated or only partially mitigated impacts should be identified. These could be divided according to the degree of severity: for example, into "less than significant impacts" and "significant unavoidable impacts".

Mitigation measures are of little or no value unless they are implemented. Hence there is a clear link between mitigation and the monitoring of outcomes, if and when a project is approved and moves to the construction and operational stages. Indeed, the incorporation of a clear monitoring programme can be one of the most important mitigation measures. Monitoring, which is discussed in Chapter 7, must include the effectiveness or otherwise of mitigation measures. The latter must therefore be devised with monitoring in mind; they must be clear enough to allow for the checking of effectiveness. The use of particular mitigation measures may also draw on previous experience of relative effectiveness, from previous monitoring activity in other relevant and comparable cases.

5.5 Summary

Impact prediction and the evaluation of the significance of impacts often constitute a "black box" in EIA studies. Intuition, often wrapped up as expert opinion, cannot provide a firm and defensible foundation for this important stage of the process. Various methods, ranging from simple to complex, are available to the analyst, and these can help to underpin analysis. Mitigation measures come into play particularly at this stage. However, the increasing sophistication of some methods does run the risk of cutting out key actors, and especially the public, from the EIA process. Chapter 6 discusses the important, but currently weak, role of public participation, the value of good presentation, and approaches to EIS review and decision-making.

References

Barde, J. P. & D. W. Pearce 1991. *Valuing the environment: six case studies*. London: Earthscan.

Bateman, I. 1991. Social discounting, monetary evaluation and practical sustainability. *Town and Country Planning* **60**(6), 174–6.

Bisset, R. 1988. Developments in EIA methods. In *Environmental impact assessment: theory and practice*, P. Wathern (ed.), 47–60. London: Unwin Hyman.

Bourdillon, N. 1996, *Limits and standards in EIA*. Oxford: Oxford Brookes University, Impact Assessment Unit, School of Planning.

Bowers, J. 1990. *Economics of the environment: the conservationists' response to the Pearce Report*. Wellington, Telford: British Association of Nature Conservationists.

Bregman, J. I. & K. M. Mackenthun 1992. *Environmental impact statements*. Boca Raton, Florida: Lewis.

CEC (Commission of the European Communities) 1985. *On the assessment of effects of certain public and private projects on the environment*, OJ L 175, 5 July. Brussels: CEC.

CEC 1993. *Environmental manual: sectoral environmental assessment sourcebook*. Brussels: CEC, DG VIII.

CEQ (Council on Environmental Quality) 1978. *National Environmental Policy Act*, Code of Federal Regulations, Title 40, Section 1508.20.

Dasgupta A. K. & D. W. Pearce 1978. *Cost-benefit analysis: theory and practice*. London: Macmillan.

De Jongh, P. E. 1988. Uncertainty in EIA. In *Environmental impact assessment: theory and practice*, P. Wathern (ed.), 62–83. London: Unwin Hyman.

DETR (Department of the Environment, Transport and the Regions) 1997. *Mitigation measures in environmental statements*. London: DETR.

157

DOE (Department of the Environment) 1989. *Environmental appraisal: a guide to the procedures*. London: HMSO.

DOE 1991. *Policy appraisal and the environment*. London: HMSO.

Edwards, W. & J. R. Newman 1982. *Multiattribute evaluation*. Los Angeles: Sage.

English Nature 1994. *Nature conservation in environmental assessment*. Peterborough: English Nature.

Fortlage, C. 1990. *Environmental assessment: a practical guide*. Aldershot: Gower.

Friend, J. K. & A. Hickling 1987. *Planning under pressure: the strategic choice approach*. Oxford: Pergamon.

Friend, J. K. & W. N. Jessop 1977. *Local government and strategic choice: an operational research approach to the processes of public planning*, 2nd edn. Oxford: Pergamon.

Glasson, J., M. J. Elson, M. Van der Wee, B. Barrett 1987. *Socio-economic impact assessment of the proposed Hinkley Point C power station*. Oxford: Oxford Polytechnic, Impacts Assessment Unit.

Glasson, J., M. Van der Wee, B. Barrett 1988. A local income and employment multiplier analysis of a proposed nuclear power station development at Hinkley Point in Somerset. *Urban Studies* **25**, 248–61.

Golden, J., R. P. Duellette, S. Saari, P. N. Cheremisinoff 1979. *Environmental impact data book*. Ann Arbor: Ann Arbor Science Publishers.

Green, H., C. Hunter, B. Moore 1989. Assessing the environmental impact of tourism development: the use of the Delphi technique. *International Journal of Environmental Studies* **35**, 51–62.

Green, H., C. Hunter, B. Moore 1990. Assessing the environmental impact of tourism development. *Tourism Management*, June, 111–20.

Hansen, P. E. & S. E. Jorgensen (eds) 1991. *Introduction to environmental management*. New York: Elsevier.

Hill, M. 1968. A goals-achievement matrix for evaluating alternative plans. *Journal of the American Institute of Planners* **34**, 19.

HMSO 1971. *Report of the Roskill Commission on the Third London Airport*. London: HMSO.

Holling, C. S. (ed.) 1978. *Adaptive environmental assessment and management*. New York: Wiley.

Institute of Environmental Assessment and Landscape Institute 1995. *Guidelines for landscape and visual assessment*. Lincoln: IEA.

Journal of American Institute of Planners 1965. Theme issue on urban development models – new tools for planners. May.

Kristensen, P., J. P. Jensen, E. Jeppesen 1990. *Eutrophication models for lakes*. Research Report C9. Copenhagen: National Agency of Environmental Protection.

Lee, N. 1987. *Environmental impact assessment: a training guide*. Occasional Paper 18. Department of Town and Country Planning, University of Manchester.

Lewis, J. A. 1988. Economic impact analysis: a UK literature survey and bibliography. *Progress in Planning* **30**(3), 161–209.

Lichfield, N. 1996. *Community impact evaluation*. London: UCL Press.

Lichfield, N., P. Kettle, M. Whitbread 1975. *Evaluation in the planning process*. Oxford: Pergamon.

Loewenstein, L. K. 1966. On the nature of analytical models. *Urban Studies* **3**.

Maclaren V. W. & J. B. Whitney (eds) 1985. *New directions in environmental impact assessment in Canada*. London: Methuen.

Mitchell, J. 1997. Mitigation in environmental assessment – furthering best practice. *Environmental Assessment* **5**/4. December.

Morris, P. & R. Therivel (eds) 1995. *Methods of environmental impact assessment*. London: UCL Press.

Munn, R. E. 1979. *Environmental impact assessment: principles and procedures*. New York: Wiley.

Parkin, J. 1992. *Judging plans and projects*. Aldershot: Avebury.

Pearce, D. 1989. Keynote speech at the 10th International Seminar on Environmental Impact Assessment and Management, University of Aberdeen, 9–22 July.

Pearce, D. & A. Markandya 1990. *Environmental policy benefits: monetary valuation*. Paris: OECD.

Pearce, D., A. Markandya, E. B. Barbier 1989. *Blueprint for a green economy*. London: Earthscan.

Rau J. G. & D. C. Wooten 1980. *Environmental impact analysis handbook*. New York: McGraw-Hill.

Richey, J. S., B. W. Mar, R. Horner 1985. The Delphi technique in environmental assessment. *Journal of Environmental Management* **21**(1), 135–46.

Suter II, G. W. 1993. *Ecological risk assessment*. Chelsea, Michigan: Lewis.

Tomlinson, P. 1989. Environmental statements: guidance for review and audit. *The Planner* **75**(28), 12–15.

US Environmental Protection Agency 1993. *Sourcebook for the environmental assessment (EA) process*.

Von Neuman, J. & O. Morgenstern 1953. *Theory of games and economic behaviour*. Princeton: Princeton University Press.

Voogd, J. H. 1983. *Multicriteria evaluation for urban and regional planning*. London: Pion.

VROM 1984. *Prediction in environmental impact assessment*. The Hague: Netherlands Ministry of Public Housing, Physical Planning and Environmental Affairs.

West Australian Environmental Protection Authority 1993. *Environmental impact assessment: administrative procedures*. Perth: EPA.

Westman, W. E. 1985. *Ecology, impact assessment and environmental planning*. New York: Wiley.

Winpenny, J. T. 1991. *Values for the environment: a guide to economic appraisal* (Overseas Development Institute). London: HMSO.

CHAPTER 6

Participation, presentation and review

6.1 Introduction

One of the aims of the EIA process is to provide information about a proposal's likely environmental impacts to the developer, public and decision-makers, so that a better decision may be made. Consultation with the public and statutory consultees in the EIA process can help to ensure the quality, comprehensiveness and effectiveness of the EIA, as well as to ensure that the various groups' views are adequately taken into consideration in the decision-making process. Consultation and participation[1] can be useful at most stages of the EIA process:

- in determining the scope of an EIA;
- in providing specialist knowledge about the site;
- in evaluating the relative significance of the likely impacts;
- in proposing mitigation measures;
- in ensuring that the EIS is objective, truthful and complete;
- in monitoring any conditions of the development agreement.

As such, how the information is presented, how the various interested parties use that information, and how the final decision incorporates the results of the EIA and the views of the various parties, are essential components in the EIA process.

Traditionally, the British system of decision-making has been characterized by administrative discretion and secrecy, with limited public input (McCormick 1991). However, there have been recent moves towards greater public participation in decision-making, and especially towards greater public access to information. In the environmental arena, the Environmental Protection Act of 1990 requires the Environment Agency and local authorities to establish public registers of information on potentially polluting processes; *This common inheritance* (DOE et al. 1990) and its annual updates compile environmental data and set forth an environmental agenda in a publicly available form; EC Directive 85/337's requirements for EIA allow greater public access to information previously not compiled, or considered confidential; and EC Directive 90/313, which requires Member States to make provisions for freedom of access to information on the environment, has been implemented in the UK through the Environmental Information Regulations 1992.

160

However, despite the positive trends towards greater consultation and participation in the EIA process and the improved communication of EIA findings, both are still underdeveloped in the UK. Few developers make a real effort to gain a sense of the public's views before presenting their applications for authorization and EISs. Few competent authorities have the time or resources to gauge public opinion adequately before making their decisions. Few EISs are truly well presented, although standards of presentation have improved rapidly since mid-1988.

This chapter discusses how consultation and participation by both the public (Section 6.2) and designated environmental consultees (Section 6.3) can be fostered, and how the results can be used to improve a proposed project and speed up its authorization process. The effective presentation of the EIS is then discussed in Section 6.4. The review of EISs and assessment of their accuracy and comprehensiveness are considered in Section 6.5. The chapter concludes with a discussion about decision-making and post-decision legal challenges.

6.2   Public consultation and participation

This section discusses how "best practice" public participation can be encouraged. It begins by considering the advantages and disadvantages of public participation. It then establishes requirements for effective public participation and reviews methods of such participation. Finally, we discuss the UK approach to public participation. The reader is also referred to Canter (1996), Jain et al. (1977), O'Riordan & Turner (1983), Westman (1985) and various contributors in Weston (1997) for further information.

Advantages and disadvantages of public participation

Developers do not usually favour public participation. It may upset a good relationship with the local planning authority. It carries the risk of giving a project a high profile, with attendant costs in time and money. It may not lead to a conclusive decision on a project, as diverse interest groups have different concerns and priorities; the decision may also represent the views of the most vocal interest groups rather than of the general public. Most developers' contact with the public comes only at the stage of planning appeals and inquiries; by this time, participation has often evolved into a systematic attempt to stop their projects. Thus, many developers never see the positive side of public participation, because they do not give it a chance.

Historically, public participation has also had connotations of extremism, confrontation, delays and blocked development. In the USA, NEPA-related lawsuits have stopped major development projects, including oil and gas developments in Wyoming, a ski resort in California, and clear-cut logging project in Alaska (Turner 1988). In Japan in the late 1960s and early 1970s, riots (so violent that six people died) delayed the construction of the Narita Airport near Tokyo by five years. In the UK, perhaps the most visible forms of public participation have been protesters wearing gas masks at nuclear power station sites, threatening to lie down in front of the bulldozers working on the M3 motorway at Twyford Down and being forcibly evicted from tunnels and tree-houses on the Newbury bypass route, which cost more than £6 million for policing before construction even began. More typically, all planners are familiar with acrimonious public meetings and "ban the project" campaigns. Public participation may provide the legal means for intentionally obstructing development; the protracted delay of a project can be an effective method of defeating it.

On the other hand, public participation can be used positively to convey information about a development, clear up misunderstandings, allow a better understanding of relevant issues and how they will be dealt with, and identify and deal with areas of controversy while a project is still in its early planning phases. The process of considering and responding to the unique contributions of local people or special interest groups may suggest measures the developer could take to avoid local opposition and environmental problems. These measures are likely to be more innovative, viable and publicly acceptable than those proposed solely by the developer. Project modifications made early in the planning process, before plans have been fully developed, are more easily and cheaply accommodated than those made later. Projects that do not have to go to inquiry are considerably cheaper than those that do. Early public participation also prevents an escalation of frustration and anger, so it helps to avoid the possibility of more forceful "participation". Finally, the implementation of a project generally proceeds more cheaply and smoothly if local residents agree with the proposal, with fewer protests, a more willing labour force, and fewer complaints about impacts such as noise and traffic.

Past experience shows that the total benefits of openness can exceed its costs, despite the expenditure and delays associated with full-scale public participation in the project planning process. The case of British Gas has already been noted (House of Lords 1981). Similarly, the conservation manager of Europe's (then) largest zinc/lead mine noted that:

> properly defined and widely used, [EIA is] an advantage rather than a deterrent. It is a mechanism for ensuring the early and orderly consideration of all relevant issues and for the involvement of affected communities. It is in this last area that its true benefit lies. We have entered an era when the people decide. It is therefore in the interests of developers to ensure that they, the people, are equipped to do so with the confidence that their concern is recognized and their future life-style protected. (Dallas 1984)

More recently, the developers of a motor-racing circuit noted:

> The [EIS] was the single most significant factor in convincing local members, residents and interested parties that measures designed to reduce existing

environmental impacts of motor racing had been uppermost in the formulation of the new proposals. The extensive environmental studies which formed the basis of the statement proved to be a robust defence against the claims from objectors and provided reassurance to independent bodies such as the Countryside Commission and the Department of the Environment. Had this not been the case, the project would undoubtedly have needed to be considered at a public inquiry. (Hancock 1992)

On the other hand, as will be seen later, many developers still see public participation as a counterproductive exercise which brings little gain at great cost.

Requirements for effective participation

The United Nations Environment Programme lists five interrelated components of effective public participation:

- identification of the groups/individuals interested in or affected by the proposed development;
- provision of accurate, understandable, pertinent and timely information;
- dialogue between those responsible for the decisions and those affected by them;
- assimilation of what the public say in the decision; and
- feedback about actions taken and how the public influenced the decision. (Clark 1994)

These points will be discussed in turn.

Although the *identification of relevant interest groups* seems superficially simple, it can be fraught with difficulty. The simple term "the public" actually refers to a complex amalgam of interest groups, which changes over time and from project to project. The public can be broadly classified into two main groups. The first consists of the voluntary groups, quasi-statutory bodies or issues-based pressure groups which are concerned with a specific aspect of the environment or with the environment as a whole. The second group consists of the people living near a proposed development who may be directly affected by it. These two groups can have very different interests and resources. The organized groups may have extensive financial and professional resources at their disposal, may concentrate on specific aspects of the development, and may see their participation as a way to gain political points or national publicity. People living locally may lack the technical, educational or financial resources, and familiarity with relevant procedures, to put their points across effectively, yet they are the ones who will be the most directly affected by the development (Mollison 1992). The people in the two groups, in turn, come from a wide range of backgrounds and have a wide variety of opinions. A multiplicity of "publics" thus exits, each of which has specific views, which may well conflict with those of other groups and those of EIA "experts".

It is debatable whether all these publics should be involved in all decisions, for instance whether "highly articulate members of the NGO, Greenpeace International, sitting in their office in Holland, also have a right to express their views on, and attempt to influence, a decision on a project which may be on the other side of the world" (Clark 1994). Participation may be tightly controlled by regulations specifying the groups and organizations that are eligible to participate or by criteria identifying those considered to be directly affected by a development (e.g. living within a certain distance of it).

Lack of information, or misinformation, about the nature of a proposed development prevents adequate public participation and causes resentment and criticism of the project. One objective of public participation is thus to *provide information* about the development and its likely impacts. Before an EIS is prepared, information may be provided at public meetings, exhibitions, or telephone hotlines. This information should be as candid and truthful as possible: people will be on their guard against evasions or biased information, and will look for confirmation of their fears. A careful balance needs to be struck between consultation that is early enough to influence decisions and consultation that is so early that there is no real information on which to base any discussions. For instance, after several experiences of problematic pre-EIS consultation, one UK developer decided to conduct

> quite elaborate consultation exercises but only after the statements were published . . . by delaying public consultation until after the initial assessment was completed, Lakewoods and their consultants were able to impart genuine information and to say what effects they thought the development would have and why. (McNab 1997)

The way information is conveyed can influence public participation. Highly technical information can be understood by only a small proportion of the public. Information in different media (e.g. newspapers, radio) will reach different sectors of the public. Ensuring the participation of groups that generally do not take part in decision-making – notably minority and low-income groups – may be a special concern, especially in light of the Brundtland Commission's emphasis on intra-generational equity and participation. A recent US study (Williams & Hill 1996) identified a number of disparities between traditional ways of communicating environmental information and the needs of minority and low-income groups; for instance:

- agencies focus on desk studies rather than working actively with these groups;
- agencies often do not understand existing power structures, so do not involve community leaders such as preachers for low-income churches, or union leaders;
- agencies hold meetings where the target groups are not represented, for instance in city centres away from where the project will be located;
- agencies hold meetings in large "fancy" places which disenfranchised groups feel are "off-limits", rather than in local churches, schools or community centres;
- agencies use newspaper notices, publication in official journals and mass mailings instead of telephone trees or leaflets handed out in schools;

- agencies prepare thick reports which confuse and overwhelm;
- agencies use formal presentation techniques such as raised platforms and slide projections.

These points suggest that a wide variety of methods for conveying information should be used, with an emphasis on techniques that would be useful for traditionally less participative groups: EIS summaries with pictures and perhaps comics as well as technical reports, meetings in schools and churches as well as in more formal venues, and contact through established community networks as well as through leaflets and newspaper notices.

Public participation in EIA also aims to *establish a dialogue* between the public and decision-makers (both the project proponent and the authorizing body) and to ensure that decision-makers *assimilate the public's views* into their decisions. Public participation can help to identify issues that concern local residents. These issues are often not the same as those of concern to the developer or outside experts. Public participation exercises should thus achieve a two-way flow of information to allow residents to voice their views. The exercises may well identify conflicts between the needs of the developer and those of various sectors of the community; but this should ideally lead to solutions of these conflicts, and to agreement on future courses of action that reflect the joint objectives of all parties.

Public participation is likely to be greatest where public comments are most likely to influence decisions. Arnstein (1971) identified "eight rungs on a ladder of citizen participation", ranging from non-participation (manipulation, therapy), through tokenism (informing, consultation, placation) to citizen power (partnership, delegated power, citizen control). Similarly, Westman (1985) has identified four levels of increasing public power in participation methods: information-feedback approaches, consultation, joint planning and delegated authority. Table 6.1 lists advantages and disadvantages of these levels.

There are many different forms of public participation. A few are listed in Table 6.2, along with an indication of how well they provide information, cater for special interests, encourage dialogue and affect decision-making. Box 6.1 gives an example from Canada, where many of these techniques have been used in practice. The effectiveness of these techniques can vary widely. One UK local authority planner gives his views:

> It is normal practice for controversial cases to be referred to a panel of committee members for a site visit. Frequently at these meetings the public attend and are invited to make comments. Often the applicant is encouraged to prepare a small exhibition of the proposals so that interested parties have the opportunity to examine the project in more detail and opinions can be exchanged. This practice gives the developer the chance to experience how local people feel about the proposal. How far this may cause a change in the details of the project is another matter. Another area of publicity is the public meeting and it is probably the least productive . . . The public meeting appears not to be the right forum for the exchange of information or opinion. It might function well

165

Table 6.1 Advantages and disadvantages of levels of increasing public influence.

Approaches	Extent of public power in decision-making	Advantages	Disadvantages
Information feedback			
Slide or film presentation, information kit, newspaper account, notices, etc.	Nil	Informative, quick	No feedback; presentation subject to bias
Consultation			
Public hearing, ombudsperson or representative, etc.	Low	Allows two-way information transfer; allows limited discussion	Does not permit ongoing communication; some-what time-consuming
Joint planning			9
Advisory committee, structured workshop, etc.	Moderate	Permits continuing input and feedback; increases education and involvement of citizens	Very time-consuming; dependent on what information is provided by planners
Delegated authority			
Citizens' review board, citizens' planning commission, etc.	High	Permits better access to relevant information; permits greater control over options and timing of decision	Long-term time commitment; difficult to include wide representation on small board

(*Source:* Westman 1985)

as a community safety valve . . . but as a contribution to environmental decision making it is often unhelpful . . .

It is becoming more usual with planning cases for them to be placed before community forums of local people and other interested parties [since] the earlier the community is involved in planning matters the better chance a project has of eventually being implemented . . . The resource implications of servicing forums are considerable and indeed risky, as the debate may go in unexpected directions. Also, importantly, such a process cannot be hurried. (Read 1997)

Finally, an essential part of effective public participation is *feedback about any decisions* and actions taken, and how the public's views affected those decisions. In the US, for instance, comments on a draft EIS are incorporated into the final EIS along with the agency's response to those comments. For example:

Table 6.2 Methods of public participation and their effectiveness.

	Provide information	Cater for special interests	Two-way communication	Impact on decision-making
Explanatory meeting, slide/film presentation	✔	$^1/_2$	$^1/_2$	–
Presentation to small groups	✔	✔	✔	$^1/_2$
Public display, exhibit, models	✔	–	–	–
Press release, legal notice	$^1/_2$	–	–	–
Written comment	–	$^1/_2$	$^1/_2$	$^1/_2$
Poll	$^1/_2$	–	✔	✔
Field office	✔	✔	$^1/_2$	–
Site visit	✔	✔	–	–
Advisory committee, task force, community representative	$^1/_2$	$^1/_2$	✔	✔
Working groups of key actors	✔	$^1/_2$	✔	✔
Citizen review board	$^1/_2$	$^1/_2$	✔	✔
Public inquiry	✔	$^1/_2$	$^1/_2$	✔/–
Litigation	$^1/_2$	–	$^1/_2$	✔/–
Demonstration protests, riots	–	–	$^1/_2$	✔/–

(Adapted from Westman 1985)

comment: I am strongly opposed to the use of herbicides in the forest. I believe in a poison-free forest!

response: Your opposition to use of herbicides was included in the content analysis of all comments received. However, evidence in the EIS indicates that low risk use of selected herbicides is assured when properly controlled – the evaluated herbicides pose minimal risk as long as mitigation measures are enforced.

Without such feedback, people are likely to question the use to which their input was put, and whether their participation had any effect at all; this could affect their approach to subsequent projects as well as their view of the one under consideration.

UK procedures

Article 6 of EC Directive 85/337 (as amended) requires that:

Member States shall ensure that:

- any request for development consent and any information gathered pursuant to Article 5 are made available to the public;

167

Box 6.1   Grande-Baleine hydropower complex, Canada

Hydro-Québec has applied for permission to build a hydropower complex in northern Québec province, which could generate 16.2 TWh of energy annually. The complex would include three dammed-up reservoirs with a total area of 3400 km^2, three generating stations, 136 dykes, a road system and three airfields. Likely environmental impacts include impacts on flora, fauna, water quality (particularly methylmercury levels which would result in restrictions on fish consumption). The project would also affect about 500 Crees, 450 Inuit and 75 people of other origins, for most of whom hunting, fishing and trapping remains central to their identity as Native Peoples of northern Québec. As part of project planning, Hydro-Québec undertook extensive public consultation and description. The following description is verbatim from a leaflet summarising Hydro-Québec's communication activities (Hydro-Québec 1993):

Local populations were regularly consulted and kept informed from the start of phase I of the feasibility study for the Grande-Baleine complex, which ran from 1977 to 1981, when work was temporarily suspended. Hydro-Québec organized regular meetings with the Native communities directly affected by the project; their views were taken into consideration in conducting studies. These communities were subsequently informed of the study results and, once again, consulted about proposed mitigative and environmental enhancement measures. Based on these consultations – to take just one example – Hydro-Québec revised its scenario for the diversion of the Petite Rivière de la Baleine to eliminate environmental impacts on the drainable basin of the Rivière Nastapoka, further north.

In 1988, with the start of phase II of the Grande-Baleine feasibility study, Hydro-Québec relaunched its information and communication initiatives. At the local level, the communication program consisted of three phases: the general information phase, designed to provide information about various components of the project; the information-feedback phase, designed to gather reactions and data to guide Hydro-Québec in its decision making; and the information-consultation phase, in which Hydro-Québec presented options to the interested parties, analyzed the opinions expressed, and explained its decisions as they were made. Shortly after the phase II of the feasibility study was under way, the Crees informed Hydro-Québec that they no longer wished to maintain dialogue and that all communications should be addressed to their legal advisors. Starting in January 1989, Hydro-Québec's local information and consultation activities were directed mainly at the Inuit, who had formed a working group in 1988.

In the general information phase, Hydro-Québec held meetings with various organizations and clarified key aspects of the project, including the project rationale, environmental studies, employment opportunities, and the overall development calendar. A bulletin summarizing this information in French, English, Cree, and Inuktitut was sent to the persons, groups, and organizations that requested it, and all interested parties were able to express their concerns.

The information-feedback phase began with a helicopter tour over the affected area given to members of the Inuit working group. It continued with numerous meetings in which the Native peoples voiced their concerns in greater detail about the project and its diverse components. Thematic workshops focused on specific subjects such as employment and training for Native peoples.

In the information-consultation phase, members of the Inuit working groups flew over the sector chosen for the new Petite Rivière de la Baleine diversion option, which had been devised in response to concerns expressed in the earlier phases. The working group also flew over the La Grande complex, where members took a close look at a section of river where the flow of water had been reduced. During workshops, specific problems, such as impacts on the beluga whales and increased mercury levels were examined in greater depth. Hydro-Québec provided detailed data on all aspects of the project and gave updates on studies then in progress.

To keep the rest of the Québec population informed, Hydro-Québec held meetings with a cross section of groups and organizations, took part in public meetings, and distributed information bulletins. Once again, interested parties were given the opportunity to express their concerns about the project. These activities were part of a national communication campaign in the print media and on radio in which the public was invited to request more information by calling a toll-free number.

In addition, following Parliamentary Commission hearings in May 1990, Hydro-Québec worked with the Québec government to establish a framework for public consultation that was designed to integrate the expectations and concerns of Québec society into its proposed development plan. The opinions expressed in 47 meetings with 75 groups were made public in November 1992, when a new development plan proposal was tabled.

Internationally, Hydro-Québec undertook information campaigns in the northeastern United States and Europe after various groups took positions based on erroneous data and the *New York Times* published a one-page advertisement in the fall of 1991 that was quite biased and took an unfavourable stance on the Grande-Baleine project. Hydro-Québec held conferences, organized visits to the La Grande complex, took part in college and university debates, set up a toll-free line in Vermont, and opened information offices in New York and Brussels. In February 1992, Hydro-Québec successfully defended the Grande-Baleine project and its assessment procedure before the International Water Tribunal in Amsterdam . . .

From the outset . . . Hydro-Québec has responded to thousands of questions from journalists and organized numerous news conferences and visits to James Bay. In its efforts to prevent the Grande-Baleine project from becoming a symbol of conflict between environmental protection and economic development Hydro-Québec has endeavoured to clarify the facts, set the record straight, and explain the complex issues involved. If the Grande-Baleine project is approved, Hydro-Québec will show the same commitment to maintaining open channels of communication and dialogue during the construction and operational phases. The utility will also remain in close contact with the communities concerned . . .

- the public concerned is given the opportunity to express an opinion before development consent is granted.

The detailed arrangements for such information and consultation shall be determined by the Member States which may in particular, depending on the particular characteristics of the projects or sites concerned:

- determine the public concerned;
- specify the places where the information can be consulted;
- specify the ways in which the public may be informed, for example by bill-posting within a certain radius, publication in local newspapers, organisation of exhibitions with plans, drawings, tables, graphs, models;
- determine the manner in which the public is to be consulted, for example by written submissions, by public enquiry;
- fix appropriate time limits for the various stages of the procedure in order to ensure that a decision is taken within a reasonable period.

169

In the UK, this has been translated by the various EIA regulations (with minor differences) into the following general requirements. Notices must be published in two local newspapers and posted at a proposed site at least seven days before the submission of the development application and EIS. These notices must describe the proposed development, state that a copy of the EIS is available for public inspection with other documents relating to the development application for at least 21 days, give an address where copies of the EIS may be obtained and the charge for the EIS, and state that written representations on the application may be made to the competent authority for at least 28 days after the notice is published. When a charge is made for an EIS, it must be reasonable, taking into account printing and distribution costs.

Environmental assessment: a guide to the procedures (DOE 1989), the government manual to developers, notes:

Developers should also consider whether to consult non-statutory bodies concerned with environmental issues, and the general public, during the preparation of the environmental statement. Bodies of these kinds may have particular knowledge and expertise to offer . . . While developers are under no obligation to publicise their proposals before submitting a planning application, consultation with local amenity groups and with the general public can be useful in identifying key environmental issues, and may put the developer in a better position to modify the project in ways which would mitigate adverse effects and recognize local environmental concerns. It will also give the developer an early indication of the issues which are likely to be important issues at the formal application stage if, for instance, the proposal goes to public inquiry.

The good practice guide on preparing EISs (DOE 1995) repeats this virtually verbatim, and adds:

It is at the scoping stage that the developer should consider the most appropriate point at which to involve members of the public. Developers may be reluctant to make a public announcement about their proposals at an early stage, perhaps because of commercial concerns . . . There may also be occasions when public disclosure of development proposals in advance of a formal planning application may cause unnecessary blight. However, early announcement of plans for prospecting and site or route selection, and the provision of opportunities for environmental/amenity groups and local people to comment on environmental issues, may channel legitimate concerns into constructive criticism.

From this it is clear that in the UK the requirements for public participation have been implemented half-heartedly at best, and developers and the competent authorities have in turn generally limited themselves to the minimal legal requirements. An environmental consultant suggests:

On the one hand assessment may be seen as a process in which all should participate; which involves the whole community in the design process and in which the statement merely becomes the statutory document required at the time the planning application is submitted. On the other hand assessment may be seen as a process in which the statement forms a critical milestone, the point at which the developer unveils his plans and gives his account of their likely environmental impacts. Discussion and debate ensue.

Both models . . . are valid. The first may be seen as an ideal where a public spirited developer has the time, resources and ability to initiate a wide ranging programme of participation. It requires all participants to take a lively and rational interest in the proposal and preferably not take up an entrenched position at the outset. It would seem most suited to public sector projects, projects initiated by the not-for-profit sector and proposals which are unlikely to provoke much opposition in principle . . . relatively few projects requiring assessment will fulfil the last criterion. The second model is more suited to the private sector developer wrestling with the problems of commercial confidentiality and time constraints. Its acceptability appears to be endorsed by the latest draft guidelines on environmental assessment published by the DOE which, whilst emphasizing the value of scoping and the need for early consultation with the LPA and statutory consultees, acknowledges the possible need for confidentiality. The guidance on public participation is similarly cautions, balancing the desirability and potential benefits of early disclosure with commercial concerns (McNab 1997).

As such the potential benefits of public participation are achieved in the uk only to a limited extent.

6.3 Consultation with statutory consultees

Some of the most useful inputs to project decision-making (Wood & Jones 1997) are comments by statutory and other relevant consultees. Article 6(1) of Directive 85/337 (as amended) states:

Member States shall take the measures necessary to ensure that the authorities likely to be concerned by the project by reasons of their specific environmental responsibilities are given an opportunity to express their opinion on the information supplied by the developer and other requests for development consent. To this end, Member States shall designate the authorities to be consulted . . . The information gathered . . . shall be forwarded to those authorities. Detailed arrangements for consultation shall be laid down by Member States.

In the UK, different statutory consultees have been designated for different types of development. For planning projects, for instance, the statutory consultees are any principal council to the area in which the land is situated (if not the LPA), the

Countryside Commission or the Countryside Council for Wales, the Environment Agency if the project is likely to have significant waste or air pollution effects, and any body the lpa would be required to consult under Article 10 of the Town and Country Planning (General Development Procedures) Order 1995, for instance the local highways authority if the project is likely to affect the road network.

In terms of best (but not mandatory) practice, the consultees should already have been consulted at the scoping stage. In addition, it is a legal requirement that the consultees should be consulted before a decision is made. Once the EIS is completed, copies can be sent to the consultees directly by the developer or by the competent authority. In practice, many competent authorities only send particular EIS chapters to the consultees, e.g. the chapter on archaeology to the archaeologist. However, this often limits the consultee's understanding of the project context and wider impacts; more recent recommendations (DOE 1996) suggest that consultees should be sent a copy of the entire EIS.

The statutory consultees have accumulated a wide range of knowledge about environmental conditions in various parts of the country, and many have published guidelines to EIA procedures (e.g. English Nature's (1992) *Environmental Assessment Handbook*). They can give valuable feedback on the appropriateness of a project and its likely impacts. However, the consultees may have their own priorities, which may prejudice their response to the EIS. In particular, the ex-HMIP (Her Majesty's Inspectorate of Pollution) was wary of commenting on EISs for fear of restricting its own negotiations on air pollution under Integrated Pollution Control; this is discussed further in Chapters 8 and 9.

6.4   EIA presentation

Although the EIA regulations specify the minimum contents required in an EIS (in Appendix 3 of the Town and Country Planning (Assessment of Environmental Effects) Regulations 1988), they do not give any standard for the presentation of this information. Past EISs have ranged from a 3-page typed and stapled EIS to glossy brochures with computer-graphics and multi-volume documents in purpose-designed binders. This section discusses the contents, organization, clarity of communication, and presentation of an EIS.

Contents and organization

An EIS should be *comprehensive*. Its contents must at least fulfil the requirements of the relevant EIA legislation. As we shall discuss in Chapter 8, past EISs have not all fulfilled these requirements; however, the situation is improving rapidly, and local planning authorities are increasingly likely to require information on topics they feel have not been adequately discussed in an EIS. A good EIS will also go further than the minimum requirements if other significant impacts are identified. Most EISs are broadly organized into four sections: a non-technical summary, a discussion of relevant methods and issues, a description of the project, and of the environmental baseline conditions, and a discussion of likely environmental impacts (which may include a discussion of baseline environmental conditions and predicted impacts, proposed mitigation measures and residual impacts). Ideally, an EIS should also include the main alternatives considered, and proposals for monitoring. It could include much or all of the information given in Appendix 4 of *Environmental assessment: a guide to the procedures* (DOE 1989). Figure 1.2 in Chapter 1 provides an example of a good EIS outline.

An EIS should *explain why some impacts are not dealt with*. The introductory chapter, or an appendix, should include a "finding of no significant impact" section to explain why some impacts may be considered insignificant. If, for instance, the development is unlikely to affect the climate, a reason should be given explaining this conclusion. An EIS should emphasize key points. These should have been identified during the scoping exercise, but additional issues may arise during the course of the EIA. The EIS should set the context of the issues. The names of the developer, relevant consultants, relevant local planning authorities and consultees should be listed, along with a contact person for further information. The main relevant planning issues and legislation should be explained. The EIS should also indicate any references used, and give a bibliography at the end. Ideally, the cost of the EIS should be given.

The preparation of a *non-technical summary* is particularly important in an EIS, as this is often the only part of the document that the public and decision-makers will read. The Dutch suggest that this summary "be such that a lay member of the public can read it and then be able to pass a considered opinion on the alternatives described and their environmental impact" (Government of the Netherlands 1991). It should thus briefly cover all relevant impacts and emphasize the most important, and should ideally contain a list or a table that allows readers to identify them at a glance. Chapter 4 gave examples of a number of techniques for identifying and summarizing impacts.

An EIS should ideally be one *unified document*, with perhaps a second volume for appendices. A common problem with the organization of EISs stems from how environmental impacts are assessed. The developer (or the consultants coordinating the EIA) often subcontracts parts of the EIA to consultancies which specialize in those fields (e.g. ecological specialists, landscape consultants). These in turn prepare reports of varying lengths and styles, making a number of (possibly different) assumptions about the project and likely future environmental conditions, and proposing different and possibly conflicting mitigation measures. One way developers have attempted to circumvent this problem has been to summarize the impact predictions in a main text, and add the full reports as appendices to the main body of the EIS. Another has been to put a "company cover" on each report and present the EIS as a multi-volume document, each volume discussing a single type of impact.

Both of these methods are problematic: the appendix method in essence discounts the great majority of findings, and the multi-volume method is cumbersome to read and carry. Neither method attempts to present findings in a cohesive manner, emphasize crucial impacts or propose a coherent package of mitigation and monitoring measures. A good EIS would incorporate the information from the subcontractors' reports into one coherent document which uses consistent assumptions and proposes consistent mitigation measures.

The EIS should be kept as *brief* as possible while still *presenting the necessary information*. The main text should include all the relevant discussion about impacts, and appendices should present only additional data and documentation. In the US, the length of an EIS is generally expected to be less than 150 pages. In the UK, the DOE (1995) recommends:

> For projects which involve a single site and relatively few areas of significant impact, it should be possible to produce a robust ES of around 50 pages. Where more complex issues arise, the main body of the statement may extend to 100 pages or so. If it exceeds 150 pages it is likely to become cumbersome and difficult to assimilate and this should generally be regarded as a maximum . . . However, the quality of an ES will not be determined by its length. What is needed is a concise, objective analysis . . .

Clarity of communication

Weiss (1989) nicely notes that an unreadable EIS is an environmental hazard:

> The issue is the quality of the document, its usefulness in support of the goals of environmental legislation, and, by implication, the quality of the environmental stewardship entrusted to the scientific community . . . An unreadable EIS not only hurts the environmental protection laws and, thus, the environment. It also turns the sincere environmental engineer into a kind of "polluter".

Weiss identifies three classes of error that mar the quality of EIS's communications:

- strategic errors, "mistakes of planning, failure to understand why the EIS is written and for whom";
- structural errors related to the EIS's organization;
- tactical errors of poor editing.

An EIS has to communicate information to many audiences, from the decision-maker, to the environmental expert, to the lay person. Although it cannot fulfil all the expectations of all its readers, it can go a long way towards being a useful document for a wide audience. It should at least be *well-written*, with good spelling and punctuation. It should have a clear structure, with easily visible titles and a logical flow of information. A table of contents, with page numbers marked, should

174

be included before the main text, allowing easy access to information. Principal points should be clearly indicated, perhaps in a table at the front or back.

An EIS should *shun technical jargon*. Any jargon it does include should be explained in the text or in footnotes. All the following examples are from actual EISS:

> *Wrong*: It is believed that the aquiclude properties of the Brithdir seams have been reduced and there is a degree of groundwater communication between the Brithdir and the underlying Rhondda beds, although . . . numerous seepages do occur on the valley flanks with the retention regime dependent upon the nature of the superficial deposits.

> *Right*: The accepted method for evaluating the importance of a site for water-fowl (i.e. waders and wildfowl) is the "1% criterion". A site is considered to be of National Importance if it regularly holds at least 1% of the estimated British population of a species of waterfowl. "Regularly" in this context means counts (usually expressed as annual peak figures), averaged over the last 5 years.

The EIS should clearly *state any assumptions* on which impact predictions are based:

> *Wrong*: As the proposed development will extend below any potential [archaeological] remains, it should be possible to establish a method of working which could allow adequate archaeological examinations to take place.

> *Right*: For each operation an assumption has been made of the type and number of plant involved. These are:

> Demolition: 2 pneumatic breakers, tracked loader
> Excavation: backacter excavator, tracked shovel . . .

The EIS should be *specific*. Although it is easier and more defensible to claim that an impact is significant or likely, the resulting EIS will be little more than a vague collection of possible future trends.

> *Wrong*: The landscape will be protected by the flexibility of the proposed [monorail] to be positioned and designed to merge in both location and scale into and with the existing environment.

> *Right*: From these [specified] sections of road, large numbers of proposed wind turbines would be visible on the skyline, where the towers would appear as either small or indistinct objects and the movement of rotors would attract the attention of road users. The change in the scenery caused by the proposals would constitute a major visual impact, mainly due to the density of visible wind turbine rotors.

Predicted impacts should be *quantified* if possible, perhaps with a range, and the use of non-quantified descriptions, such as severe or minimal, should be explained:

> *Wrong*: The effect on residential properties will be minimal with the nearest properties . . . at least 200 m from the closest area of filling.

175

Table 6.3 Presentation of environmental effects.

Feature	Effects	Affected	Timescale	Magnitude	Controversial	Probability	Mitigation	Significance
Pigeon House Road	Reduced risk of HGV traffic	Residents	Short term permanent	Local	No	High	None	Minor beneficial
	Perceived severance due to elevated structure	Residents	Short term reducing with time	Local	Potentially	Low	None	Minor adverse
Bremen Grove	Reductions in traffic flow by about 80%	Residents and children using the park	Short term permanent	Local	No	High	None	Minor beneficial
Beach Road	Reductions in traffic flow by about 80%	School children	Short term permanent	Local	No	High	None	Minor beneficial

(*Source:* P. Tomlinson, Ove Arup)

Right: Without the bypass, traffic in the town centre can be expected to increase by about 50–75% by the year 2008. With the bypass, however, the overall reduction to 65–75% of the 1986 level can be achieved.

Even better, predictions should give an *indication of the probability* that an impact will occur, and the degree of confidence with which the prediction can be made (see Ch. 9 for a good example). In cases of uncertainty, the EIS should propose worst case scenarios:

Right: In terms of traffic generation, the "worst case" scenario would be for 100% usage of the car park . . . For a more realistic analysis, a redistribution of 50% has been assumed.

Finally, an EIS should be *honest and unbiased*. A review of local authorities noted that "[a] number of respondents felt that the Environmental Statement concentrated too much on supporting the proposal rather than focusing on its impacts and was therefore not sufficiently objective" (Kenyan 1991). Developers cannot be expected to conclude that their projects have such major environmental impacts that they should be stopped. However, it is unlikely that all major environmental issues will have been resolved by the time the statement is written.

Wrong: The proposed site lies adjacent to lagoons, mud and sands which form four regional Special Sites of Scientific Interest [*sic*]. The loss of habitat for birds, is unlikely to be significant, owing to the availability of similar habitats in the vicinity.

Table 6.3 provides a simple example of a clear presentation of the environmental effects of a road development on adjacent areas. Table 6.4 provides an example of a useful summary table of environmental impacts.

Presentation

Although it would be good to report that EISs are read only for their contents and clarity, in reality, as for prime ministers and presidents, presentation can have a great influence on how they are received. EIAs are, indirectly, public relations exercises, and an EIS can be seen as a publicity document for the developer. Good presentation can convey a concern for the environment, a rigorous approach to the impact analysis and a positive attitude to the public. Bad presentation, in turn, suggests a lack of care, and perhaps a lack of financial backing. Similarly, good presentation can help to convey information clearly, whereas bad presentation can negatively affect even a well-organized EIS.

The presentation of an EIS will say much about the developer. The type of paper used – recycled or not, glossy or not, heavy or light-weight – will affect the image projected, as will the choice of coloured or black-and-white diagrams and the use of dividers between chapters. The ultra-green company will opt for double-sided printing on recycled paper, while the luxury developer will use glossy, heavy-weight

Table 6.4 Example of ES summary table showing relative weights given to significance of impacts (note: only a selection of key issues given).

Topic area	Description of impact	Geographical level of Importance of Issue					Impact	Nature	Significance
		I	N	R	D	L			
Human beings	Disturbance to existing properties from traffic and noise				*		Adverse	St, R	Major
	Coalescence of existing settlements			*			Adverse	Lt, IR	Major
Flora and fauna	Loss of grassland of local nature conservation value					*	Adverse	Lt, IR	Minor
	Creation of new habitats					*	Beneficial	Lt, R	Minor
	Increased recreation pressure on SSSI		*				Adverse	Lt, R	Minor
Soil and geology	Loss of 300 acres agricultural soils (grade 3B)			*			Adverse	Lt, IR	Minor
Water	Increased rates of surface water run-off				*		Adverse	Lt, IR	Minor
	Reduction in groundwater recharge			*			Adverse	Lt, R	Minor

Key: I International   St Short-term
      N National   Lt Long-term
      R Regional   R Reversible
      D District   IR Irreversible
      L Local

(*Source:* DOE 1995)

paper with a distinctive binder. Generally, a strong binder that stands up well under heavy handling is most suitable for EISs. Unless the document is very thin, a spiral binder is likely to snap or bend open with continued handling; similarly, stapled documents are likely to tear. Multi-volume documents are difficult to keep together unless a box is provided.

Finally, the use of maps, graphs, photo-montages, diagrams and other forms of visual communication can greatly help the EIS presentation. As we noted in Chapter 4, a location map, a site layout of the project and a process diagram are virtually essential to a proper description of the development. Maps showing, for example, the extent of visual impacts, the location of designated areas or classes of agricultural land are a succinct and clear way of presenting such information. Graphs are often much more effective than tables or figures in conveying numerical information. Forms of visual communication break up the page, and add interest to an EIS.

6.5   Review of EISs

The comprehensiveness and accuracy of EISs are matters of concern. As will be shown in Chapter 8, many EISs do not meet even the minimum regulatory requirements, much less provide adequate information on which to base decisions. In some countries, for example the Netherlands, Canada, Malaysia and Indonesia, EIA Commissions have been established to review EISs and act as a quality assurance process. However, in the UK there are no mandatory requirements regarding the pre-decision review of EISs to ensure that they are comprehensive and accurate. A planning application cannot be judged invalid just because it is accompanied by an inadequate or incomplete EIS: a competent authority may only request further information, or refuse permission and risk an appeal.[2]

Many competent authorities do not have the full range of technical expertise needed to assess the adequacy and comprehensiveness of an EIS. Some authorities, especially those which receive few EISs, have consequently had difficulties in dealing with the technical complexities of EISs. In about 10–20 per cent of cases, consultants have been brought in to review the EISs. Other authorities have joined the Institute of Environmental Assessment, which reviews one EIS at no cost for member organizations. Others have been reluctant to buy outside expertise, especially at a time of restrictions on local spending (McDonic 1992, Fuller 1992). A technique advocated by the International Association for Impact Assessment (Partidario 1996), although not seen often in practice, is to involve parties other than just the competent authority in EIS review, especially the public.

In an attempt to fill the void previously left by the national government, several non-mandatory review criteria have been established. Effective review criteria should allow a competent authority to:

- ensure that all relevant information has been analysed and presented,
- assess the validity and accuracy of information contained in the EIS,
- quickly become familiar with the proposed project and consider whether additional information is needed,
- assess the significance of the project's environmental effects,
- evaluate the need for mitigation and monitoring of environmental impacts, and
- advise on whether a project should be allowed to proceed (Tomlinson 1989).

To fulfil these criteria, Tomlinson proposed review criteria in the form of yes/no questions concerning nine main issues: administration/procedural requirements, effective communication, impact identification, alternatives, information assembly, baseline description, impact prediction, mitigation measures and monitoring/audits.[3]

Lee and Colley (1990) in turn proposed a hierarchical review framework. At the top of the hierarchy is a comprehensive mark (A = well-performed and complete, through to F = very unsatisfactory) for the entire report. This mark is based on marks given to four broad sub-headings: description of the development, local environment and baseline conditions; identification and evaluation of key impacts;

alternatives and mitigation of impacts; communication of results. Each of these, in turn, is based on two further layers of increasingly specific topics or questions. Lee and Colley's criteria have been used either directly or in a modified form (e.g. by the Institute of Environmental Assessment) to review a range of EISs in the UK. It is the most commonly-used review method in the UK. Appendix 3 gives the Lee & Colley framework.

In 1994, the EC (1993) also published recommended review criteria. These are similar to Lee & Colley's, but use eight subheadings instead of four, include a longer list of specific questions, and judge the information based on relevance to the project context and importance for decision-making as well as presence/absence in the EIS.

The review criteria given in Appendix 4 are an amalgamation and extension of Lee and Colley's and the EC's criteria, developed by the Impact Assessment Unit (IAU) at Oxford Brookes University. It is unlikely that any EIS will fulfil all the criteria. Similarly, some criteria may not apply to all projects. However, they should act as a checklist of good practice for both those preparing and those reviewing EISs. Table 6.5 shows a number of possible ways of using these criteria. Example (a), which relates to minimum requirements, amplifies the presence or otherwise of key information. Example (b) includes a simple grading, which could be on the A–F scale used by Lee and Colley, for each criterion (only one of which is shown here). Example (c) takes the format of the EC criteria, which appraise the relevance of the information and then judge whether it is complete, adequate (not complete but need not prevent decision-making from proceeding) or inadequate for decision-making.

6.6 Decisions on projects

EIA and project authorization

Decisions to authorize or reject projects are made at several levels:

> At the top of the tree are the relevant Secretaries of State (Environment, Wales, Scotland and Northern Ireland); below them are a host of Inspectors, sometimes called Reporters (Scotland); further down the list come Councillors, the elected members of district, county, unitary or metropolitan borough councils; and at the very bottom are chief or senior planning officers who deal with "delegated decisions". . . . [as] a rough guide, the larger the project the higher up the pyramid of decision makers the decision is made. (Weston 1997)

Where required by the competent authority, an EIS must be submitted with the application for authorization.[4] The decision on an application with an EIS must be made within a specified period (e.g. 16 weeks for a planning application), unless

Table 6.5 Examples of possible uses for EIS review criteria.

(a)

Criterion	Presence/absence (page number)	Information	Key information absent
Describes the proposed development, including its design, and size or scale	✔ (p. 5)	Location (in plans), existing operations, access	Working method, vehicle movements, restoration plans
Indicates the physical presence of the development	X		Site buildings (location, size), restoration

(b)

Criterion	Presence/absence (page number)	Comments	Grade
Explains the purposes and objectives of the development	✔ (p. 11)	Briefly in introduction, more details in Sec. 2	A
Gives the estimated duration of construction etc. phases	✔ (p. 12)	Not decommissioning	B

(c)

Criterion	Relevant? (Y/N)	Judgement (C/A/I)*	Comment
Considers the "no action" alternative, alternative processes, etc.	Y	A	Alternative sites discussed, but not alternative processes
If unexpectedly severe adverse impacts are identified, alternatives are reappraised	N		Impacts of sand/gravel working well understood

*C complete   A adequate   I inadequate

the developer agrees to a longer period. As we noted in Section 6.5, it is at this stage that the EIS review is undertaken. When making a decision, the competent authority is required to have regard to all the environmental information, i.e. "the information contained in the environmental statement and any comment made by the statutory consultees and representations from members of the public, as well as to other material considerations" (Circular 15/88). By any standards, making decisions on development projects is a complex undertaking. Decisions for projects requiring EIAs tend to be even more complex, because by definition they deal with larger, more complex projects, and probably a greater range of interest groups: "The competition of interests is not simply between the developer and the consultees. It can also be a conflict between consultees, with the developer stuck in the middle hardly able to satisfy all parties and the 'competent authority' left to establish a planning balance where no such balance can be struck" (Weston 1997).

Whereas in the early years the decision-making process for projects with EIA was accepted as being basically a black box, more recently attempts have been made to make the process more rigorous and transparent. Research by the University of Manchester (see Wood & Jones 1997) and Oxford Brookes University (see Weston et al. 1997) has focused on how environmental information is used in UK decision-making; this is discussed further in Chapter 8. Similar work carried out by Land Use Consultants resulted in a research report (DOE 1994a) and a good practice guide (DOE 1994b) on the evaluation of environmental information for planning projects; the advice, however, could relate equally well to other types of project. The good practice guide begins with a definition of evaluation:

> ... in the context of environmental assessment, there are a number of different stages or levels of evaluation. These are concerned with:
>
> - checking the adequacy of the information supplied as part of the ES, or contributed from other sources;
> - examining the magnitude, importance and significance of individual environmental impacts and their effects on specific areas of concern ... ;
> - preparing an overall "weighing" of environmental and other material considerations in order to arrive at a basis for the planning decision.

The guide suggests that, after vetting the application and EIS, advertising the proposals and EIS, and relevant consultation, the LPA should carry out two stages of decision-making: an evaluation of the individual environmental impacts and their effects, and weighing the information to reach a decision. The evaluation of impacts and effects first involves verifying any factual statements in the EIS, perhaps by highlighting any statements of concern and discussing these with the developer. The nature and character of particular impacts can then be examined; either the EIS will already have provided such an analysis (e.g. in the form of Table 6.3 or 6.4) or the case-work officer could prepare such a table. Finally, the significance and importance of the impacts can be weighed up, taking into consideration such issues as the extent of the area affected, the scale and probability of the effects, the scope for mitigation and the importance of the issue.

Weighing up the information to reach a decision involves not only considering the views of different interest groups and the importance of the environmental issues, but also determining whether the proposed project is in accordance with the development plan: "all development control decisions are required to accord with the development plan unless material considerations indicate otherwise and proposals should be consistent with policy" (DOE 1994b). The guide suggests that environmental impacts can be divided into three groups: those which by themselves

182

provide grounds for refusal or approval, those which in conjunction with others influence the decision, and those which are unlikely to influence the outcome of the decision. Then, "decision-makers will usually be faced with a choice. The planning merits will depend upon a comparison of the advantages and disadvantages arising from the construction and operation of the development, with the consequences of maintaining the status quo – or 'do-nothing' option" (DOE 1994b). In the case of a planning application, the planning officer's recommendations will then go to the planning committee, which makes the final decision.

The range of decision options are as for any application for project authorization: the competent authority can grant permission for the project (with or without conditions) or refuse permission. It can also suggest further mitigation measures following consultations, and will seek to negotiate these with the developer. If the development is refused, the developer can appeal against the decision. If the development is permitted, people or organizations can challenge the permission. The relevant Secretary of State can also "call in" an application, for a variety of reasons. A public inquiry may result.[5]

EIA and public inquiries

Compellingly Weston (1997) discusses why all parties involved in EIA try to avoid public inquiries:

> By the time a project becomes the subject of a public inquiry the sides are drawn and the hearing becomes a focus for adversarial debate between opposing, expensive, experts directed and spurred on by advocates schooled in the art of cajoling witnesses into submission and contradictions. Such debates are seldom rational or in any other way related to the systematic, iterative and co-operative characteristics of good practice EIA. By the time the inquiry comes around, and all the investment has been made in expert witnesses and smooth talking barristers, it is far too late for all that. (Weston 1997)

Nevertheless, by 1996 more than a hundred projects involving EIA had gone to inquiry.

The environmental impact of proposals, especially traffic, landscape and amenity issues, will certainly be examined in detail during any inquiry. The EIA regulations allow inquiry inspectors and the secretary of state to require (a) the submission of an EIS before a public inquiry, if they regard this as appropriate, and (b) further information from the developer if they consider the EIS is inadequate as it stands. In practice, before public inquiries involving EIAs the inspector generally receives a case file (including the EIS) which is examined to determine whether any further information is required. Pre-inquiry meetings may be held where the inspector may seek further information; these meetings may also assist the developer and competent authority to arrive at a list of agreed matters before the start of the inquiry; this can avoid unnecessary delays during it. At the inquiry, inspectors often ask for further information, and they may adjourn the inquiry if the information cannot be produced within the available time. The information contained in the EIS will be among the material considerations taken into account. However, an inadequate EIS is not a valid reason for preventing authorization, or even for delaying an inquiry.[6]

An analysis of ten public inquiries involving projects for which EISs had been prepared (Jones & Wood 1995) suggested that in their recommendations most inspectors give "moderate" or "considerable" weight to the EIS and consultations on the EIS, and that environmental information is of "reasonable" importance to the decision whether to grant consent. However, a study of 54 decision letters from inspectors (Weston 1997) suggests that EIA has had little influence on the inquiry process: in about two-thirds of the cases, national or local land-use policies were the determining issues identified by the inspectors and the secretary of state, and in the remaining cases other traditional planning matters predominated:

> Other issues which are more directly related to the introduction of the Regulations [were] not, or [were] rarely, discussed under the headings given in Schedule 3. Climate was not an issue in a single case and cultural heritage and material assets, although discussed, were not debated under those headings. The headings which dominate the decision letters of the Inspectors and Secretaries of State are the traditional planning material considerations such as amenity, various forms of risk, traffic and need, although some factors such as flora and fauna, noise and landscape do tend to be discussed separately. (Weston 1997)

Challenging a decision: judicial review

The UK planning system has no official provisions for an appeal against development consent. However, if permission is granted, a third party may wish to challenge that decision on the grounds, for example, that no EIA was prepared when it should have been, or that the competent authority did not adequately consider the relevant environmental information. The only way to do this is through judicial review proceedings in the courts, or through the European Community.

Judicial review proceedings in the UK courts first require that the third party shows it has "standing" to bring in the application, namely sufficient interest in the project by virtue of attributes specific to it or circumstances which differentiate it from all other parties (e.g. a financial or health interest). Establishing standing is one of the main difficulties in applying for judicial review.[7] If standing is established, the third party must then convince the court that the competent authority did not act according to the relevant EIA procedures. The court does not make its own decision about the merits of the case, but only reviews the way in which the competent authority arrived at its decision:

> The court will only quash a decision of the [competent authority] where it acted without jurisdiction or exceeded its jurisdiction or failed to comply with

the rules of natural justice in a case where those rules apply or where there is an error of law on the face of the record or the decision is so unreasonable that no [competent authority] could have made it. (Atkinson and Ainsworth 1992)

Various possible scenarios emerge. A competent authority may fail to require an EIA for a Schedule 1 project, or may grant permission for such a project without considering the environmental information. In such a case, its decision would be void.

A competent authority may decide that a project does not require EIA because it is not in Schedule 1 or Schedule 2 with significant environmental effects. This was the issue in the case of *R v. Swale Borough Council and Medways Port Authority ex parte RSPB* (1991, 1 P.L.R. 6) concerning the construction of a storage area for cargo, which would require the infill of Lappel Bank, a mudflat important for its wading birds. The appellants argued that the project fell either within Schedule 1 or Schedule 2 with significant environmental effects; the local planning authority felt that an EIA was unnecessary. The courts held that a project's falling within a Schedule is a decision for the local authority to make, and open for review only if no reasonable local authority could have made it.

A competent authority may make a decision in the absence of a formal EIA, but with environmental information available in other forms. This was the case in *R v. Poole Borough Council, ex parte Beebee and others* (1991, J.P.L. 643) concerning a decision to develop part of Canford Heath. In this case the courts ruled that, despite the lack of an EIS and the attendant rigour and publicity, enough environmental information was available for the council to make an informed decision. In a similar Scottish appeal case against a LPA decision to refuse planning permission for an opencast coal mine, the Reporter felt that an EIA would not have raised issues that would not have been raised by other means (Weston 1997). A different judgement may have been made if the competent authority had been shown to have made its decision before it had received all the relevant environmental information.[8]

In several cases (e.g. the M3 at Twyford Down and a large afforestation scheme in Glen Dye (1992, J.E.L. 289)) a competent authority approved a project, in the absence of an EIA, after 3 July 1988 – when Directive 85/337 should have been implemented – but before the relevant UK regulation came into effect. These cases have questioned how far the Directive applies to transitional projects proposed before 3 July 1988 but decided after that date. Macrory (1992) summarizes the courts' judgement: "The Directive is intended to influence the 'process at every state', and in the absence of clear transitional measures it would be against the aims of the Directive to attempt to retrospectively impose such requirements on decision-making procedures already commenced." The cases have also focused on whether Directive 85/337 can have a direct effect in the Member States. However, a legal judgement of 1994 (*Wychavon D.C. v. Secretary of State for the Environment and Velcourt Ltd., Times* 7.1.94) concluded that since various of the Directive's articles were not certain and unconditional the Directive is incapable of having direct effect.

In summary, judicial reviews of competent authority decisions have to date been severely limited by the issue of standing, and by the courts' narrow interpretation of the duties of competent authorities under the EIA regulations. Future court cases may widen this interpretation, but it is very unlikely that the UK courts will play as active a role as those in the US did in relation to the NEPA.

Challenging a decision: the European Commission

Another avenue by which third parties can challenge a competent authority's decision to permit development, or not to require EIA, is the European Community. Such cases need to show that the UK failed to fulfil its obligations as a Member State under the Treaty of Rome by not properly implementing EC legislation, in this case Directive 85/337. In such a case, Article 169 of the Treaty allows a declaration of non-compliance to be sought from the European Court of Justice. The issue of standing is not a problem here, since the European Commission can begin proceedings either on its own initiative or based on the written complaint of any person. To use this mechanism, the Commission must first state its case to the Member State and seek its observations. The Commission may then issue a "reasoned opinion". If the Member State fails to comply within the specified time, the case proceeds to the European Court of Justice.

This mechanism was first used in October 1991, by the EC Commissioner for the Environment against the UK government in the case of seven projects:

(a) The extension of the M3 bisecting Twyford Down near Winchester.
(b) The extension of the M11 to link it to the Blackwall Tunnel.
(c) The construction of a clinical waste incinerator at South Warwick Hospital in the West Midlands.
(d) The construction of a soft drinks and can manufacturing plant at Brackmills in Northamptonshire.
(e) The construction of a high-speed train link between the Channel Tunnel and London, and a London rail terminal.
(f) The extension of British Petroleum's gas separation plant at Kinneil near Falkirk.
(g) The East London River Crossing linking Becton in the Docklands to Greenwich, and passing through the ancient woodland of Oxleas Wood.

The EC dropped legal proceedings against the M3 proposal in late July 1992, after being satisfied with the UK's response that the Directive had been complied with (the road has since been built). The EC accepted that consent for the M11 had been granted before the Directive came into effect. The waste incinerator has been refused on appeal, a second EIA has been prepared for the Brackmills plant, and the EC was satisfied that the Channel Tunnel rail link and terminal could be considered separate projects (*Simmons and Simmons Environmental Law Newsletter* 15). However, in late 1992 the EC was still considering issuing a "reasoned opinion" over the case of the Kinneil gas separation plant, arguing that no public consultation had taken place, and that the competent authority had not considered the information the developer had prepared (*Planning* 980); and in May 1993 it issued a "reasoned opinion" on the East London River Crossing, alleging that the procedural requirements of the Directive had not been complied with, and that no non-technical summary had been prepared (*Simmons and Simmons Environmental Law Newsletter* 18).

Under Article 171 of the Treaty of Rome, if the European Court of Justice finds that a Member State has failed to fulfil an obligation under the Treaty, it may require the Member State to take the necessary measures to comply with the Court's judgement. Under Article 186, the EC may take interim measures to require a Member State to desist from certain actions until a decision is taken on the main action. However, to do so the Commission must show the need for urgent relief, and that irreparable damage to Community interests would result if these measures were not taken. Suggested amendments to the Treaty of Maastricht would enable the European Court of Justice to impose fines on Member States in the future. We refer readers to Atkinson & Ainsworth (1992), Buxton (1992), and Salter (1992a, b, c) for further information.

6.7 Summary

Active public participation, thorough consultation with relevant consultees, and good presentation are important aspects of a successful EIA process. All have been undervalued to date. The presentation of environmental information has improved, and statutory consultees are becoming increasingly familiar with the EIA process, but public participation is likely to remain a weak aspect of EIA in the UK until developers and competent authorities see the benefits exceeding the costs.

A formal review of EIA is also rarely carried out, despite the availability of several non-mandatory review guidelines and recent government advice on the use of environmental information for decision-making. Such review procedures can contribute to the processing of the EIS as part of the decision-making stage. The impact of the EIS on the quality of the outcome is discussed in Chapter 8.

Several appeals against development consents or against competent authorities' failure to require EIA have been brought to the UK courts or the EC. The UK courts have been unwilling to overturn the decisions of competent authorities, and have generally given a narrow interpretation of the duties of competent authorities under the EIA regulations. The EC, by contrast, has proved willing to challenge the UK government on its implementation of Directive 85/337 and on a number of specific decisions resulting from this implementation.

More positively, the next step in a good EIA procedure is the monitoring of the development's actual impacts and the comparison of actual and predicted impacts. This is discussed in the next chapter.

References

Arnstein, S. R. 1971. A ladder of public participation in the USA. *Journal of the Royal Town Planning Institute*, April, 216–24.

Atkinson, N. and R. Ainsworth 1992. Environmental assessment and the local authority: facing the European imperative. *Environmental Policy and Practice* 2(2), 111–28.

Buxton, R. 1992. Scope for legal challenge. In *Environmental assessment and audit: a user's guide*, Ambit (ed.), 43–44. Gloucester: Ambit.

Canter, L. W. 1996. *Environmental impact assessment*, 2nd edn. New York: McGraw-Hill.

Clark, B. 1994. Improving public participation in environmental impact assessment. *Built Environment* 20(4), 294–308.

Commission of European Communities 1993. *Checklist for review of environmental information submitted under EIA procedures*. DG XI. Brussels: CEC.

Council for the European Community 1990. Directive 90/313, *Official Journal* L158, 23 June, 56.

Council for the Protection of Rural England 1991. *The Environmental Assessment Directive – five years on*. London: CPRE.

Dallas, W. G. 1984. Experiences of environmental impact assessment procedures in Ireland. In *Planning and Ecology*, R. D. Roberts and T. M. Roberts (eds), 389–95. London: Chapman and Hall.

DOE (Department of the Environment) 1989. *Environmental assessment: a guide to the procedures*. London: HMSO.

DOE 1991. *Monitoring environmental assessment and planning*. Report by the EIA Centre, Department of Planning and Landscape, University of Manchester, HMSO.

DOE 1994a. *Good practice on the evaluation of environmental information for planning projects: research report*. London: HMSO.

DOE 1994b. *Evaluation of environmental information for planning projects: a good practice guide*. London: HMSO.

DOE 1995. *Preparation of environmental statements for planning projects that require environmental assessment*. London: HMSO.

DOE 1996. *Changes in the quality of environmental statements for planning projects*. London: HMSO.

DOE et al. 1990. *This common inheritance: Britain's environmental strategy* (Cmnd 1200). London: HMSO.

Elkin, T. J. & P. G. R. Smith 1988. What is a good environmental impact statement? Reviewing screening reports from Canada's National Parks. *Journal of Environmental Management* 26, 71–89.

Fuller, K. 1992. Working with assessment. In *Environmental assessment and audit: a user's guide*, Ambit (ed.), 14–15. Gloucester: Ambit.

Government of the Netherlands 1991. *Environmental impact assessment*. The Hague: Ministry of Housing, Physical Planning and the Environment.

Hancock, T. 1992. Statement as an aid to consent. In *Environmental assessment and audit: a user's guide*, Ambit (ed.), 34–5. Gloucester: Ambit.

House of Lords 1981. *Environmental assessment of projects*, Select Committee on the European Communities, 11th Report, Session 1980–81, HMSO.

Hydro-Québec 1993. *Overview 21: communication and environmental assessment*. Quebec: Hydro-Québec.

Jain, R. K., L. V. Urban & G. S. Stacey 1977. *Environmental impact analysis*. New York: Van Nostrand Reinhold.

Jones, C. E. & C. Wood 1995. The impact of environmental assessment in public inquiry decisions. *Journal of Planning and Environment Law*, October, 890–904.

Kenyan, R. C. 1991. Environmental assessment: an overview on behalf of the R.I.C.S. *Journal of Planning and Environment Law*, 419–22.

Lee, N. & R. Colley 1990. *Reviewing the quality of environmental statements*, Occasional Paper No. 24. Manchester: University of Manchester, EIA Centre.

McCormick, J. 1991. *British politics and the environment*. London: Earthscan.

McDonic, G. 1992. The first four years. In *Environmental assessment and audit: a user's guide*, Ambit (ed.), 2–3. Gloucester: Ambit.

McNab, A. 1997. Scoping and public participation. In *Planning and EIA in practice*, J. Weston (ed.), 60–77. Harlow: Longman.

Macrory, R. 1992. Analysis of the petition of the Kincardine and Deeside District Council. *Journal of Environmental Law* 4(2), 298–304.

Mollison, K. 1992. A discussion of public consultation in the EIA process with reference to Holland and Ireland (written for MSc Diploma course in Environmental Assessment and Management). Oxford: Oxford Polytechnic.

O'Riordan, T. & R. K. Turner 1983. *An annotated reader in environmental planning and management*. Oxford: Pergamon.

Partidario, M. R. 1996. *16th Annual Meeting, International Association for Impact Assessment: Synthesis of Workshop Conclusions*. Estoril: IAIA.

Read, R. 1997. Planning authority review. In *Planning and EIA in practice*, J. Weston (ed.). Harlow: Longman.

Salter, J. R. 1992a. Environmental assessment: the challenge from Brussels. *Journal of Planning and Environment Law*, January, 14–20.

Salter, J. R. 1992b. Environmental assessment – the need for transparency. *Journal of Planning and Environment Law*, March, 214–21.

Salter, J. R. 1992c. Environmental assessment – the question of implementation. *Journal of Planning and Environment Law*, April, 313–18.

Tomlinson, P. 1989. Environmental statements: guidance for review and audit. *The Planner* 75(28), 12–15.

Turner, T. 1988. The legal eagles. *Amicus Journal*, Winter, 25–37.

Weiss, E. H. 1989. An unreadable EIS is an environmental hazard. *Environmental Professional* 11, 236–240.

Westman, W. E. 1985. *Ecology, impact assessment and environmental planning*. New York: Wiley.

Weston, J. (ed.) 1997. *Planning and EIA in practice*. Harlow: Longman.

Weston, J., J. Glasson, R. Therivel, E. Weston R. Frost 1997. *Environmental information and planning projects*, Working Paper No. 170. Oxford: Oxford Brookes University, School of Planning.

Williams, G. & A. Hill 1996. Are we failing at environment justice? How minority and low income populations are kept out of the public involvement process. *Proceedings of the 16th Annual Meeting of the International Association for Impact Assessment*. Estoril: IAIA.

Winter, P. 1994. Planning and sustainability: an examination of the role of the planning system as an instrument for the delivery of sustainable development. *Journal of Planning and Environment Law*, October, 883–900.

Wood, C. M. & C. Jones 1997. The effect of environmental assessment on UK local authority planning decisions. *Urban Studies* **34**(8), 1237–57.

Notes

1   Although this section refers to public consultation and participation together as "public participation", the two are in fact separate. Consultation is in essence an exercise concerning a passive audience: views are solicited, but respondents have little active influence over any resulting decisions. In contrast, public participation involves an active role for the public, with some influence over any modifications to the project and over the ultimate decision.

2   Weston (1997) notes that LPAs need to be aware that they have the power to ask for further information, and that failure to use it could later be seen as tacit acceptance of the information provided. For instance, when deciding on an appeal for a Scottish quarry extension, the Reporter noted that it was significant that the LPA had not requested further information when they were processing the application, and had not objected to the EIS until the development came to appeal.

3   These are based on a Canadian framework initially established by Elkin & Smith (1988).

4   Where the project has already been built without authorization, the competent authority considers the environmental information when determining whether the project will be demolished or not.

5   For instance, in 1993–4, about 1,200 public inquiries were held, only a few of which involved EIAs. About a hundred of the 1,200, mostly the high-profile cases, were called in by the SOS.

6   For instance, in the case of a Scottish appeal regarding a proposed quarry extension (Scottish Office, P/PPA/SQ/336, 6 January 1992), the Reporter noted that: "The ES has been strongly criticised . . . [it] does not demonstrate that a proper analysis of environmental impacts has been made . . . Despite its shortcomings, the ES appears to me to comply broadly with the statutory requirements of the EA regulations."

7   A recent EC court case, for instance, ruled that Greenpeace had insufficient individual concerns to contest a decision to use regional funds to help build power stations in the Canary Islands (*Greenpeace v. Commission of the European Communities* (1996) 8 Journal of Environmental Law 139). Similar judgements have been made in the UK context.

8   The UK is not alone in this. A 1994 German Federal Administrative Court ruling held that it was necessary for a plaintiff to demonstrate that a decision would have to be different had an EIA been carried out, before that decision could be quashed (Weston 1997).

CHAPTER 7
Monitoring and auditing: after the decision

7.1 Introduction

Major projects, such as roads, airports, power stations, petrochemical plants, mineral developments and holiday villages, have a life-cycle, with a number of stages (see Fig. 1.5). It may cover a very long period (e.g. 50–60 years for the planning, construction, operation and decommissioning of a fossil-fuelled power station). EIA, as it is currently practised in the UK and in many other countries, relates primarily to the period *before* the decision. At its worst, it is a partial linear exercise related to one site, produced in-house by a developer, without any public participation. There is a danger of a short-sighted "build it and forget it" approach (Culhane 1993). However, EIA should not stop at the decision. It should be more than an auxiliary to the procedures to obtain a planning permission; rather it should be a means to obtain good environmental management *over the life* of the project. This means including monitoring and auditing in the EIA process.

The first section clarifies the definitions of and differences between monitoring and auditing, and outlines their potentially important roles in EIA. An approach to the better integration of monitoring into the process, drawing in particular on Californian practice, is then outlined. We then discuss approaches to environmental impact auditing, including a review of recent attempts to audit a range of EISs in a number of countries. The final section draws briefly on detailed monitoring and auditing studies of the local socio-economic impacts of the construction of the Sizewell B PWR nuclear power station in the UK.

7.2 The importance of monitoring and auditing in the EIA process

Monitoring involves the measuring and recording of physical, social and economic variables associated with development impacts (e.g. traffic flows, air quality, noise, employment levels). The activity seeks to provide information on the characteristics and functioning of variables in time and space, and in particular on the occurrence and magnitude of impacts. Monitoring can improve project management. It can be used, for example, as an early warning system, to identify harmful trends in a locality before it is too late to take remedial action. It can help to identify and correct unanticipated impacts. Monitoring can also provide an accepted data base, which can be useful in mediation between interested parties. Thus, monitoring of the origins, pathways and destinations of, for example, dust in an industrial area may clarify where the responsibilities lie. Monitoring is also essential for successful environmental impact auditing, and can be one of the most effective guarantees of commitment to undertakings and to mitigation measures.

As noted by Buckley (1991), the term environmental auditing is currently used in two main ways. *Environmental impact auditing*, which is covered in this chapter, involves comparing the impacts predicted in an EIS with those that actually occur after implementation, in order to assess whether the impact prediction performs satisfactorily. The audit can be of both impact predictions (how good were the predictions?) and of mitigation measures and conditions attached to the development (is the mitigation effective, are the conditions being honoured?). This approach to auditing contrasts with *environmental management auditing*, which focuses on public and private corporate structures and programmes for environmental management and the associated risks and liabilities. We discuss this latter approach further in Chapter 12.

In total, monitoring and auditing can make important contributions to the better planning and EIA of future projects (see Fig. 7.1). Sadler (1988) writes of the need to introduce feedback in order to learn from experience; we must avoid the constant "reinventing of the wheel" in EIA. Monitoring and auditing of outcomes can contribute to an improvement in all aspects of the EIA process, from understanding baseline conditions to the framing of effective mitigating measures. In addition Greene et al. (1985) note that monitoring and auditing should reduce time and resource commitments to EIA by allowing all participants to learn from past experience; they should also contribute to a general enhancing of the credibility of proponents, regulatory agencies and EIA processes. We are learning, and there is a considerable growth of interest in examining the effectiveness of the EIA process in practice. Unfortunately there are a number of significant issues that have greatly limited the use of monitoring and auditing to date. These issues and possible ways forward for monitoring and auditing in practice are now discussed.

7.3 Monitoring in practice

Key elements

Monitoring implies the systematic collection of a potentially large quantity of information over a long period of time. Such information should include not only the

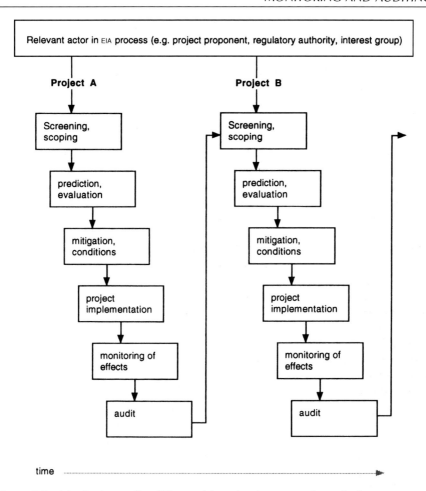

Figure 7.1 Monitoring and auditing and learning from experience in the EIA process. (Adapted from Bisset & Tomlinson 1988, Sadler 1988)

traditional *indicators* (e.g. ambient air quality, noise levels, the size of a workforce) but also *causal underlying factors* (e.g. the *decisions and policies* of the local authority and developer). The causal factors determine the impacts and may have to be changed if there is a wish to modify impacts. *Opinions* about impacts are also important. Individual and group "social constructions of reality" (IAIA 1994) are often sidelined as "mere perceptions, or emotions", not to be weighted as heavily as facts. But such opinions can be very influential in determining the response to a project. To ignore or undervalue them may not be methodologically defensible and is likely to raise hostility. Monitoring should also analyse impact equity. The distribution of impacts will vary between groups and locations; major projects may be more vulnerable than others, as a result of factors such as age, race, gender and

193

income. So a systematic attempt to identify opinions can be an important input into a monitoring study.

The information collected needs to be stored, analysed and communicated to relevant participants in the EIA process. A primary requirement, therefore, is to focus monitoring activity only on "those environmental parameters expected to experience a significant impact, together with those parameters for which the assessment methodology or basic data were not so well established as desired" (Lee & Wood 1980).

Monitoring is an integral part of EIA; baseline data, project descriptions, impact predictions and mitigation measures should be developed with monitoring implications in mind. An EIS should include a *monitoring programme* which has clear objectives, temporal and spatial controls, an adequate duration (e.g. covering the main stages of the project's implementation), practical methodologies, sufficient funding, clear responsibilities and open and regular reporting. Ideally, the monitoring activity should include a partnership between the parties involved; for example, the collection of information could involve the developer, local authority and local community. Monitoring programmes should also be adapted to the dynamic nature of the environment (Holling 1978).

Mandatory or discretionary

Unfortunately monitoring is not a mandatory step in many EIA procedures, including those current in the UK. European Commission regulations do not specifically require monitoring. This omission was recognized in the review of Directive 85/337 (CEC 1993). The Commission is a strong advocate for the inclusion of a formal monitoring programme in an EIS, but EU Member States are normally more defensive and reactive. In consequence, the amended Directive does not include a mandatory monitoring requirement. However, this has not deterred some Member States. For example, in the Netherlands the competent authority is required to monitor project implementation, based on information provided by the developer, and to make the monitoring information publicly available. If actual impacts exceed those predicted, the competent authority must take measures to reduce or mitigate these impacts.

In other Member States, in the absence of mandatory procedures, it is usually difficult to persuade developers that it is in their interest to have a continuing approach to EIA. This is particularly the case where the proponent has a one-off project, and has less interest in learning from experience for application to future projects. Fortunately, we can turn to some examples of good practice in a few other countries. A brief summary of monitoring procedures in Canada is included in Chapter 11. In Hong Kong, a systematic, comprehensive environmental monitoring and auditing system was introduced in 1990 for major projects. The environmental monitoring and audit manual includes three stages of an event-action plan: (1) trigger level, to provide an early warning; (2) action level, at which action is to be taken before an upper limit of impacts is reached; (3) target level, beyond which a predetermined plan response is initiated to avoid or rectify any problems. The approach does build monitoring much more into project decision-making, requiring proponents to agree monitoring and audit protocols and event-action plans in advance; however, enforcement is still a problem (Au & Sanvicens 1996).

The case of California

The monitoring procedures used in California, for projects subject to the California Environmental Quality Act (CEQA), are of particular interest (California Resources Agency 1988). Since January 1989, state and local agencies in California have been required to adopt a monitoring and/or reporting programme for mitigation measures and project changes which have been imposed as conditions to address significant environmental impacts. The aim is to provide a mechanism which will help to ensure that mitigation measures will be implemented in a timely manner in accordance with the terms of the project's approval. Monitoring refers to the observation and oversight of mitigation activities at a project site, whereas reporting refers to the communication of the monitoring results to the agency and public. If the implementation of a project is to be phased, the mitigation and subsequent reporting and monitoring may also have to be phased. If monitoring reveals that mitigation measures are ignored or are not completed, sanctions could be imposed; these can include, for example, "stop work" orders, fines and restitution. The components of a monitoring programme would normally include the following:

- a summary of the significant impacts identified in the Environmental Impact Report;
- the mitigation measures recommended for each significant impact;
- the monitoring requirements for each mitigation measure;
- the person or agency responsible for the monitoring of the mitigation measure;
- the timing and/or frequency of the monitoring;
- the agency responsible for ensuring compliance with the monitoring programme;
- the reporting requirements.

Figure 7.2 provides an extract from a monitoring programme for a woodwaste conversion facility at West Berkeley in California.

UK experience

Although monitoring is not a mandatory requirement under UK EIA regulations, there is monitoring activity. A research study at Oxford Brookes University (see Glasson 1994; Frost 1997) has sought to provide an initial estimate of the extent of

195

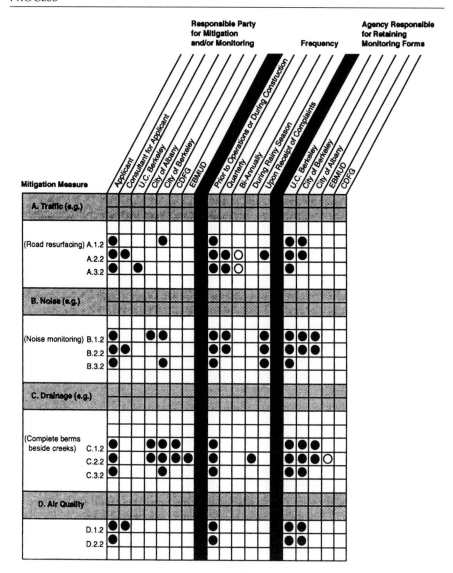

Figure 7.2 Example of Californian monitoring programme. (*Source:* Baseline Environmental Consulting 1989)

such activity using a "contents analysis" and a "practice analysis". The contents analysis of references to monitoring intentions uses a representative sample of almost 700 EISs and summaries of EISs (taken from the Institute of Environmental Assessment's *Digest of Environmental Statements*) (IEA 1993). For some EISs there was a clearly indicated monitoring section; for others monitoring was covered in sections related to mitigation. In several cases there were generic monitoring

Table 7.1 Types of impact monitoring in UK EISS.

Type	% of total monitoring proposals
Water quality	16
Air emissions	15
Aqueous emissions	13
Noise	12
General	9
Others	7
Ecological	7
Archaeological	6
Air quality	5
Structural survey	4
Liaison group	3
Water levels	3
	100

(*Source:* Glasson 1994)

proposals with, for example, a proposal to check that contractors are in compliance with contract specifications. Overall, approximately 30 per cent of the cases included at least one reference to impact monitoring. The maximum number of monitoring types was six, suggesting that impact monitoring is unlikely to be approaching comprehensiveness in even a select few cases. Table 7.1 shows the types of monitoring in EISS. Water quality monitoring was more frequently cited than air quality monitoring. Point of origin monitoring of air and aqueous emissions was also frequently cited. There was only very limited reference to the monitoring of non-biophysical (i.e. socio-economic) impacts. The type of monitoring varied between project types. For Combined Cycle Gas Turbine (CCGT) power stations, proposals were often made for monitoring air emissions, air quality and construction noise; for landfill projects, the proposals were skewed towards the monitoring of leachate, landfill gas and water quality.

The practice analysis used a small representative sample of 17 projects, with EIS monitoring proposals, which had started. The LPAS were contacted to clarify monitoring arrangements, including, for example, whether monitoring arrangements had been made operational under the terms of various consents (e.g. planning conditions, S106 agreements, Integrated Pollution Control (IPC) conditions, site licence conditions), or whether monitoring was being carried out voluntarily. The findings revealed that overall EISS tended to understate, on average by about 30 per cent, the amount of monitoring actually undertaken. This may be a response to planning conditions and agreements resulting from the decision-making process; it may also relate to other relevant licensing procedures, such as IPC. Whatever the case, the findings do suggest that monitoring proposals in EISS are carried out and are often more extensive than the, admittedly often limited, coverage in EIAS. The findings do not, of course, provide any information on the quality of the monitoring or about the accuracy of the predictions.

197

7.4 Auditing in practice

Auditing is already developing a considerable variety of types. Tomlinson & Atkinson (1987a, 1987b) have attempted to standardize *definitions* with a set of terms for seven different points of audit in the "standard" EIA process, as follows:

- decision point audit (draft EIS) – by regulatory authority in the planning approval process;
- decision point audit (final EIS) – also by regulatory authority in the planning approval process;
- implementation audit – to cover start up; it could include scrutiny by the government and the public and focus on the proponent's compliance with mitigation and other imposed conditions;
- performance audit – to cover full operation; it could also include government and public scrutiny;
- predictive techniques audit – to compare actual with predicted impacts as a means of comparing the value of different predictive techniques;
- project impact audits – also to compare actual with predicted impacts and to provide feedback for improving project management and for future projects;
- procedures audit – external review (e.g. by the public) of the procedures used by the government and industry during the EIA processes.

These terms can and do overlap. The focus here is on project, performance and implementation audits. Whatever the focus, auditing faces a number of major *problems*. Buckley (1991) identifies the following:

- EISs often contain very few testable predictions, which may only relate to relatively minor impacts;
- environmental parameters that are monitored may not correspond with those for which predictions were made;
- monitoring techniques may not enable predictions to be tested, because, *inter alia*, time periods and locations do not match, there are too few samples etc;
- projects are almost always modified between the design used for the EIA and in practice;
- monitoring data provided by the developers or project operators may possibly be biased towards their interests.

Such problems may partly explain the dismal record of the Canadian EISs examined, from an ecological perspective, by Beanlands and Duinker (1983), for which accurate predictions appeared to be the exception rather than the rule. There are several examples, also from Canada, of situations where an EIA has failed to predict significant impacts. Berkes (1988) indicated how an EIA on the James Bay HEP megaproject (1971–85) failed to pick up a sequence of interlinked impacts, which resulted in a significant increase in the mercury contamination of fish and in the mercury poisoning of native people. Dickman (1991) identified the failings of an EIA to pick up the impacts of increased lead and zinc mine tailings on the fish population in Garrow Lake, Canada's most northerly hypersaline lake. Such outcomes are not unique to Canada. Canada is a leader in monitoring, and the incidence of such research may result in improved and better predictions than in most countries.

Findings from the limited auditing activity in the UK are not too encouraging. A study of four major developments – the Sullom Voe (Shetlands) and Flotta (Orkneys) oil terminals, the Cow Green reservoir and the Redcar steelworks – suggested that 88 per cent of the predictions were not auditable. Of those that were auditable, fewer than half were accurate (Bisset 1984). Mills's (1992) monitoring study of the visual impacts of five recent UK major project developments (a trunk road, two windfarms, a power station and an opencast coalmine) revealed that there were often significant differences between what was stated in an EIS and what actually happened. Project descriptions changed fundamentally in some cases, landscape descriptions were restricted to land immediately surrounding the site, and aesthetic considerations were often omitted. However, mitigation measures were generally carried out well.

More recent examples of auditing include the Toyota plant study (Ecotech Research and Consulting Ltd 1994), and various wind farm studies (Blandford, C. Associates 1994, ETSU 1994). The Toyota study took a wide perspective on environmental impacts; auditing revealed some underestimate of the impacts of employment and emissions, some overestimate of housing impacts and a reasonable identification of the impacts of construction traffic. The study by Blandford, C. Associates of the construction stage of three windfarms in Wales confirmed the predictions of low ecological impacts, but suggested that the visual impacts were greater than predicted, with visibility distance greater than the predicted 15 km. However the latter finding related to a winter audit; visibility may be less in the haze of summer.

One of the most comprehensive nationwide auditing studies of the precision and accuracy of environmental impact predictions has been carried out by Buckley (1991) in Australia. At the time of his study, he found that adequate monitoring data to test predictions were available for only 3 per cent of the up to 1,000 EISS produced between 1974 and 1982. In general, he found that testable predictions and monitoring data were available only for large complex projects, which had often been the subject of public controversy, and whose monitoring was aimed primarily at testing compliance with standards rather than with impact predictions. Some examples of over 300 major and subsidiary predictions tested are illustrated in Table 7.2.

Overall, Buckley found the average accuracy of quantified, critical, testable predictions was 44% ± 5% standard error. The more severe the impact, the lower the accuracy. Inaccuracy was highest for predictions of groundwater seepage. Accuracy assessments are of course influenced by the degree of precision applied to a prediction in the first place. In this respect, the use of ranges, reflecting the probabilistic nature of many impact predictions, may be a sensible way forward and would certainly make compliance monitoring more straightforward and less subject to dispute.

199

Table 7.2 Examples of auditing of environmental impact predictions.

Component/ parameter	Type of development	Predicted impact	Actual impact	Accuracy/ precision
Surface water quality: salts, pH	Bauxite mine	No detectable increase in stream salinity	None detected	Correct
Noise	Bauxite mine	Blast noise < 115 dBA	Only 90% < 115 dBA	Incorrect: 90% accurate, worse
Workforce	Aluminium smelter	1,500 during construction	Up to 2,500	Incorrect: 60% accurate, worse

(*Source:* Buckley 1991)

Buckley's national survey, showing less than 50 per cent accuracy, does not provide grounds for complacency. Indeed, as it was based on monitoring data provided by the operating corporations concerned, it may present a better result than would be generated from a wider trawl of EISs. On the other hand, we are learning from experience, and more recent EISs may contain better and more accurate predictions.

There has not, until recently, been much emphasis in auditing studies on the important area of predictive techniques audit, and on the value of particular predictive techniques. Where there have been studies, they have tended to focus on identifying errors associated with predictive methods rather than on explaining the errors. There is a need to develop appropriate audit methodologies, and as more projects are implemented there should be more scope for such studies. The pioneering study by Wood on visibility, noise and air quality impacts, using GIS to audit and model EIA errors, provides an example of a way forward for such work (Wood 1997).

7.5 A UK case study: monitoring and auditing the local socio-economic impacts of the Sizewell B PWR construction project

Background to the case study

Although monitoring and auditing impacts are not mandatory in EIA procedures in the UK, the physical and socio-economic effects of developments are not completely ignored. For example, a number of public agencies monitor particular pollutants. Local planning authorities monitor some of the conditions attached to development permissions. However, there is no systematic approach to the monitoring and auditing of impact predictions and mitigation measures. This case study reports on one attempt to introduce a more systematic, although still very partial, approach to the subject.

In the 1970s and early 1980s, Britain had an active programme of nuclear power station construction. This included a commitment, since revised, to build a family of pressurized water reactor (PWR) stations. The first such station to be approved was Sizewell B in East Anglia. The approval was controversial, and followed the longest public inquiry in UK history. Construction started in 1987, and the project was completed in 1995. The Impact Assessment Unit (IAU) in the School of Planning at Oxford Brookes University had studied the impacts of a number of power stations and made contributions to EISs, with a focus on the socio-economic impacts. A proposal was made to the relevant public utility, the Central Electricity Generating Board (CEGB), that the construction of Sizewell B provided an invaluable opportunity to monitor in detail the project construction stage, and to check on the predictions made at the public inquiry and on the mitigating conditions attached to the project's approval. Although the predictions were not formally packaged in an EIS, but rather as a series of reports based on the inquiry, the research was extensive and comprehensive (DOEn 1986). The CEGB supported a monitoring study, which began in 1988. To the credit of the utility, which is now Nuclear Electric/British Energy following privatization, there has been a continuing commitment to the monitoring study – despite the uncertainty about further PWR developments in Britain. Monitoring reports for the whole construction period and into the project's operation have now been completed (Glasson et al. 1989–1997).

Operational characteristics of the monitoring study

It is important to clarify the *objectives of the monitoring study*, otherwise irrelevant information may be collected and resources wasted. Figure 7.3 outlines the scope of the study. The development under consideration is the construction stage of the Sizewell B PWR 1,200 MW nuclear power station. The focus is on the socio-economic impacts of the development, although with some limited consideration of physical impacts. The socio-economic element of EIA involves "the systematic advanced appraisal of the impacts on the day to day quality of life of people and communities when the environment is affected by development or policy change" (Bowles 1981). This involves a consideration of the impacts on employment, social structure, expenditure, services etc. Although to date socio-economic studies have often been the poor relation in impact assessment studies, meriting no more than a chapter or two in EISs, they are important, not least because they consider the impacts of developments on people, who can answer back and object to developments.

The highest priority in the study has been to identify the impacts of the development on local employment; this emphasis reflects the pivotal role of employment impacts in the generation of other local impacts, particularly accommodation and local services. In addition to providing an updated and improved data base to inform future assessments, assisting project management of the Sizewell B project in the local community and auditing impact predictions, the study is also monitoring and auditing some of the conditions and undertakings associated with permission to

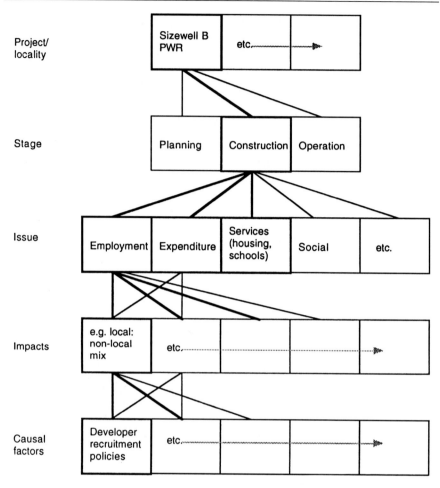

Figure 7.3 Scope of study and data base organization – Sizewell B monitoring study. (*Source:* Glasson et al. 1989–97)

proceed with the construction of the power station. These include undertakings on the use of rail and the routeing of road construction traffic, as well as conditions on the use of local labour and local firms, local liaison arrangements and (traffic) noise (DoEn 1986).

The monitoring study includes the collection of a range of information, including statistical data (e.g. the mixture of local and non-local construction stage workers, the housing tenure status and expenditure patterns of workers), decisions, opinions and perceptions of impacts. The spatial scope of the study extends to the commuting zone for construction workers (Fig. 7.4). The study includes information from the developer and the main contractors on site, from the relevant local authorities and other public agencies, from the local community and from the construction

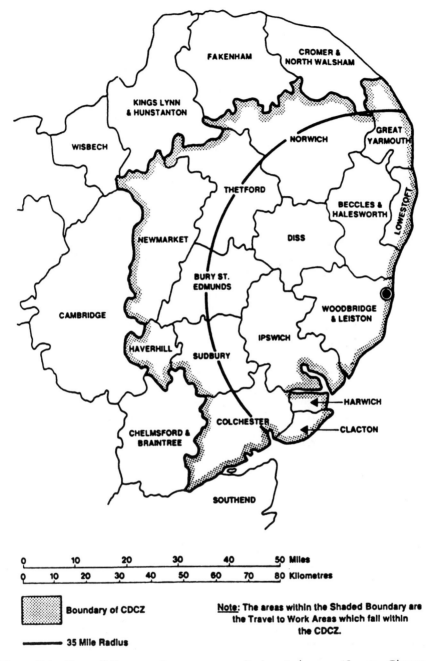

Figure 7.4 Sizewell B commuting zone – monitoring study area. (*Source:* Glasson et al. 1989–97)

203

workers. The local upper school Geography A-level students helped to collect data on the local perceptions of impacts via biennial questionnaire surveys in the town of Leiston, which is adjacent to the construction site. A major survey of the socio-economic characteristics and activities of a 20 per cent sample of the project workforce was also carried out every two years. The Impacts Assessment Unit team operated as the catalyst to bring the data together. There has been a high level of support for the study, and the results are openly available in published annual monitoring reports and in summary broadsheets, which are available free to the local community (Glasson et al. 1989 to 1997).

The study has highlighted a number of *methodological difficulties with monitoring and auditing*. The first relates to the disaggregation of project-related impacts from baseline trends. Data are available that indicate local trends in a number of variables, such as unemployment levels, traffic volumes and crime levels. But problems are encountered when we attempt to explain these local trends. To what extent are they due to (a) the construction project itself, (b) national and regional factors, or (c) other local changes independent of the construction project? It is straightforward to isolate the role of national and regional factors, but the relative roles of the construction project and other local changes are very difficult to determine. "Controls" are used where possible to isolate the project-related impacts.

A second problem relates to the identification of the indirect, knock-on effects of a construction project. Indirect impacts – particularly on employment – may well be significant, but they are not easily observed or measured. For example, indirect employment effects may result from the replacement of employees leaving local employment to take up work on site. Are these local recruits replaced by their previous employers? If so, do these replacements come from other local employees, the local unemployed or in-migrant workers? It has not been feasible to obtain this sort of information. Further indirect employment impacts may stem from local businesses gaining work as suppliers or contractors at Sizewell B. They may need to take on additional labour to meet their extra workload. The extent to which this has occurred is again very difficult to estimate, although surveys of local companies have provided some useful information on these issues.

Some findings from the studies

A very brief summary of a number of the findings are outlined below and in Figures 7.5a and b.

Employment
An important prediction and condition was that at least 50 per cent of construction employment should go to local people (within daily commuting distance of the site). This has been the case, although, predictably in a rural area, local people have the largely semi-skilled or unskilled jobs. As the employment on site has increased, with a shift from civil engineering to mechanical and electrical engineering trades, so the pressure on maintaining the 50 per cent proportion has increased. In 1989, a

Employment impacts

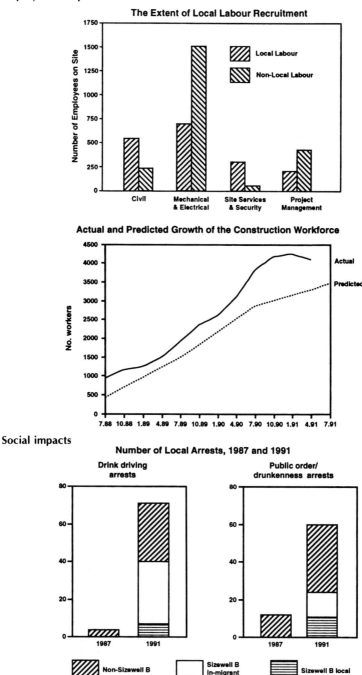

Figure 7.5a Brief summary of some findings from the Sizewell B PWR construction project monitoring and auditing study. (*Source:* Glasson et al. 1989–97)

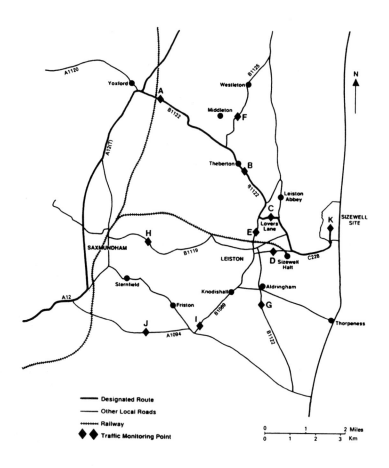

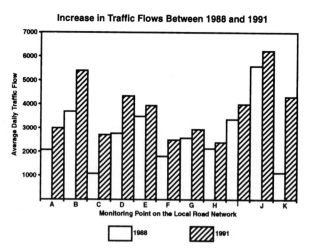

Figure 7.5b More findings from the Sizewell B PWR construction project monitoring and auditing study. (*Source:* Glasson et al. 1989–97)

training centre was opened in the nearest local town, Leiston, to supply between 80 and 120 trainees from the local unemployed.

Local economy

A major project has an economic multiplier effect on a local economy. By the end of 1991, Sizewell B workers were spending about £500,000 per week in Suffolk and Norfolk, Nuclear Electric had placed orders worth over £40 million with local companies, and a "good neighbour" policy was funding a range of community projects (including £1.9 million for a swimming pool in Leiston).

Housing

A major project, with a large in-migrant workforce, can also distort the local housing market. One mitigating measure at Sizewell B was the requirement of the developer to provide a large site hostel. A 600-bed hostel (subsequently increased to 900) has been provided. It has been very well used, accommodating in 1991 over 40 per cent of the in-migrants to the development, at an average occupancy rate of over 85 per cent, and has helped to reduce demand for accommodation in the locality.

Traffic and noise

The traffic generated by a large construction project can badly affect local towns and villages. To mitigate such impacts, there is a designated construction route to Sizewell B. The monitoring of traffic flows on designated and non-designated (control) routes indicated that this mitigation measure was working. Between 1988 and 1989, the amount of traffic rose substantially at the four monitoring points on the designated route, but much less so at most of the seven points not on that route. Construction noise on site has been a local issue. Monitoring has led to modifications in some construction methods, notably improvements to the railway sidings and changes in the piling methods used.

Crime

An increase in local crime is normally associated with the construction stage of major projects. The Leiston police division did see a significant increase in the number of arrests in certain offence categories after the start of the project. However, local people not employed on the project were involved in most of the arrests, and in the increase in arrests, with the exception of drink-driving, for which Sizewell B employees (mainly in-migrants) accounted for most arrests and for most of the increase. However, the early diagnosis of the problems facilitated remedial action, including the introduction of a shuttle minibus service for workers, the provision of a large bar in the site hostel, the stressing at site-workers' induction courses of the problems of drink-driving, and the exclusion from the site (effectively the exclusion from Sizewell B jobs) of workers found guilty of serious misconduct or crime. Since the early stages of the project, worker-related crime has fallen substantially, and the police have considered the project workforce to be relatively trouble-free, with fewer serious offences than anticipated.

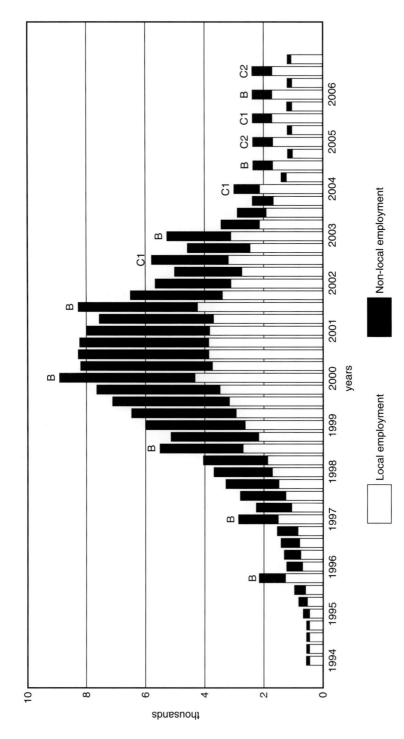

Figure 7.6 Predicted local employment impacts of Sizewell B operational station and Sizewell C construction project (with reactors C1 and C2). (*Source:* Nuclear Electric 1993, 1994)

Residents' perceptions
Surveys of local residents in 1989 and 1991 revealed more negative than positive perceived impacts, increased traffic and disturbance by workers being seen as the main negative impacts. The main positive impacts of the project were seen to be the employment, additional trade and ameliorative measures associated with the project. The monitoring of complaints about the development revealed substantially fewer complaints over time, despite the rapid build-up of the project.

Learning from monitoring: Sizewell B and Sizewell C
The monitoring of impacts and the auditing of the predictions and mitigation measures revealed that many of the predictions used in the Sizewell B public inquiry were reasonably accurate – although there was an underestimate of the build-up of construction employment and an overestimate of the secondary effects on the local economy. Predictions of traffic impacts, and on the local proportion of the construction workforce, were very close to the actual outcomes. Mitigation measures also appeared to have some effect. Other local issues have been revealed by monitoring, allowing some modifications to manage the project better in the community. Unfortunately, such systematic monitoring is still discretionary in the UK and very much dependent on the goodwill of developers.

Information gained from monitoring can also provide vital intelligence for the planning and assessment of future projects. This is particularly so when the subsequent project is of the same type, and in the same location, as that which has been monitored. Nuclear Electric applied for consent to build and operate a replica of Sizewell B, to be known as Sizewell C. A full EIS was produced for the project (Nuclear Electric 1993). Its prediction of the socio-economic impacts drew directly on the findings from the Sizewell B monitoring study. Figure 7.6 provides an overview of the cumulative employment impacts of the operational Sizewell B plus the construction of Sizewell C (with two reactors, C1 and C2). The regular peaks in the figure are the refuelling intervals. However, this proposed follow-on project fell victim to the abandonment of the UK nuclear power station programme.

7.6 Summary

A mediation of the relationship between a project and its environment is needed throughout the life of a project. Environmental impact assessment is meant to establish the terms and conditions for project implementation; yet there is often little follow-through to this stage and even less follow-up after it. Some projects have very long lives, and their impacts need to be monitored on a regular basis. Such monitoring can improve project management and contribute to the auditing of both impact predictions and mitigating measures. Monitoring and auditing can provide essential feedback to improve the EIA process, yet this is probably the weakest step of the process in many countries. Discretionary measures are not enough; monitoring and auditing need to be more fully integrated into EIA procedures on a mandatory basis.

References

Au, E. & G. Sanvicens 1996. EIA follow up monitoring and management. *EIA Process Strengthening*. Canberra: Environmental Protection Agency.

Baseline Environmental Consulting 1989. *Mitigation monitoring and reporting plan for a woodwaste conversion facility, West Berkeley, California*. Emeryville, California: Baseline Environmental Consulting.

Beanlands, G. E. & P. Duinker 1983. *An ecological framework for environmental impact assessment*. Halifax, Nova Scotia: Dalhousie University, Institute for Resource and Environmental Studies.

Berkes, F. 1988. The intrinsic difficulty of predicting impacts: lessons from the James Bay hydro project. *Environmental Impact Assessment Review* **8**, 201–20.

Bisset, R. 1984. Post development audits to investigate the accuracy of environmental impact predictions. *Zeitschift für Umweltpolitik* **7**, 463–84.

Bisset, R. & P. Tomlinson 1988. Monitoring and auditing of impacts. In *Environmental impact assessment*, P. Wathern (ed.). London: Unwin Hyman.

Blandford, C. Associates 1994. *Wind turbine power station construction monitoring study*, Gwynedd: Countryside for Wales.

Bowles, R. T. 1981. *Social impact assessment in small communities*. London: Butterworth.

Buckley, R. 1991. Auditing the precision and accuracy of environmental impact predictions in Australia. *Environmental Impact Assessment Review* **11**, 1–23.

California Resources Agency 1988. *California EIA monitor*. State of California.

Culhane, P. J. 1993. Post-EIS environmental auditing: a first step to making rational environmental assessment a reality. *Environmental Professional*, **5**.

Dickman, M. 1991. Failure of environmental impact assessment to predict the impact of mine tailings on Canada's most northerly hypersaline lake. *Environmental Impact Assessment Review* **11**, 171–80.

DOEn (Department of Energy) 1986. *Sizewell B Public Inquiry: Report by Sir Frank Layfield*. London: HMSO.

CEC (Commission of the European Communities) 1993. *Report from the Commission of the Implementation of Directive 85/337/EEC on the assessment of the effects of certain public and private projects on the environment*, vol. 13, *Annexes for all Member States*, COM (93)28 final. Brussels: EC, Directorate-General XI.

Ecotech Research and Consulting Ltd 1994. *Toyota impact study summary*, unpublished.

ETSU (Energy Technology Support Unit) 1994. *Cemmaes Wind Farm: sociological impact study*. Market Research Associates and Dulas Engineering, ETSU.

Frost, R. 1997. EIA monitoring and audit. In Planning and EIA in practice, J. Weston (ed.), 141–64. Harlow: Longman.

Glasson, J. (1994). Life after the decision: the importance of monitoring in EIA. *Built Environment* **20**(4), 309–20.

Glasson, J. & A. Chadwick 1989–97. *Local socio-economic impacts of the Sizewell B PWR Construction Project*. Oxford: Oxford Polytechnic/Oxford Brookes University, Impacts Assessment Unit.

Glasson, J. & D. Heaney 1993. Socio-economic impacts: the poor relations in British environmental impact statements. *Journal of Environmental Planning and Management* **36**(3), 335–43.

Greene, G., J. W. MacLaren & B. Sadler 1985. Workshop summary. In *Audit and evaluation in environmental assessment and management: Canadian and international experience*, 301–21. Banff: The Banff Centre.

Holling, C. S. (ed.) 1978. *Adaptive environmental assessment and management*. Chichester: Wiley.

IAIA (International Association for Impact Assessment) 1994. Report of special committee on guidelines on principles for social impact assessment. *Impact Assessment*, **12**(2).

IEA 1993. *Digest of Environmental Statements*. London: Sweet & Maxwell.

Lee, N. & C. Wood 1980. *Methods of environmental impact assessment for use in project appraisal and physical planning, Occasional Paper no. 7*. University of Manchester, Department of Town and Country Planning.

Mills, J. 1992. *Monitoring the visual impacts of major projects* (MSc dissertation in Environmental Assessment and Management, School of Planning, Oxford Brookes University).

Nuclear Electric 1993. *Proposed Sizewell C PWR Power Station – environmental statement*. Gloucester: Nuclear Electric.

Nuclear Electric 1994. *Suffolk coastal economy with and without Sizewell C*. Gloucester: Nuclear Electric.

Sadler, B. 1988. The evaluation of assessment: post-EIS research and process. In *Environmental Impact Assessment* P. Wathern (ed.). London: Unwin Hyman.

Tomlinson, P. & S. F. Atkinson 1987a. Environmental audits: proposed terminology. *Environmental Monitoring and Assessment* **8**, 187–98.

Tomlinson, P. & S. F. Atkinson 1987b. Environmental audits: a literature review. *Environmental Monitoring and Assessment* **8**, 239.

Wood, G. 1997. *Auditing and modelling environmental impact assessment errors using GIS* (unpublished PhD thesis, School of Planning, Oxford Brookes University).

Part 3

Practice

LET'S MAKE SURE I'VE GOT THIS RIGHT. WE GET TO KEEP SOME LIZARDS AND BLUE BUTTERFLIES ON OUR HEATHLAND, AND IN RETURN YOU GET TO BUILD 3,200 NEW HOUSES ON OUR GREEN BELT . . . .

CHAPTER 8

An overview of UK practice to date

8.1 Introduction

Part 3 considers EIA practice: what is done rather than what should be done. Chapter 8 provides an overview of the first eight years or so of UK practice since EC Directive 85/337 became operational. We develop this further, with reference to particular sectors and their associated legislation, in Chapters 9 and 10: Chapter 9 focuses on new settlements and waste disposal facilities, which both come under the Town and Country Planning (Assessment of Environmental Effects) Regulations. Chapter 10 deals with projects that fall under different regulations, focusing on trunk roads and power stations. Each of these two chapters includes case studies that provide examples of current practice in various aspects of the EIA process. Chapter 11 discusses international practice in terms of "best practice" systems, emerging EIA systems and the role of development agencies in EIA.

These chapters can be set in the context of the recently completed international study on EIA effectiveness, a major three-year study, whose results have been written up by Sadler (1996). Sadler suggests that EIA effectiveness can be tested at different stages in a cycle of EIA systems: (1) whether a given EIA policy is effectively translated into practice through the application of relevant processes and procedures, (2) whether the practice results in effective EIA performance through contributions to decision-making, and (3) whether this performance then effectively feeds back into changes in the EIA policy by examining whether EIA realizes its purpose.

Sadler also notes that these questions and the attendant techniques for investigating them must be seen in the context of the decision-making framework in which the relevant EIA system operates. As was discussed in previous chapters, EIA in the UK can broadly be described as having been

- imposed on a reluctant government by the EC;
- implemented since then relatively punctually and thoroughly;
- based on a strong pre-existing planning system, but with inelegant "patching" where Directive 85/337 has required EIA for projects covered by other authorization systems, and where regulations have since been amended;

- often implemented through negotiations rather than through direct confrontations between the relevant interest groups, with the attendant weakening of many decisions but also relatively good implementation;
- focused on qualitative rather than quantitative techniques, eschewing high-tech methods and leading to short, quite readable EISs.

Chapter 8 broadly addresses Sadler's first two points in sequence. Section 8.2 considers the number, type and location of projects for which EIAS have been carried out in the UK since mid-1988, as well as where the resulting EISs can be found. Section 8.3 discusses the stages of EIA before the submission of the EIS and application for authorization. Section 8.4 addresses what has, to date, been the most heavily studied aspect of EIA practice, the quality of EISs. Section 8.5 considers the post-submission stages of EIA, and how environmental information is used in decision-making by LPAS and inspectors. Finally, Section 8.6 discusses the costs and benefits of EIA as seen from various perspectives. Sadler's third point is partially addressed by recently published DOE good-practice guides on EIA preparation and review (DOE 1994, 1995), which reflect a first cycle of limited policy changes by the UK government in response to early research findings regarding EIS and EIA effectiveness.

The information in this chapter was correct at the time of writing in 1997; it will obviously change as more EIAS are carried out.

8.2 Number, type and location of EISs and projects

In the absence of a central government lead in maintaining a comprehensive database of EISs, several organizations have begun to establish such databases (e.g. IEA 1993, Frost & Wenham 1996). This section considers how many EISs have been produced, for which projects and developers, and where. It concludes with a brief review of where collections of EISs are kept.

This analysis is complicated by several problems. First, some projects fall under more than one schedule classification, for example mineral extraction schemes (Sched. 2.2) that are later filled in with waste (Sched. 2.11), or industrial/residential developments (Sched. 2.10) that also have a leisure component (Sched. 2.11). Secondly, the mere description of a project is often not enough to identify the regulations under which its EIA was carried out. For instance, power stations may fall under Schedule 1.2 or 2.3a depending on size. Roads may come under highways or planning regulations depending on whether they are trunk roads or local highways. Thirdly, many EISs do not mention when, by whom or for whom they were prepared. Fourthly, locational analysis after 1995 is complicated by local government reorganization and many changes in the nature and boundaries of authorities in England, Scotland and Wales. All these factors affect the analysis. This chapter is based primarily on information from Frost & Wenham (1996), but their findings are very similar to others' (e.g. Wood 1996).

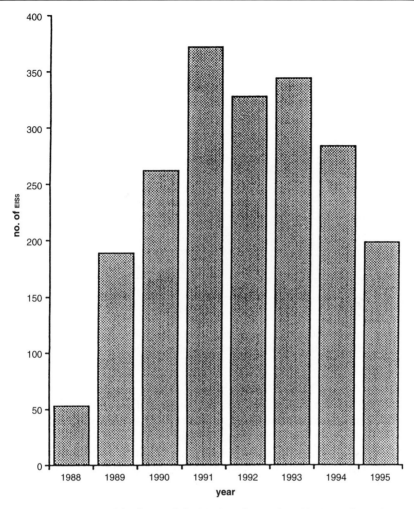

Figure 8.1 EISS prepared in the UK, July 1988 to September 1995: total number received. (*Sources:* Frost & Wenham 1996, Wood 1996)

Number of EISs

Between the mid-1970s and the mid-1980s, approximately twenty EISs were prepared annually in the UK (Petts & Hills 1982). After the implementation of Directive 85/337, this number rose dramatically and, despite the recession, a peak of about 350 EISs per year were produced in the early 1990s. However, as can be seen from Figure 8.1, this number began to drop in the mid-1990s. By late 1995, a total of about 2,500 EISs had been produced in the UK, of which about 78 per cent were in England, 12 per cent in Scotland, 8 per cent in Wales, and 2 per cent in Northern Ireland.

217

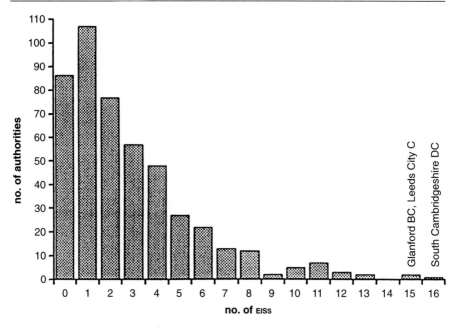

Figure 8.2 EISS prepared in the UK, July 1988 to September 1995: number received by local level authorities. (*Source:* Frost & Wenham 1996)

In parallel with the gradual increase in the number of EIAS, the participants in the EIA process have become increasingly familiar with EIA. Surveys of UK local authorities carried out by Oxford Brookes University have showed that of 502 (97 per cent of the total) authorities for which some information on EIAS was known by late 1995 412 (82 per cent) had received at least one EIS: Figures 8.2 and 8.3 show the number of EISs known to have been received by, respectively, local level authorities (district, borough, metropolitan borough and city councils, and development corporations) and strategic level authorities (county and regional councils and national park authorities). On average, strategic level authorities had received 12 EIS and the other authorities four. Surveys of environmental consultants (Weston 1995, Radcliff & Edward-Jones 1995) have found that about one-third of the consultancies surveyed had prepared ten or more EISS.

Types of projects

Figure 8.4 shows the regulations under which the EISs were prepared: the reduction during the period in EISS prepared under the planning regulations is noticeable, as is the increase in EISS from Northern Ireland. Figure 8.5 shows the types of project for which EISs have been prepared. The largest project types were waste disposal (Scheds

218

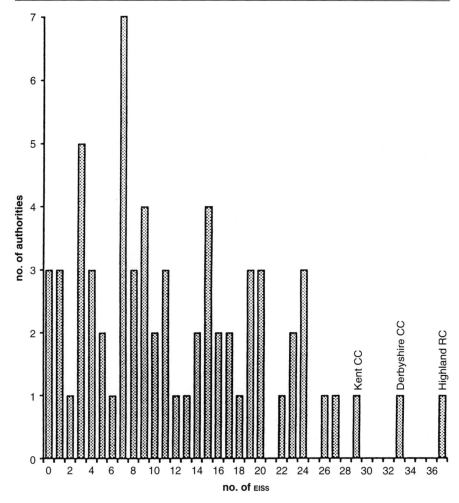

Figure 8.3 EISS prepared in the UK, July 1988 to September 1995: number received by strategic level authorities. (*Source:* Frost & Wenham 1996)

1.9, 2.11c&d)[1], industrial and urban developments (Scheds 2.4, 2.6, 2.8, 2.10a&b), roads (Scheds 1.7, 2.10d), extraction schemes (Sched. 2.2c&d), and energy projects (Scheds 1.2, 2.3, 2.10h) (Frost & Wenham 1996). About 10 per cent of these projects were Schedule 1 projects, primarily toxic waste disposal installations, power stations and motorways (Wood 1996).

Although these ratios have remained broadly steady over the years, some project types show clear trends. For instance, in response to privatization and the "dash for gas" the number of EISS for combined-cycle gas turbine power stations peaked at about eight per year (1989–91), falling back to about three per year (1992–95). EISS for roads increased steadily from about 30 in 1989 to about 80 in 1993 as a result

219

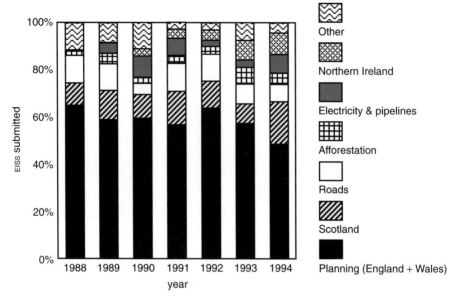

Figure 8.4 EISS prepared in the UK, July 1988 to September 1994: regulations under which prepared. (*Source:* Wood 1996)

of the roads programme, and can be expected to fall back substantially, given restrictions on government funding of road construction and policy trends towards traffic management. EISS for incinerators and waste-water treatment plants grew rapidly until 1992, as the government worked to meet EC water quality standards, and have since dropped. The number of business park EISS dropped sharply after 1990, as the recession affected speculative development (Frost & Wenham 1996).

In the first few years following the implementation of Directive 85/337, 40 per cent of EISS were produced for the public sector and 60 per cent for the private sector (Wood 1991). This has remained broadly the same: the percentage of private sector projects has increased slightly owing to privatization, but much of this has been balanced by the heavy government investment in – and consequently EISS for – new roads. A particularly interesting subset is that of the 10 per cent of EIAS for which one agency acts as both the project proponent and the competent authority (e.g. the Highway Agency for roads, the Forestry Authority for afforestation).

Location of projects

Figure 8.6 shows the distribution of known EISS in England, Scotland and Wales by county or region. Generally, more is known about English and Scottish than about

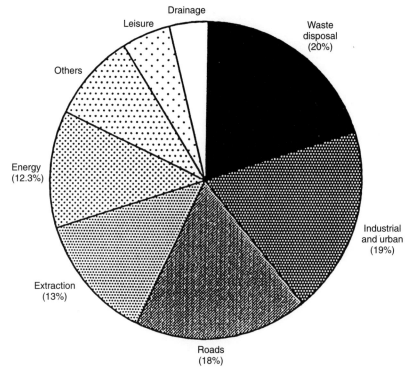

Figure 8.5 EISS prepared in the UK, July 1988 to September 1995: types of project. (*Source:* Frost & Wenham 1996)

Welsh and Northern Irish EISs, so the figure may understate the situation in Wales and Ireland. By the end of 1996, the most EISs – about 130 – had been prepared for projects in Kent: the Channel Tunnel, the M20 "missing link" and the Dartford Crossing have spawned proposals for major secondary projects, including mixed-use, road and waste-disposal projects. The authorities in the major conurbations have also received a relatively large number of EISs. In Greater London (about 80) many of them were for road and rail schemes. In Greater Manchester (60) and the West Midlands (60), light rail and waste disposal also feature prominently. Many EISs (110) have been prepared in Strathclyde, largely for extraction, incinerator and leisure proposals. Recently several wind-farm EISs have also been submitted. The areas with the fewest EISs are in southern Scotland, mid-Wales, the northern Home Counties and Somerset.

The types of development vary considerably between regions, reflecting differences in the local economic bases. Thus certain areas show concentrations of particular project types, for instance opencast coal schemes in Derbyshire and the northern English counties, power stations in Greater London and Humberside,

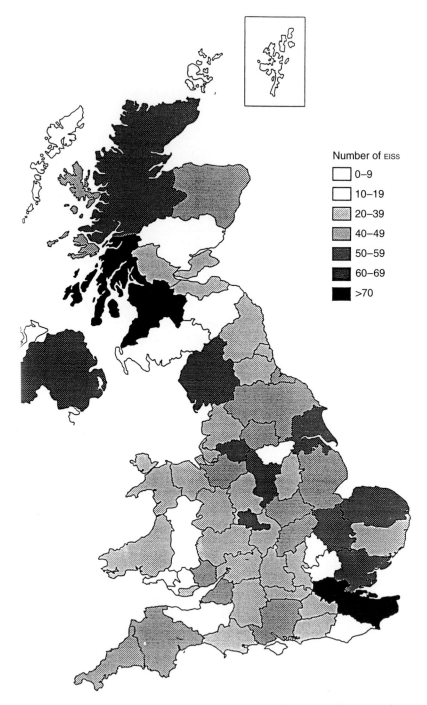

Figure 8.6 EISS prepared in the UK, July 1988 to September 1995: location of projects. (*Source:* Frost & Wenham 1996)

wind-farms in Wales and Cornwall, afforestation schemes in the Scottish Highlands and agricultural projects in Lincolnshire and Shropshire.

Sources of EISs

When a local planning authority receives an EIS, it is required to send a copy to the regional office of the DOE, which then forwards it to the DOE (now DETR) library in London once the application has been dealt with. However, this process is a long one: in early 1996, the DOE library in London held about 550 EISs (including a large number not falling under the town and country planning system). The library is open to the public by appointment; photocopies can be made off the premises. In Wales, planning EISs are forwarded to the Welsh Office (holding about 160 EISs). In Scotland, all EISs are sent to the Scottish Office (about 80), while in Northern Ireland they are sent to the Northern Ireland DOE (100). Other government agencies, such as the Ministry of Agriculture, Fisheries and Food, and the Forestry Authority also hold collections and lists of the EISs that fall under their jurisdiction. These collections are, however, generally not publicly available, although limited access for research purposes may be allowed.

The Institute of Environmental Assessment, based in Lincoln, has a collection of about a thousand EISs, which are available by pre-arrangement with institute staff and can also be mailed on a one-week loan basis to members. It has also published a regularly updated *Digest of Environmental Statements* (IEA 1993), which provides comprehensive summaries of 1,800 EISs. The EIA Centre at the University of Manchester keeps a large database of EISs and EIA-related literature: its collection of over 500 EISs is, like its database, open to the public, by appointment. Oxford Brookes University's collection of approximately 650 EISs is open to the public, by appointment, and photocopies can be made on the premises. The University's Impacts Assessment Unit publishes an annual directory of EISs, the latest being by Bellanger and Frost (1997). The addresses of these and other organizations are given in Appendix 7. Other organizations, such as the Royal Society for the Protection of Birds, the Institute of Terrestrial Ecology, English Nature and the Council for the Protection of Rural England, as well as many environmental consultancies, also have limited collections of EISs, but these are generally kept by individuals within the organization for in-house use only, and are not available to the public.

The difficulty of finding out which EISs exist and their often prohibitive cost make the acquisition and analysis of EISs very arduous. Various organizations, e.g. the Institute of Environmental Assessment, the University of Manchester and Oxford Brookes University, have called for one central repository for all EISs in the UK.

8.3   The pre-submission EIA process

This is the first of three sections which discuss how EIAS are carried out in practice in the UK. It focuses on some of the pre-EIS submission stages of EIA, namely screening, scoping, and pre-submission consultation.

Screening

Underpinning any analysis of the implementation of EIA in the UK are the requirements of the EC and UK government legislation. With respect to screening, Directive 85/337 has a number of limitations, including its coverage of types of project (e.g. EIA for small projects which themselves do not cause a significant impact but which, cumulatively with others, do) and the narrow focus on the project level only. These limitations are reinforced by UK legislation. Indeed, some projects that require EIA under Directive 85/337, for instance afforestation projects that do not require Forestry Authority grants, are not covered in UK EIA legislation.

Competent authorities in the UK are given wide discretion to determine which Schedule 2 projects require EIA within a framework of varying criteria and thresholds established by the 40-plus regulations and additional guidance. Generally this screening process works well (CEC 1993). However, some specific problems arise regarding screening in the UK. First, because of the largely discretionary system for screening, LPAS often – about half of the time – require an EIS to be submitted only after they receive a planning application (DOE 1996). For the same reason, screening requirements vary considerably between competent authorities. For instance, a 1991 search of 24 LPA returns, registers and files revealed 30 projects in 12 authorities for which EIAs could have been required but were not (DOE 1991). Most of these types of development (e.g. six mineral extraction schemes, two landfills, 17 mixed-use developments) had been subject to EIA in other local authorities. The decision not to require an EIA had mostly been taken by junior members of staff who had never considered the need for an EIA, or who thought (incorrectly) that no EIA was required if the land was designated for the type of use specified in the development plan, or if the site was being extended or redeveloped rather than newly developed. The new screening criteria established by the amendments to the Directive are likely to reduce these problems in the future (see Sections 3.4 and 4.3).

Similarly, different DOE regional offices have given different decisions on appeals for what are essentially very similar developments. For instance:

> . . . there have been two applications for major out-of-town regional shopping centres, both well above the threshold given in (Circular 15/88) of 10,000 sq. metres. The request for the smaller of the two (The Richings at Iver, South Buckinghamshire DC) was successfully challenged by the developer. The request for the larger of the two (at Lea Cross, Newham borough) was also

challenged and in this case an assessment was required. The direction letters came from two different DOE regional offices. (Gosling 1990)

The DOE regional offices' decisions have also differed substantially from the thresholds indicated in Circular 15/88:

The threshold for sand and gravel cases is set as "sites of more than 50 ha may well require EIA and significantly smaller sites could require EIA if they are in a sensitive area or if subjected to particularly obtrusive operations" . . . Two of the applications in Cambridgeshire are well over the threshold. The Barleycroft Farm (117 ha) proposal is not identified within the minerals local plan, . . . is part in an Area of Best Landscape and is close to residential development. On being challenged a direction was received that assessment was not required. The Fenstanton case (74 ha) is entirely within an Area of Best Landscape and the application has been refused without waiting on the receipt of the environmental statement. (Gosling 1990)

Finally, EIA in the UK is an all-or-nothing process: either an EIA is needed or it is not. This is in contrast with some other countries (e.g. Peru and China, see Section 11.3), where a brief environmental study is carried out to determine whether a full-scale EIA is needed. In the UK, where the provision of environmental information with planning applications was the norm before the implementation of Directive 85/337, many developers still voluntarily submit environmental documents without specifying whether these are EISs or not. Some competent authorities treat these documents as EISs, with the attendant requirements for consultation and publicity, but in other cases they simply treat them as additional information (Hughes & Wood 1996).

Scoping and pre-submission consultation

Competent authorities also have much discretion to determine the scope of EIAS. As we discussed in Chapter 3, Directive 85/337's Annex III was interpreted in UK legislation as being in part mandatory and in part discretionary. Table 8.1 shows the type of information included in EISs, based on a survey of 100 prepared before 1990 (Jones et al. 1991). It shows that, although the mandatory requirements of the legislation are generally carried out, the discretionary elements (e.g. the consideration of alternatives, forecasting methods, secondary and indirect impacts, scoping) have, understandably, been carried out less often. Since EIS quality is improving, it is likely that a more up-to-date analysis would show higher proportions of EISs including the relevant information.

Although early scoping discussions between the developer, the consultants carrying out the EIA work, the competent authority and relevant consultees are advised in government guidance and are increasingly considered vital for effective EIA (Jones 1995, Sadler 1996), in practice pre-submission consultation is carried out sporadically. For instance, a survey of environmental consultants (Weston 1995) showed that

225

Table 8.1 EISS prepared in the UK July 1988 to December 1989: comprehensiveness.

Type of information	EISS including information (%)
Specified information	
Description of proposed development	93
Data to identify and assess the main environmental effects	76
Description of likely significant effects	88
With reference to:	
• human beings	75
• flora/fauna	85
• soil/geology	51
• water	65
• air	54
• climate	24
• landscape	91
• interaction between them	14
• material assets/cultural heritage	48
Description of mitigating measures	75
Non-technical summary	67
Additional information	
Physical characteristics of proposed development:	
• construction	51
• operation	74
Residues and emissions from the development	
With reference to:	
• water	63
• air	54
• soil	29
• noise	68
• vibration	17
• light	9
• heat	2
• radiation	1
Outline of main alternatives studied	34
Forecasting methods used	45
Difficulties in compiling information	4

(*Source:* Jones et al. 1991)

only 3 per cent had been asked to prepare their EISs before site identification, and 28 per cent before detailed design. Local planning authorities are consulted by the developer before EIS submission in between 30 and 70 per cent of cases, although this seems to be increasing (DOE 1996, Lee et al. 1994, Leu et al. 1993, Radcliff & Edward-Jones 1995, Weston 1995). A survey by Weston (1995) showed that English Nature was consulted before EIS submission in about half the cases, the (then) National Rivers Authority in about 40 per cent, the Countryside Commission in about one-quarter, and HMIP only rarely. Other studies (DOE 1996, Pritchard et al. 1995) also showed that very limited consultation with statutory or non-statutory consultees or the public occurred at this stage, although where extensive consultation had been carried out project design was often modified significantly before the submission of the planning application (Pritchard et al. 1995).

However, even early consultation does not necessarily mean that the consultees will be satisfied with the outcome. For instance, in some cases groups have lodged objections to planning applications despite having been consulted. In particular, consultees from whom the developer has requested information before EIS submission may expect the EIS to cover more than just the data that they have provided (DOE 1996). Similarly, consultation may be widespread but may avoid organizations that could be hostile to the project (Pritchard et al. 1995).

Considerable experience has been gained with screening and scoping. After initial hiccups, the screening process now seems to be relatively well accepted. However, despite the fact that scoping is generally considered to be a very valuable and cost-effective part of EIA by all those concerned, it is carried out in only about half the cases, and then generally only in a limited manner. It is to be hoped that the changes following the amended EC Directive (see Chs 2 and 3) will help to raise the profile of this important part of the EIA process.

8.4 EIS quality

As we mentioned in Section 8.1, the preparation of high-quality EISs is one component of an effective translation of EIA policy into practice. Two schools of thought exist about the standards that should be required of an initial EIS. Some argue that developers should be encouraged to submit EISs of the highest standard from the outset. This reduces the need for costly interaction between developer and competent authority (Ferrary 1994), provides a better basis for public participation (Sheate 1994), places the onus appropriately on the developer and increases the chance of effective EIA overall. Others argue that it is the entirety of environmental information that is important, and that the advice of statutory consultees, the comments of the general public and the expertise of the competent authority can substantially overcome the limitations of a poor EIS (Braun 1993). This view is also supported by planning inspectors at appeal and judicial review cases.

EIS quality in the UK is affected by the limited legal basis for EIA and by these facts: that planning applications cannot be rejected if the EIS is inadequate, that (to date) some crucial steps of the EIA process (e.g. public participation, the consideration of alternatives, monitoring) are not mandatory, and that developers undertake EIAs for their own projects. This section first considers the quality of EISs produced in the UK, based on several academic studies. It continues with a brief discussion of other perceptions of EIS quality, since competent authorities, statutory consultees and developers require different things from EIA and may thus have different views of EIS quality. It concludes with a discussion of factors that may influence EIS quality.

Table 8.2 Aggregated EIS quality, % satisfactory*.

	Authors and year of study**				
	Lee & Colley 1990	Wood & Jones 1991	Lee & Brown 1992	Lee et al. 1994	Jones 1995
Year(s) EISs were prepared:					
1988–89	25	37	34	17	
1989–90			48		
1990–91			60	47	
1988–93					"just over half"

* Satisfactory means marks of A, B or C based on the Lee & Colley criteria (1990 or 1992).
** No. of EISs analysed: Lee & Colley, 12; Wood & Jones, 24; Lee & Brown, 83; Lee et al., 47; Jones, 40.

Academic studies of EIS quality

Academic studies of EIS quality can broadly be classified as aggregated or disaggregated. Aggregated studies consider the quality of a number of EISs overall, where the EISs either represent the total population of EISs or a specific subgroup (e.g. type of project). Disaggregated approaches focus on the quality of the treatment of individual EIS topic areas (e.g. landscape or noise), or performance with respect to certain EIS components (e.g. baseline data, the consideration of alternatives) or their presentation (Lee et al. 1994).

Researchers from the University of Manchester have studied aggregated approaches to EIS quality over the years, using the Lee & Colley criteria (1992) (see Appendix 3). Based on these criteria, EISs have been divided into "satisfactory" (i.e. marks of A, B or C) and "unsatisfactory" (D or below). Table 8.2 summarizes some of the findings. It shows EIS quality to be increasing after dismal beginnings, about half of recent EISs being satisfactory.

A recent study carried out by Oxford Brookes University's Impacts Assessment Unit (IAU) for the DOE compared 25 EISs prepared before 1991 with matched[2] EISs prepared after 1991, on the basis of four sets of criteria, including simple "regulatory requirements" and comprehensive criteria devised by the IAU. For the simple regulatory requirements, 44 per cent of the post-1991 EISs fulfilled all the nine criteria used, compared with 36 per cent of the pre-1991 EISs. A more detailed analysis indicated that 92 per cent of the post-1991 EISs fulfilled six or more of the criteria, compared with 64 per cent of the pre-1991 criteria. However, this review framework (see Table 8.3), with simple yes/no grading and a very limited list of criteria, could be regarded as providing a crude and perhaps over-harsh review of quality. Using the more comprehensive range of criteria established by the IAU (see Table 8.4, and Section 6.5), the quality of EISs rose from just unsatisfactory (D) before 1991 to just satisfactory (C) after 1991. The percentage of satisfactory EISs[3] increased from 36 to 60 per cent.

Table 8.3 Disaggregated EIS quality based on simple "regulatory requirements", % covered.

Criterion	25 pre-1991 EISs	25 post-1991 EISs
1. Describes the proposed development, including its design and size or scale.	76	84
2. Defines the land areas taken up by the development site and any associated works, and shows their location on a map.	76	92
3. Describes the uses to which this land will be put and demarcates the land use areas.	68	92
4. Considers direct and indirect effects of the project and any consequential development.	60	80
5. Investigates these impacts in so far as they affect human beings, flora, fauna, soil, water, air, climate, landscape, interactions between the above, material assets and cultural heritage.	56	76
6. Considers the mitigation of all significant negative impacts.	68	92
7. Mitigation measures include the modification of the project, the replacement of facilities and the creation of new resources.	60	92
8. There is a non-technical summary, which contains at least a brief description of the project and environment, the main mitigation measures and a description of any remaining impacts.	64	80
9. The summary presents the main findings of the assessment and covers all the main issues raised.	52	72
All criteria	36	44

(*Source:* DOE 1996)

Table 8.4 Disaggregated EIS quality based on IAU criteria (average marks).

Criterion	25 pre-1991 EISs	25 post-1991 EISs
1. Description of the development	C/D	C
2. Description of the environment	C/D	C
3. Scoping, consultation and impact identification	D	C/D
4. Prediction and evaluation of impacts	D	C
5. Alternatives	E	D
6. Mitigation and monitoring	C/D	C/D
7. Non-technical summary	D	C/D
8. Organization and presentation of information	C/D	C
Overall mark	D	C
% satisfactory (A–C)	36	60
% marginal (C–D)	12	4
% unsatisfactory (D–F)	52	36

(*Source:* DOE 1996)

Other studies have focused on specific project types: for instance Kobus & Lee (1993) and Pritchard et al. (1995) reviewed EISs for extractive industry projects, Prenton-Jones for pig and poultry developments (Weston 1996), Radcliff & Edward-Jones (1995) for clinical waste incinerators (see Section 9.3) and Davison (1992) and Zambellas (1995) for roads. These studies also broadly suggest that EIS quality is not very good, but improving.

In terms of disaggregated approaches, Lee & Dancey (1993) analysed 83 EISs and found 60 per cent to be satisfactory in terms of their description of the development, local environment and baseline conditions, 36 per cent in terms of identification of key impacts, 47 per cent in terms of alternatives and mitigation, and 49 per cent in terms of communication and presentation of results. Nelson (1994), Pritchard et al. (1995) and Jones (1995) made broadly similar findings. Tables 8.3 and 8.4 show the results of an analysis of the 25 matched pairs of EISs carried out for the DOE (1996) based on, respectively, simple "regulatory requirements" and IAU criteria. Coverage of each of the regulatory requirement criteria improved over time, from an average of about two-thirds before 1991 to more than 80 per cent since 1991. Based on the IAU criteria, quality in general also rose significantly between 1988 and 1990 and between 1992 and 1994, with improvements in each of the eight main categories of assessment. The EISs' description of monitoring and mitigation improved only marginally, but the other categories generally improved by about half of a mark (e.g. from D to C/D). A particular improvement was seen in the approach to alternatives. Of the 25 EIS pairs, 15 showed an improvement in quality, while nine became worse (DOE 1996).

Other studies have analysed the quality of specific EIS environmental components, for instance landscape/visual impacts (e.g. Mills 1994) and socio-economic impacts (e.g. Hall 1994). These show similar findings to those discussed above.

In sum, although both aggregated and disaggregated studies by academics show a continued and pleasing improvement in EIS quality, there must still be concern that many of the most recent EISs, from between one-third to one-half depending on the criteria used, are still not satisfactory, and in several cases poor.

Quality for whom?

These findings must, however, be considered in the wider context of "quality for whom?" Academics may find that an EIS is of a certain quality, but the relevant planners or consultees may perceive it quite differently. For instance, the DOE (1996) study, Radcliff & Edward-Jones (1995), and Jones (1995) found little agreement about EIS quality between planners, consultees and the researchers; the only consistent trend was that consultees were more critical of EIS quality than planners were.

In interviews conducted by the IAU (DOE 1996), planning officers thought EISs were intended to gain planning permission and minimize the implication of impacts. Just over 40 per cent felt that EIS quality had improved, although this improvement was usually only marginal. Most of the others felt that this was difficult

Table 8.5 Changing perceptions of participants of quality of EISs in relation to particular criteria (%).

	Good		Marginal		Poor	
	pre-1991	post-1991	pre-1991	post-1991	pre-1991	post-1991
Comprehensiveness	31	55	31	27	38	18
Objectivity	18	41	37	35	45	24
Clarity of information	25	55	56	38	19	7

(*Source:* DOE 1996)

to assess when individual officers see so few EISs and when those they do see tend to be for different types of project, which raises different issues. A lack of adequate scoping and discussion of alternatives were felt to be major problems. EISs were seen to be getting "better but also bigger". Some officers linked EIS quality with the reputation of the consultants producing them, and believed that the use of experienced and reputable consultants is the best way to achieve good quality EISs.

Statutory consultees differed about whether EIS quality is improving. Statutory consultees generally felt that an EIS's objectivity and clear presentation were important, improving, yet still wanting. Table 8.5 from the DOE study (1996) indicates that participants (LPAs, developers, consultants and consultees) generally thought the key EIS criteria of comprehensiveness, objectivity and clear information were improving. Yet it is interesting to note that only about 40 per cent of interviewees regarded the objectivity of recent EISs as good.

Developers and consultants link EIS quality with ability to achieve planning permission. Consultants felt that developers are increasingly recognizing the need for environmental protection and are starting to bring in consultants early in project planning, so that a project can be designed around that need. One reason for this improvement may be that pressure groups are becoming more experienced with EIA, and thus have higher expectations of the process (DOE 1996).

Determinants of EIS quality

Several factors affect EIS quality, including the type and size of a project, and the nature and experience of various participants in the EIA process. Certain types of *project* have been associated with higher quality EISs. For instance, Schedule 1 projects, which generally have a high profile and attract substantial attention and resources, are likely to have better EISs. Better EISs have been linked with projects coming under the electricity and pipeline EIA regulations, the Scottish EIA regulations (Lee & Brown 1992) and the post-1993 highways regulations (Zambellas 1995) and, within the planning regulations, with wind-farms, recent waste-disposal and treatment plants, sand and gravel extraction schemes and opencast coal projects

Table 8.6 Project size v. EIS quality, % satisfactory*.

	Lee & Brown 1992**	DOE 1996**
Project size***:		
small	20	50
medium	35	54
large, very large	50–65	64

* Satisfactory means marks of A, B or C based on the Lee & Colley
(1990 or 1992) criteria.
** Lee & Brown reviewed 83 EISs, DOE 50.
*** Small is defined as < 75% of the threshold size used to determine
whether EIA is needed (DOE 1989), large as > 125% of the threshold size,
and medium between the two.

(DOE 1996). Larger projects generally have more satisfactory EISs than smaller projects, as is shown in Table 8.6.

Regarding the *nature and experience of the participants* in the EIA process, EISs produced in-house by developers are generally of poorer quality than those produced by outside consultants: the DOE (1996) study, for instance, showed that EISs prepared in-house had an average mark of D/E, while those prepared by consultants averaged C/D, and those prepared by both B/C. Lee & Brown's (1992) analysis of 83 EISs concluded that 57 per cent of those prepared by environmental consultants were satisfactory, compared with only 17 per cent of those prepared in-house. Similarly, EISs prepared by independent applicants tend to be better (C/D) than those prepared by local authorities for their own projects (D/E) (DOE 1996).

The experience of the developer, consultant, and competent authority also affects EIS quality. For instance, Lee & Brown (1992) showed that of EISs prepared by developers (without consultants) who had already submitted at least one EIS 27 per cent were satisfactory, compared with 8 per cent of those prepared by developers with no prior experience; Kobus & Lee (1993) cited 43 and 14 per cent respectively. A study by Lee & Dancey (1993) showed that of EISs prepared by authors with prior experience of four or more 68 per cent were satisfactory compared with 24 per cent of those with no prior experience. The DOE (1996) study showed that of the EISs prepared by consultants with experience of five or fewer about half were satisfactory, compared with about 85 per cent of those prepared by consultants with experience of eight or more. EISs prepared for local authorities with no prior EIS experience were just over one-third satisfactory, compared with two-thirds for local authorities with experience of eight or more (DOE 1996).

Other determinants of EIS quality include: the availability of EIA guidance and legislation, more guidance (e.g. DOE 1995, DOT 1993, local authority guides such as those of Kent and Essex) leading to better EISs; the stage in project planning at which the development application and EIA are submitted, EISs for detailed planning applications generally being better than those for outline applications; and issues related to the interaction between the parties involved in the EIA process, including commitment to EIA, the resources allocated to the EIA, and communication between the parties. There are also more EISs in the public domain to provide evidence of good practice.

EIS length also shows some correlation with EIS quality. For instance, Lee & Brown (1992) showed that the percentage of satisfactory EISs rose from 10 per cent of EISs less than 25 pages long to 78 per cent of those more than 100 pages long. In the DOE (1996) study, quality was shown to rise from an average of E/F for EISs of less than 20 pages to C for those of over 50 pages. However, as EISs became much longer than 150 pages, quality became more variable: although the very large EISs may contain more information, their length seems to be a symptom of poor organization and coordination.

8.5 The post-submission EIA process

After a competent authority receives an EIS and application for project authorization, it must review it, consult with statutory consultees and the public, and come to a decision about the project. This section covers these points in turn.

Review

Interviews with local authority planners show that planning officers see little difference between projects subject to EIA and other projects of similar complexity and controversy: once an application is lodged, the development process takes over. Competent authorities usually review EISs using their own knowledge and experience to pinpoint limitations and errors: the review is carried out primarily by reading through the EIS, consulting with other officers in the competent authority, consulting externally and comparing the EIS with the relevant regulations.

Despite the ready availability of the Lee & Colley (1992) review criteria, only about one-third of local authorities use any form of review methods at all, and then usually as indicative criteria, to identity areas for further investigation, rather than in a formal way. About 10–20 per cent of EISs are sent for review by external consultants or by the Institute of Environmental Assessment; but even when outside consultants are hired to appraise an EIS, it is doubtful whether the appraisal will be wholly unbiased if the consultants might otherwise be in competition with each other. There are also problems involved in getting feedback from the reviewing consultants quickly enough, given the tight timetable for making a project determination. An innovative approach being used by some developers requires consultants who are bidding to carry out an EIA to include as part of their bid an "independent" peer reviewer who will guarantee the quality of the consultants' work.

Various studies (e.g. Jones 1995, Kobus & Lee 1993, Lee et al. 1994, Weston 1995) suggest that LPAS require additional EIA information in about two-thirds of cases. This is usually done informally, without invoking the regulations.

Post-submission consultation and public participation

Competent authorities generally rely heavily on statutory and non-statutory consultees to review the different elements of an EIS, and in most cases the comments received are "substantial or very substantial" (Kobus & Lee 1993). Where the EIS contains insufficient information about a specific environmental component, competent authorities often put the developer and consultee in direct contact with each other rather than formally require further information themselves (DOE 1996).

Although there were early problems when EISs were not sent to the consultees (DOE 1991, Lee & Brown 1992), these seem to have mostly been ironed out (Jones 1995). In particular, English Nature and the Countryside Commission seem to participate quite actively at this stage of EIA, as well as local interest groups. However, some LPAS are not consulting all statutory consultees despite regulatory requirements to do so, possibly because they feel that a proposed project does not affect their area of interest (Pritchard et al. 1995). The Environment Agency also has little reason to carry out extensive consultation as part of the EIS process:

> HMIP, the NRA and the waste regulatory authorities (which have since been merged to form the Environment Agency) require impact assessments to be supplied with pollution permit applications. Therefore in their role as statutory planning EIS consultees, HMIP and the NRA are unlikely to waste time complaining about the poorly detailed designs given in a planning EIS, if they will be receiving another type of EIA document which precisely covers their area of concern. The Didcot B case study showed that even though HMIP considered the EIS to be satisfactory, they later demanded major design changes. (In the case of the Hamilton Oil gas terminal project in Liverpool Bay) HMIP raised no objectives to the EIS, but then rejected the IPC authorization on the grounds that design neither met the requirements of BATNEEC nor represented the BPEO. (Bird 1996)

This problem of duplicate authorization procedures and the lack of discussion between EIA participants will be discussed further in Chapter 10. The CEC (1993) feels that the UK situation:

> is satisfactory concerning the publication of ESs and their availability for consultation once they have been submitted. Copies can, in most cases, be obtained from either the developer or the competent authority concerned. Where the information was available to the EIA Centre, just under half of 290 ESs were available free of charge, with 18% available for purchase at £20 or

less, and the remaining 33% available at more than £20. In most cases copies of ESs are available, particularly in the specific locality where an application for consent is submitted. However, in a few cases copies of ESs are only available for consultation, but not for purchase by the public.

In general, however, public participation in the UK EIA system is often very partial, limited to a short period following EIS submission. For other wider aspects of public consultation and participation the reader is referred to Section 6.2.

Decision-making

As we noted in Chapter 1, one of the main purposes of EIA is to help to make better decisions, and it is therefore important to assess the performance of EIA to date in relation to this purpose. It is also important to remember that all decisions involve trade-offs. Wood (1995) identifies some of these, including trade-offs in the EIA process between simplification and complexity, urgency and the need for better information, facts and values, forecasts and evaluation, certainty and uncertainty. There are also trade-offs of a more substantive nature, in particular between the socio-economic and biophysical impacts of projects – sometimes reduced to the "jobs versus the environment" dilemma. Box 8.1 illustrates the trade-off issue in relation to the UK's Newbury by pass, which generated direct action by aggrieved parties, who sought to influence the project decision.

Some impacts may be more tradable in decision-making than others. Sippe (1994) provides an illustration, for both socio-economic and biophysical categories, of negotiable and non-negotiable impacts (see Table 8.7). Sadler (1996) identifies such trade-offs as the core of decision-making for sustainable development.

In the UK there is an important decision-making stage linked most normally to a planning approval process by the competent authority, and involving the consideration of the EIS and associated information. The EIS may have an impact on a planning officer's report, on a planning committee's decision, and on modifications and conditions to the project before and after submission. But the impact of EIA on decision-making may be much wider than this, influencing, for example, the alternatives under consideration, project design and redesign, and the range of mitigation measures and monitoring procedures. Indeed, the very presence of an effective EIA system may lead to the withdrawal of unsound projects and the deterrence of the initiation of environmentally damaging projects.

In Chapter 3 the various participants in the EIA process were identified. These participants will have varying perspectives on EIA in decision-making. A local planning officer may be concerned with the *centrality* of EIA in decision-making (does it make a difference?), central government might be concerned about *consistency* in application to development proposals across the country; pressure groups may also be concerned with these criteria, but also with *fairness* (in providing opportunities for participation) and *integration* in the project cycle and

235

Box 8.1 Socio-economic and biophysical impact trade-offs – the example of Newbury Bypass, UK

MIXED FEELINGS GREET GO-AHEAD TO NEWBURY

The Government's shock decision to approve the A34 Newbury Bypass has been greeted with relief by district and county planners, who feel that traffic congestion is paralysing the Berkshire town. Environmental campaigners, however, have reacted angrily to ex-transport secretary Brian Mawhinney's announcement last Wednesday – his final one before being replaced by Sir George Young. Protests on a similar scale to those at Twyford Down are now expected at the site, where construction work could begin before the end of the year.

Last December, Mawhinney said he would delay any decision on the controversial proposal for a year to consider alternatives, and his sudden announcement took both local authorities and environmentalists by surprise. "I had no doubt that the current situation on the A34 was intolerable and that there was strong economic justification for a bypass" he explained. "But I wanted to be sure of the environmental balance between the principal route alternatives and to confirm that the route proposed was the best solution to the problems of congestion in Newbury."

Peter Gilmour, community officer at Newbury Borough Council, called the decision "a triumph for common sense and local democracy". Council planning officers and members had backed the scheme "root and branch", he said – while 13,000 of the town's 27,000 population had signed a petition of support. According to council research 50,000 vehicles a day travel through Newbury, an estimated two thirds of which are "simply passing through the town", said Gilmour. Incidents of asthma are growing among local people, while traffic delays of half-an-hour are common. "People are reluctant to come into the town to shop, while some local industries have moved away because they are unable to transport goods effectively," Gilmour continued. "It's all very well protesters taking the moral high ground, but we have a moral duty to the welfare of people in the town."

Berkshire County Council also welcomed the announcement, which follows years of uncertainty and two public inquiries to decide on the western route. The council's environment committee has been lobbying Mawhinney to back the scheme – arguing that the delay was making transport planning very difficult. "The council's transport strategy generally has been to move away from major roads but we feel this is one of the exceptions", said county environment officer Keith Reed.

He stressed that the council is concerned about potential environmental damage from the scheme, which protesters say will destroy parts of four SSSIs, partly cross a civil war battlefield and pass near to the 14th century Grade 1 listed Donnington Castle and the Watermill Theatre at Bagnor. "We will be pressing the DoT to carry out more archaeological studies, and calling for the use of porous asphalt to cut down on traffic noise levels", said Reed. Environmental campaigners remain furious, nonetheless. Tony Juniper of Friends of the Earth said Mawhinney's decision "makes a mockery of the 'great transport debate'. Local people have been preparing to start local dialogue this month on alternatives to the plan, but will now have their efforts thrown back in their face". "This road must be stopped at all costs. It is one of the most destructive schemes in the national roads programme and will mobilise massive countrywide opposition", Juniper said.

Graham Wynne, RSPB conservation director, said the bypass will destroy part of Snelsmore Common – home to a special community of plants and animals including the rare nightjar. "This decision flies in the face of the Government's stated commitment to a sustainable transport policy. No-one denies that Newbury needs a solution to its traffic problems, but it should not be regardless of the environmental cost." The European Commission is also considering legal action to stop the scheme, which protesters say will break various environmental directives.

(*Source: Planning Week*, July 1995)

Table 8.7 Judging environmental acceptability – trade offs.

	Non-negotiable impacts	Negotiable impacts*
Ecological (physical and biological components)	Degrades essential life support systems	No degradation beyond carrying capacity
	Degrades conservation estate	No degradation of productive systems
	Adversely affects ecological integrity	Wise use of natural resources
	Loss of biodiversity	
Social (humans as individuals or in social groupings)	Loss of human life	Community benefits and costs and where they are borne
	Reduces public health and safety unacceptably	Reasonable apportionment of costs and benefits
	Unreasonably degrades quality of life where people live	Reasonable apportionment of inter-generational equity
		Compatibility with defined environmental policy goals

* In terms of net environmental benefits.
(*Source:* Sippe 1994)

approval process (to what extent is EIA easily bypassed?). A number of studies have attempted to determine whether EIA and associated consultations have influenced decisions about whether and how to authorize a project.

Early surveys of local planning officers (Kobus & Lee 1993, Lee et al. 1994) suggest that EISs were important in the decision in about half of the cases. More recent interviews with a wider range of interest groups (DOE 1996) found that about 20 per cent of respondents felt that the EIS had "much" influence on the decision, more than 50 per cent felt that it had "some" influence, and the remaining 20–30 per cent felt that it had little or no influence. Jones (1995) found that about one-third of planning officers, developers and public interest groups felt that the EIS influenced the decision, compared with almost half of environmental consultants and only a very small proportion of consultees. For planning decisions, it is the members of the planning committees who make the final decision. Interviews suggest that they are not generally interested in reading the EIS, but instead rely on the officer's report to summarize the main issues (DOE 1996). According to Wood & Jones (1997), planning committees followed officers' recommendations in 97 per cent of the cases they studied.

The consultations related to the EIS are generally seen to be at least as important as the EIS itself (Jones 1995, Kobus & Lee 1993, Lee et al. 1994, Wood & Jones 1997). Figure 8.7 clearly illustrates this point. On the other hand, many interviewees from non-statutory bodies felt excluded from the decision-making process, and one national non-statutory wildlife body complained that if the Nature Conservancy

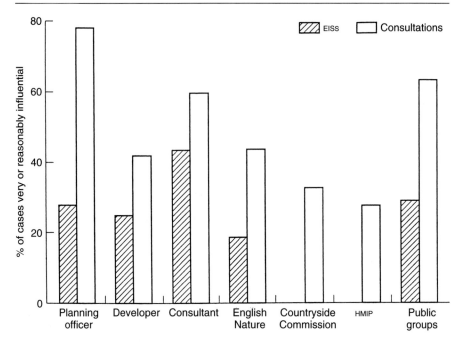

Figure 8.7 Opinions about the influence of EISs and consultations on decisions. Note: consultees include equivalents in Scotland and Wales, in each case. (*Source:* Wood & Jones 1997)

Council or the Countryside Commission did not object then their own objections went largely ignored (DOE 1996).

While studies of early EISs (e.g. Kobus & Lee 1993, Lee et al. 1994) suggest that material considerations were slightly more important than environmental considerations in the final decision on a project's authorization, a recent study (Jones 1995) suggests that the environment was the principal factor influencing the decision, with planning policies given slightly less weight. Wood & Jones (1997) report that the environment was seen to be the overriding factor influencing the decisions in 37 per cent of the cases they studied. However, only in a very few cases would the final decision have been different in the absence of an EIS.

Project applications with EISs are not treated much differently from those without EISs. Although environmental issues are addressed more formally, in a discrete document, the final decision-making process is not changed much by EIA. The main procedural difference brought about by EIA is the need to consult people about the EIS, and the broader scope for public participation (not often used in practice) that it brings. However, the result of the entire EIA process is a modification of projects due to EIA, possibly additional or different conditions on the project, and perhaps a more comprehensive consideration of environmental issues by the competent authority.

8.6 Costs and benefits of EIA

Much of the early resistance to the imposition of EIA was based on the idea that it would cause additional expense and delay in the planning process. EIA proponents refuted this by claiming that the benefits of EIA would well outweigh its costs. This chapter concludes with a discussion of the costs and benefits of EIA to various parties in the UK.

Costs of EIA

Generally, EIA has probably slightly increased the cost to *developers* of obtaining planning permission. An EIS generally costs between 0.000025 and 5 per cent of project costs (Coles et al. 1992). Weston's (1995) survey of consultants showed that consultancies received on average £34,000 for preparing a whole EIS, £40,000 for several EIS sections, and £14,750 for one section: this itself highlights the variability of the costs involved. Another study of 20 EISs showed EIS preparation to vary from 22 person-days at a cost of £5,000 to 3–4 person-months with additional work contracted out (DOE 1996). Pritchard et al.'s (1995) study of eight EIAs found that developers felt that "the preparation of the ES had cost them too much time and money, and that the large amounts of work involved in EA often yielded few tangible benefits in terms of the actual planning decision reached".

In terms of the delay caused to planning decisions, various studies (e.g. DOE 1991, Tarling 1991, Lee et al. 1994) have shown that the mean time to decide planning applications with EISs is about 40 weeks; but there are wide variations. This is considerably more than for an applications without an EIS (DOE 1996), but then the projects with an EIS also tend to be larger, more complex and more politically sensitive. An early study (Coles et al. 1992) found that, on average, the entire EIA process, from the notification of intent for the project to the decision, took 62 weeks, the EIS preparation taking 25 weeks. Although some consultants feel that EIA slows down the decision-making process, imposes additional costs on developers and is a means through which LPAs can make unreasonable demands on developers to provide detailed information on issues "which are not strictly relevant to the planning decision" (Weston 1995), others feel that EIA does not necessarily slow things down: "The more organised approach makes it more efficient and in some cases it allows issues to be picked up earlier. The EIS can thus speed up the system" (DOE 1996). An EIA may well shorten the planning application stage but lengthen the period before the EIS is submitted.

There has been some concern that competition and cost-cutting by consultancies, an increase in "cowboy" consultancies and the tendency for developers to accept the lowest bid for preparing an EIS, may affect the quality of the resulting EIAs by limiting the consultants' time, expertise or equipment. Consultants note that "on all but the largest developments there is always a limited budget – an EA expands to fill the available budget, and then some" (Radcliff & Edward-Jones 1995). However, Fuller (1992) argues that this may not be helpful to a developer in the long run:

> A poor-quality statement is often a major contributory factor to delays in the system, as additional information has to be sought on issues not addressed, or only poorly addressed, in the original . . . Therefore, reducing the cost of an environmental assessment below the level required for a thorough job is often a false economy.

The cost of EIA to *competent authorities* is much more difficult to measure and has until now been based on interviews rather than on a more systematic methodology. An early study (Lee & Brown 1992) found that about half the officers interviewed felt that the EIS had not influenced how long it took to reach a decision; the rest were about evenly split between those who felt that the EIA had speeded up or slowed down the process. In more recent interviews (DOE 1996), many planning officers felt that dealing with the EIS and the planning application were one and the same and "just part of the job". Estimates for reviewing the EIS and associated consultation ranged from five hours to 6–8 months of staff time. Planning officers handling EIS cases tend to be development control team leaders and above, so staff costs would generally be higher than for standard planning applications. Where LPAs had engaged consultants to help them appraise an EIS, the cost of such review was between £1,000 and £10,000 for half the cases, the remaining being broadly split evenly between more than £10,000 and less than £1,000 (Leu et al. 1993).

In 20 case studies, the time spent by *consultees* on EIA ranged from four hours to one-and-a-half days for statutory consultees, and from one hour to two weeks for non-statutory consultees. Although some consultees, like planning officers, argued that "this is what we are here for", others suggested that they needed to prioritize what developments they got involved in because of time and resource constraints (DOE 1996).

Benefits of EIA

The benefits of EIA are mostly unquantifiable, so a direct comparison with the costs of EIA is not possible. Perhaps the clearest way to gauge whether EIA helps to reduce a project's environmental impacts is to determine whether a project was modified as a result of EIA. Early studies on EIA effectiveness (e.g. Tarling 1991, Kobus & Lee 1993) showed that modifications to the project as a result of the EIA process were required in almost half the cases, with most modifications regarded as significant. Jones's (1995) study of 40 EISs prepared before March 1993 showed that modifications before EIS submission were made in one-third of cases, modification after EIS submission in about 15 per cent of cases, modifications before and after the EIS submission in one-fifth of cases, and no modifications in just under one-third of cases.

240

EIA can have other benefits in addition to project modification. A recent survey of *environmental consultants* (Weston 1995) showed that about three-quarters of them felt that EIA had brought about at least some improvements in environmental protection, primarily through the incorporation of mitigation measures early in project design and the higher regard given to environmental issues. However, other consultants felt that the system is "often a sham with ESs full of platitudes". Jones (1995) found that only one-fifth of developers and consultants felt that there had been no benefits associated with EIA, and two-fifths felt that EIA had conferred no disadvantages.

Competent authorities generally feel that projects and the environment benefit greatly from EIA (Jones 1995, Lee et al. 1994). EIA is seen as a way to focus the mind, highlight important issues, reduce uncertainty, consider environmental impacts in a systematic manner, save time by removing the need for planning officers to collect the information themselves, and identify problems early and direct them to the right people (DOE 1996, Jones 1995, Pritchard et al. 1995). One planning officer noted: "when the system first appeared I was rather sceptical because I believed we had always taken all these matters into account. Now I am a big fan of the process. It enables me to focus on the detail of individual aspects at an early stage" (DOE 1996).

Consultees broadly agree that EIA creates a more structured approach to handling planning applications, and that an EIS gives them "something to work from rather than having to dig around for information ourselves". However, when issues are not covered in the EIS, consultees are left in the same position as with non-EIA applications: some of their objections are not because the impacts are bad but because they have not been given any information on the impacts or any explanation of why a particular impact has been left out of the assessment. Consultees feel that an EIA can give them data on sites that they would not otherwise be able to afford to collect themselves, and that it can help parties involved in an otherwise too often confrontational planning system to reach common ground (DOE 1996).

8.7 Summary

All the parties involved agree that EIA as practised in the UK helps to improve projects and protect the environment, although the system could be much stronger: EIA is thus at least partly achieving its main aims. There are time and money costs involved, but there are also tangible benefits in the form of project modifications and more informed decision-making. When asked whether EIA was a net benefit or cost, "the overwhelming response from both planning officers and developers/consultants was that it had been a benefit. Only a small percentage of both respondents felt that EIA had been a drawback" (Jones 1995). Some stages in EIA – particularly early scoping, good consultation of all the relevant parties and the preparation of a clear and unbiased EIS – are consistently cited as leading to particularly clear and cost-effective benefits (Kobus & Lee 1993, DOE 1996).

Chapters 9 and 10 examine more specifically how EIA is used in practice in the UK. Suggestions for future directions in EIA in the UK and beyond are discussed in Chapter 12.

References

Bellanger, C. & R. Frost 1997. *Directory of environmental impact statements, July 1988–January 1997*. Oxford: Oxford Brookes University School of Planning.

Bird, A. 1996. *Auditing environmental impact statements using information held in public registers of environmental information*, Working Paper 165. Oxford: Oxford Brookes University, School of Planning.

Braun, C. 1993. EIA in the 1990s: the view from Marsham Street. EIA in the UK: *Evaluation and prospect conference*, 6 April. Manchester: University of Manchester.

CEC (Commission of the European Communities) 1993. *Report from the Commission of the Implementation of Directive 85/337/EEC on the assessment of the effects of certain public and private projects on the environment*, vol. 13, *Annexes for all Member States*, COM(93)28 final. Brussels: EC, Directorate-General XI.

Coles, T. F., K. G. Fuller, M. Slater 1992. *Practical experience of environmental assessment in the UK*, Proceedings of Advances in Environmental Assessment Conference. London: IBC Technical Services.

Davison, J. B. R. 1992. *An evaluation of the quality of Department of Transport environmental statements* (MSc dissertation, Oxford Brookes University).

DOE (Department of the Environment) 1989. *Environmental assessment: a good practice guide*. London: HMSO.

DOE 1991. *Monitoring environmental assessment and planning*. London: HMSO.

DOE 1994. *Evaluation of environmental information for planning projects: a good practice guide*. London: HMSO.

DOE 1995. *Preparation of environmental statements for planning projects that require environmental assessment: a good practice guide*. London: HMSO.

DOE 1996. *Changes in the quality of environmental impact statements*. London: HMSO.

Ferrary, C. 1994. Environmental assessment: our client's perspective. Environmental Assessment: RTPI Conference, 20 April. Andover.

Frost, R. & D. Wenham 1996. *Directory of environmental impact statements July 1988–Sept. 1995*. Oxford: Oxford Brookes University, School of Planning.

Fuller, K. 1992. Working with assessment. In *Environmental assessment and audit: a user's guide*, 14–15. Gloucester: Ambit.

Gosling, J. 1990. *The Town and Country (Assessment of Environmental Effects) Regulations 1988: the first year of application*. Proposal for a working paper, Department of Land Management and Development, University of Reading.

Hall, E. 1994. *The environment versus people? A study of the treatment of social effects in environmental impact assessment* (MSc dissertation, Oxford Brookes University).

Hughes, J. & C. Wood 1996. Formal and informal environmental assessment reports, *Land Use Policy* **13**(2), 101–13.

IEA (Institute of Environmental Assessment) 1993. *Digest of environmental statements.* London: Sweet & Maxwell.

Jones, C. E. 1995. The effect of environmental assessment on planning decisions, *Report*, special edition (October), 5–7.

Jones, C. E., N. Lee, C. Wood 1991. *UK environmental statements 1988–1990: an analysis.* Occasional Paper no. 29, EIA Centre, University of Manchester.

Kobus, D. & N. Lee 1993. The role of environmental assessment in the planning and authorisation of extractive industry projects. *Project Appraisal* **8**(3), 147–56.

Lee, N. & D. Brown 1992. Quality control in environmental assessment. *Project Appraisal* **7**(1) 41–5.

Lee, N. & R. Colley 1990 (updated 1992). *Reviewing the quality of environmental statements.* Occasional Paper no. 24, University of Manchester.

Lee, N. & R. Dancey 1993. The quality of environmental impact statements in Ireland and the United Kingdom: a comparative analysis. *Project Appraisal* **8**(1), 31–6.

Lee, N., F. Walsh, G. Reeder 1994. Assessing the performance of the EA process. *Project Appraisal* **9**(3), 161–72.

Leu, W.-S., W. P. Williams, A. W. Bark 1993. An evaluation of the implementation of environmental assessment by UK local authorities. *Project Appraisal* **10**(2), 91–102.

McMahon, N. 1996. Quality of environmental statements submitted in Northern Ireland in relation to the disposal of waste on land. *Project Appraisal* **11**(2), 85–94.

Mills, J. 1994. The adequacy of visual impact assessments in environmental impact statements. In *Issues in environmental impact assessment.* Working Paper no. 144, School of Planning, Oxford Brookes University, 4–16.

Nelson, P. 1994. Better guidance for better EIA. *Built Environment* **20**(4), 280–93.

Petts, J. & P. Hills 1982. *Environmental assessment in the UK.* Nottingham: University of Nottingham, Institute of Planning Studies.

Pritchard, G., C. Wood, C. E. Jones 1995. The effect of environmental assessment on extractive industry planning decisions. *Mineral Planning* **65** (December), 14–16.

Radcliff, A. & G. Edward-Jones 1995. The quality of the environmental assessment process: a case study on clinical waste incinerators in the UK. *Project Appraisal* **10**(1), 31–8.

Sadler, B. 1996. *Environmental assessment in a changing world: evaluating practice to improve performance.* Final report of the international study on the effectiveness of environmental assessment, Canadian Environmental Assessment Agency.

Sheate, W. 1994. *Making an impact: a guide to EIA law and policy.* London: Cameron May.

Sippe, R. 1994. Policy and environmental assessment in Western Australia: objectives, options, operations and outcomes. Paper for International Workshop, Directorate General for Environmental Protection, Ministry of Housing, Spatial Planning and the Environment, The Hague, Netherlands.

Tarling, J. P. 1991. A comparison of environmental assessment procedures and experience in the UK and the Netherlands (MSc dissertation, University of Stirling).

Weston, J. 1995. Consultants in the EIA process. *Environmental Policy and Practice* **5**(3), 131–4.

Weston, J. 1996. Quality of statement is down on the farm. *Planning* **1182**, 6–7.

Wood, C. 1991. Environmental impact assessment in the United Kingdom. Paper presented at the ACSP-AESOP Joint International Planning Congress, Oxford Polytechnic, Oxford, July.

Wood, C. 1995. *Environmental impact assessment: a comparative review*. Harlow: Longman Scientific and Technical.

Wood, C. 1996. Progress on EIA since 1985 – a UK overview. In the proceedings of the IBC Conference on Advances in Environmental Impact Assessment, 9 July. London: IBC UK Conferences.

Wood, C. & C. Jones 1991. *Monitoring environmental assessment and planning*, DOE Planning and Research Programme. London: HMSO.

Wood, C. & C. Jones 1997. The effect of environmental assessment on local planning authorities. *Urban Studies* **34**(8), pp. 1237–57.

Zambellas, L. 1995. *Changes in the quality of environmental statements for roads* (MSc dissertation, Oxford Brookes University).

Notes

1 This was made up largely of landfill/raise projects (10 per cent), waste-water or sewage treatment schemes (4 per cent) and incinerators (3 per cent).

2 In terms of type and size of project, location, local planning authority and developer.

3 "Satisfactory" for the IAU criteria is very similar to "satisfactory" for the Lee & Colley criteria.

CHAPTER 9

Environmental impact assessment and projects requiring planning permission

9.1   Introduction

This chapter examines the application of the UK Town & Country Planning (Assessment of Environmental Effects) Regulations in two contrasting case study sectors. Section 9.2 looks at the use of EIA in proposals for new settlements in the countryside. New settlements include a variety of activities and land-uses and provide some of the most comprehensive projects for the EIA procedures. The section illustrates some of the issues arising from the implementation of the Regulations, including ambiguity over the need for EIAs for certain projects, the appropriate timing of the submission of EISs and the important role of the planning inquiry in the EIA process. In Section 9.3, two detailed case studies of specific new settlement proposals are examined, to highlight certain features of current good EIA practice.

Section 9.4 then examines the application of EIA to waste treatment and disposal projects. Such projects account for roughly one fifth of all environmental statements submitted in the UK since the implementation of Directive 85/337 – they therefore represent a significant part of EIA activity in the UK to date. Issues highlighted in Section 9.4 include the variable quality of waste-disposal EISs, concerns about the adequacy of the wider EIA process and the problems of overlap between planning and pollution controls.

9.2   EIA and new settlements

The nature of new settlements

New settlements are not a new concept. Their development can be traced back to the garden city movement of Ebenezer Howard, followed by the new towns of the post-war years (Ward 1992). In the 1960s and 1970s, a small number of privately

Table 9.1 Location of free-standing new settlement proposals submitted July 1988 to December 1992.

Region	Applications submitted	Applications expected
East Anglia	15	2
East Midlands	7	1
South-east	5	7
West Midlands	3	1
Yorkshire & Humberside	1	3
South-west	1	0
North	0	1
TOTAL	32	15

(*Sources: Journal of Planning and Environment Law,* Therivel (1991), planning press, personal communications with local planning authority officers.)

Note: The figures show the position at the end of 1992. Only planning applications submitted since the implementation of the EC Directive in July 1988 are included. Expected applications are those for schemes awaiting the outcome of development plan work or the outcome of other neighbouring applications. Only free-standing schemes are included. Breheny et al. (1993) list a total of 125 schemes proposed between 1989 and mid-1992.

funded new settlements, such as New Ash Green, Bar Hill and South Woodham Ferrers, were also developed. More recently, a series of proposals for "new country towns" was initiated by Consortium Developments Ltd in the mid-1980s (Breheny et al. 1993). Other developers took up the concept in response to the opportunities presented by the round of structure plan reviews in the late 1980s and early 1990s. The number of new settlement schemes promoted since the late 1980s was considerable, despite the lukewarm support for the concept in the DOE's revised Planning Policy Guidance Note 3 (PPG3) published in March 1992 (DOE 1992a) and a lengthening list of appeal refusals in recent years. However, since 1993 the flow of applications has reduced substantially.

Between the implementation of EC Directive 85/337 in July 1988 and the end of 1992, planning applications were submitted for at least 32 free-standing new settlement schemes in England and Wales. Two-thirds of these were schemes located in East Anglia and the East Midlands, most of the remainder were in south-east England and the West Midlands (see Table 9.1). Just over half the schemes were located in two counties. Cambridgeshire experienced the largest number of applications (13), reflecting the A10 and A45 new settlement policies in the replacement structure plan of 1989; Leicestershire also saw a large number of applications (four), although there no lead was provided by the county's structure plan policies.

The applications submitted between 1988 and 1992 ranged in size from 200 dwellings to over 4,000; the average was just over 1,900. Very few proposals were submitted during this period for new settlements of more than 3,000 dwellings, half the schemes proposing 1,500 or fewer (see Table 9.2). Almost all the schemes included village centres with shopping and community facilities, and most also incorporated elements of commercial and/or industrial development.

Table 9.2 Size of new settlement proposals submitted between July 1988 and December 1992.

No. of dwellings proposed	Applications submitted	Applications expected	All proposals
Up to 500	5	2	7
501–1,000	2	3	5
1,001–1,500	10	2	12
1,501–2,000	0	1	1
2,001–2,500	3	1	4
2,501–3,000	8	0	8
Over 3,000	4	2	6
Not known	0	4	4
TOTAL	32	15	47
Average no. of dwellings	1,939	1,700	1,863

(*Sources:* As for Table 9.1)

The need for EIA for new settlements

Guidance on need for EIA

Free-standing new settlement schemes are not specifically identified in either Schedule 1 or 2 of the EIA regulations. This has led to some confusion over the need for formal EIA for such schemes. An early ruling by the DOE that one of the new settlement proposals near Cambridge (for up to 3,000 houses and a business park) was neither a Schedule 1 nor a Schedule 2 project, and that EIA was therefore not required, only added to this confusion.[1] In a further twist, the developers in this case eventually submitted an EIS for the scheme voluntarily, despite the Secretary of State's ruling.

It could reasonably be argued that new settlements are embraced within the term "urban development projects" (Sched. 2.10b). Given that most proposals include some commercial, industrial or retail development, they may also contain an element of "industrial estate development" (Sched. 2.10a). Granted that this *is* the case, which schemes are likely to require formal EIA? The DOE's indicative criteria and thresholds unfortunately provide little guidance on this matter. Most of the criteria for urban development and industrial estate projects appear to have been designed primarily for projects located within existing urban areas, rather than in free-standing locations – for example, close proximity to a significant number of dwellings is a possible factor in determining the need for EIA, which is clearly of no use in the case of free-standing schemes. A floorspace threshold of 10,000 m^2 (for retail and commercial development) and a site area threshold of 20 ha (for industrial estate development) are suggested, but it is not clear whether these are applicable to both free-standing and urban-area proposals. Rather confusingly, a higher floorspace threshold of 20,000 m^2 is suggested for out-of-town retail developments. No guidance is provided for housing schemes, whether free-standing or not. As one of the DOE's own commissioned research studies concluded:

Several indicative criteria and thresholds in . . . the circular are ambiguous, especially those relating to "urban development" schemes and "redevelopment" projects, there being an absence of criteria against which to determine whether or not new settlements in the countryside should be subject to EA. (DOE 1991)

Clearly, therefore, local planning authorities have been left with a large amount of discretion in determining the need for EIA for such schemes, with little guidance on how this discretion might be exercised.

The interpretation of need in practice

In practice, EISs have been submitted in many cases for new settlement proposals. Table 9.3 provides further details on the 32 applications submitted between 1988 and 1992. EISs were submitted in two-thirds of these cases, either voluntarily or in response to requests by local planning authorities. The remaining one-third of cases were not subject to formal EIA and were treated as normal planning applications. Breheny et al. (1993) note that of 28 EISs for new settlement schemes submitted before mid-1992, almost all (26) were submitted voluntarily by the developers. In only two cases was an EIS requested by the local planning authority after the submission of the planning application. Of the 28 EISs identified by Breheny et al., only eight (less than 30 per cent) were definitely of statutory status. Over half (16) of the statements were informal and therefore did not trigger the statutory procedures under the Town & Country Planning (AEE) Regulations.

One reason local planning authorities did not request EIA in these cases is that the size of the new settlement was too small to justify the use of the formal EIA procedures. The number of dwellings proposed is an obvious indicator of the size of a new settlement, although the total site area and the scale of any commercial, industrial or retail development proposed will also be important. The average number of dwellings proposed in schemes subject to EIA was approximately 2,300, which is well above the average of 1,200 dwellings in schemes not subject to EIA. Indeed, half the projects not subject to EIA were small schemes of 500 dwellings or fewer (see Table 9.3). Nevertheless, there are examples of larger schemes (of more than 2,000 dwellings) that escaped the need for EIA, as well as of much smaller schemes (of fewer than 1,000 dwellings) that *were* subject to formal EIA. Factors other than the size of a proposal are clearly at work in determining the need for EIA.

A second reason local planning authorities did not require EIA is that a proposed site was already allocated for residential development in the relevant local plan. To date, this appears to have been the case for very few proposals. However, an example is provided by a district council scheme for up to 1,150 houses on a redundant hospital site near Newark in Nottinghamshire. This application did not result in a request for EIA by the LPA, apparently because 700 dwellings were already allocated on the site in the deposit version of the Newark Area Local Plan. Bulleid (1997) notes a further example in the case of a town expansion scheme in Aberdeen. As discussed below, this type of situation may become more common in future as more new settlement proposals are pursued through the local plan process (Breheny et al. 1993).

Table 9.3 Size of new settlement proposals for which planning applications were submitted July 1988 to December 1992.

No. of dwellings proposed	EIS submitted	EIS not submitted	Total
Up to 500	0	5	5
500–100	1	1	2
1,001–1,500	9	1	10
1,501–2,000	0	0	0
2,001–2,500	2	1	3
2,501–3,000	6	2	8
Over 3,000	4	0	4
TOTAL	22	10	32
Average no. of dwellings	2,292	1,186	1,939

(*Sources:* As for Table 9.1)

A study commissioned by the DOE (1991) suggests other reasons why EIA is not requested: (a) the developer may have already provided a substantial amount of supporting information with the planning application; (b) the local planning authority may intend to refuse the application and regard any subsequent appeal inquiry as being likely to deal adequately with the environmental implications of the development; (c) if the application is for outline planning permission, the design of the proposal may not be far enough advanced to allow EIA to be carried out; (d) there may have been no formal consideration by the local planning authority of the need for EIA. It is not known whether any of these factors were applicable in any of the cases shown in Table 9.4.

EIA and the development plan process

Of the 32 planning applications submitted between 1988 and 1992 for new settlement proposals, 19 had already been determined by the end of 1992, and a further five were withdrawn before determination. Of those determined, the great majority (17) were either called in by the Secretary of State or taken to appeal following refusal by the local planning authority. The role of EIA and the EIS in resultant public inquiries has been discussed in Section 6.5. All the appeals and called-in applications for the new settlements shown in Table 9.3 were dismissed by the Secretary of State, although one of these schemes was resubmitted in revised form and subsequently granted outline planning permission by the end of 1992.

The lack of success at appeal and the revised PPG3 indicate that proposals are likely to be successful only when promoted through the development plan process, and this now appears to be the dominant approach adopted by promoters of schemes (Breheny et al. 1993). The promotion of schemes through the local plan process raises issues about the role of EIA and the appropriate timing of the submission of the EIS in such cases. If a proposal has been incorporated as an allocation in the

249

Table 9.4 Free-standing new settlement schemes for which planning applications were submitted July 1988 to December 1992.

Name of scheme and location	Local planning authority	No. of dwellings	EIS submitted?	Planning decision
Hare Park A45 east of Cambridge	East Cambridgeshire DC	3,000	no	1
Allington A45 east of Cambridge	East Cambridgeshire DC	3,370	yes	1
Kennett North-east of Newmarket	East Cambridgeshire DC	1,500	yes	5
Westmere A10 north of Cambridge	East Cambridgeshire DC	1,500	yes	1
Waterfenton A10 north of Cambridge	South Cambridgeshire DC*	1,500	yes	1
Scotland Park A45 west of Cambridge	South Cambridgeshire DC	3,000	yes	1
Highfields A45 west of Cambridge	South Cambridgeshire DC	3,300	yes	1
Great Common Farm A45 west of Cambridge	South Cambridgeshire DC	3,000	yes	2
Bourn Airfield A45 west of Cambridge	South Cambridgeshire DC	3,000	yes	1
Swansley Wood A45 west of Cambridge	South Cambridgeshire DC	3,300	yes	2
Belham Hill A45 west of Cambridge	South Cambridgeshire DC	3,000	yes	1
Crow Green A45 west of Cambridge	South Cambridgeshire DC	3,000	yes	5
Mangreen South of A47 Norwich Southern Bypass	South Norfolk DC	1,500	yes	5
Leziate East of King's Lynn	Kings Lynn & West Norfolk BC	450	no	6
Hilton Near Burnaston Toyota plant, A516 west of Derby	South Derbyshire DC	1,120	yes	3
Kettleby Magna Great Dalby airfield, south of Melton Mowbray	Melton BC	1,200	yes	4
Six Hills Village Alongside A46, between Melton Mowbray & Loughborough	Melton BC	1,400	yes	5
Stretton Magna East of Leicester, between A6 and A47	Harborough DC	2,400	yes	5
Wymeswold airfield East of Loughborough	Charnwood DC	2,200	no	1
Bilsthorpe Village Expansion Between Mansfield & Newark	Newark & Sherwood DC	990	yes	6

Table 9.4   (cont'd)

Name of scheme and location	Local planning authority	No. of dwellings	EIS submitted?	Planning decision
Balderton hospital South-east of Newark	Newark & Sherwood DC	1,150	no	6
Marston Park Marston Moretaine, between Bedford & Milton Keynes	Mid Bedfordshire DC	800	no	1
Upper Donnington North of Newbury	Newbury BC	300	no	1
Chiltern Acres Near Stoke Mandeville, south of Aylesbury	Aylesbury Vale DC and Wycombe DC	400	no	1
Northwick Village Canvey Island	Castle Point DC	4,300	yes	1
Otterham Quay East of Gillingham	Gillingham BC	200	no	1
Strensham upon Avon M5 between Cheltenham & Worcester	Wychavon DC	1,250–1,750	yes	1
Brockhill Near Redditch	Redditch BC	1,300	yes	8
Aston Prior Near Shifnal, east of Telford	Bridgnorth DC	360	no	1
Acaster Malbis South of A64 York southern bypass	Selby DC	2,250	yes	7
Poundbury Near Dorchester	West Dorset DC	2,500–3,000	no	3

(*Sources:* as for Figure 9.1)

Note: Planning status (as at end of 1992) is coded as follows:

1 – Appeal or called-in application dismissed by Secretary of State after public inquiry.
2 – As with 1, but application resubmitted in modified form and yet to be determined.
3 – Outline planning permission granted by local planning authority.
4 – Grant of outline planning permission by local planning authority, subject to completion of Section 106 agreement.
5 – Application or appeal withdrawn prior to determination.
6 – Decision by local planning authority awaited – dependent on progress with local plan.
7 – Holding direction issued by Secretary of State, preventing local planning authority determining application until completion of structure plan and Green Belt review.
8 – Appeal outcome awaited.
* Two schemes were submitted: one for 1,500, one for 3,000. Planning decision – 1 in both cases.

local plan after extensive consultation and the submission of considerable informa-tion by the developer, the local planning authority may decide that requesting an EIS with any subsequent application would be superfluous and might prejudice relations with the developer. There is therefore a possibility that such proposals might escape the need for formal EIA.

It could be argued that for schemes promoted through the local plan process formal EIA should be carried out at a stage earlier than the submission of a planning application. The resulting EIS could then be taken into account by the local planning authority and made available for public comment, either during the consultation period on the draft plan or at the local plan inquiry. An alternative approach would be for the LPA to carry out an EIA of the local plan itself, which would include an assessment of the environmental impacts of any major residential allocations in the plan (see Ch. 13). To date, there is no evidence that local planning authorities have requested EISs from developers during the local plan consultation process and *before* the submission of planning applications. However, there *are* examples of developers voluntarily submitting EISs in such circumstances. One example is provided by a village expansion scheme at Lighthorne Heath, adjacent to the M40 in Warwickshire. The site involved had already been allocated for housing in the draft Stratford-on-Avon District Local Plan, and the proposal had been worked up in considerable detail. The developers agreed to submit a voluntary EIS as part of their representations on the draft plan, "in view of the complex nature of the project and its possible impact on the area". The required scope of the EIA was determined largely by the district council. Statutory and other consultees were contacted during the preparation of the EIS, in accordance with the EIA regulations. However, as no planning application had been submitted, the requirements relating to publicity in the regulations were not regarded as applicable.

The submission of such voluntary EISs, before the planning application and outside the formal EIA regulations, could be expected to become more common as an increasing number of proposals are pursued through the local plan process. Whether a further or updated EIS would be submitted or requested at the time of the planning application in such cases is not certain. If not, then the requirements relating to publicity of the EIS and consultation with statutory and other bodies could be bypassed.

9.3 New settlement case studies

Introduction to the case studies

This section now presents detailed case studies of EIA for two new settlement proposals. The aim is not to provide a comprehensive review of the entire EIA process, but rather to highlight certain features of current good practice. The first case study examines the use of EIA for a substantial village expansion scheme in Nottinghamshire. Interesting features include the scope of the EIA, the important role of pre-application consultation, the approach to prediction and assessment of significance, and the treatment of mitigation and monitoring. The second case study concerns one of the many new settlement proposals in the Cambridge area. The way in which the scope of the EIA was determined and the methods used to assess the importance and significance of impacts are of particular interest.

A case study of EIA for an expanded village: Bilsthorpe, Nottinghamshire

Introduction

This first case study is about the proposal to expand the existing village of Bilsthorpe in Nottinghamshire. The scheme was proposed by Nottinghamshire-based environmental consultants David Tyldesley and Associates. An outline planning application was submitted in December 1990, with funding provided by the owner of the site.

The village of Bilsthorpe is located in Newark and Sherwood District, approximately seven miles east of Mansfield. The existing settlement lies immediately to the north and east of the application site, and the A614 trunk road formed the site's western boundary. The site comprised approximately 125 ha of land, almost entirely given over to arable farming (Fig. 9.1). The existing village had a population of about 3,100 at the time of the application, with a housing stock of about 1,100. No significant housing development had taken place since 1970, mainly because of the limited capacity of Bilsthorpe sewage works. Since the early 1970s, the village's population had declined slightly, with an increasing proportion in the older age groups. Community facilities, including the primary school, were concentrated in the northern half of the village. Bilsthorpe colliery was the dominant employer, accounting for 60 per cent of jobs. There were more jobs in Bilsthorpe than economically active residents, although there was substantial commuting both into and out of the village. The future of the colliery was thrown into doubt by British Coal's announcement in 1992 of the proposed closure of up to 31 pits (including Bilsthorpe).

The proposed scheme, application and planning context

The proposal envisaged the construction of just under a thousand houses, almost doubling the size of the existing settlement. A total of 10 ha on the western edge of the site would be allocated for industrial development, potentially creating a total of 500 jobs. The total site area would be 125 ha, and 50 ha of this would be established as three new areas of woodland, alongside the A614, along a ridge line leading into the centre of the site and on a prominent slope in the northern part of the site. The planting and subsequent management of these areas would attempt to recreate the natural habitats associated with Sherwood Forest. A new junction with the A614 was proposed, with the closure of the existing junction. A new village centre would be provided, trebling the amount of retail floorspace and enabling the provision of services that the existing village could not support. Sites were reserved for a new primary school and a community/sports centre. Development was to be phased over a 15-year period, resulting in an annual average building rate of about 65 houses.

An outline planning application for the proposal was lodged with Newark and Sherwood District Council in December 1990. The Nottinghamshire Structure Plan Review Examination in Public had been held earlier in the year. The review had

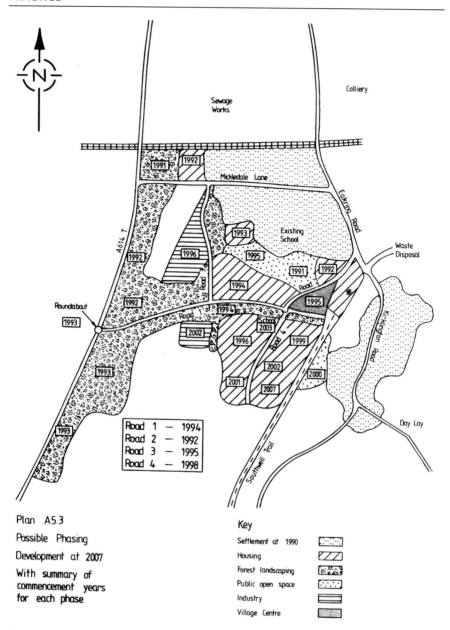

Plan A5.3

Possible Phasing

Development at 2007

With summary of
commencement years
for each phase

Key

Settlement at 1990

Housing

Forest landscaping

Public open space

Industry

Village Centre

Figure 9.1 Proposed expanded village at Bilsthorpe. (*Source:* David Tyldesley & Associates 1990)

identified Bilsthorpe as one of several villages suitable for "limited" residential and employment development, although no definition of what was meant by limited development had been provided. At the time of the planning application, the Bilsthorpe area was not covered by a local plan. However, the district council was in the process of preparing three separate local plans. One of these covered the western part of the district, which includes the Bilsthorpe area. Public consultation on the draft version of this Western Area Local Plan did not take place until *after* the submission of the planning application, during the spring of 1991. The Western Area Local Plan did not proceed to the Local Plan Inquiry stage, but was incorporated into a draft district-wide local plan.

At the end of July 1990, the local planning authority, Newark and Sherwood District Council, requested that an EIA of the proposal should be undertaken. Accordingly, an EIS was prepared by David Tyldesey and Associates (1990) and submitted with the outline planning application in December 1990.

The scope of the EIS

The effects considered in the EIS were determined partly by reference to the "specified information" in Schedule 3 of the EIA regulations. However, certain effects included in the specified information were not considered relevant for this proposal and were therefore not included in the EIS – these included the effects on air quality and climate. Appendix 4 of the DOE publication, *Environmental assessment: a guide to the procedures*, contains a more detailed checklist of matters that may need to be considered in an EIS, although it is recognized that not all of these will be relevant to every project. The EIS identifies which of these matters were regarded as irrelevant or insignificant, and provides a brief justification of this conclusion. For example:

- The EIS does not address the effects of waste disposal, the effects of pollutants on water courses and the effects on hydrology outside the site boundaries. It argues that such effects would be adequately handled by the normal requirements of the National Rivers Authority (NRA) and Severn Trent Water, which would incorporate any necessary mitigation measures.
- The possible production processes and operational features of industrial land-users, and their potential effects, are not addressed. The EIS argues that the use of the proposed industrial sites should be limited to occupiers in Classes B1, B2 and B8, and that this could be controlled by a condition attached to any planning consent.
- The EIS considers that the construction phase would not involve any unusual or unacceptable methods of construction. It assumes that mitigation of the effects of construction would not need to be considered, because of the physical separation of the site from residential areas and the speed and type of construction activities anticipated. Despite this, certain effects (e.g. the effects of construction traffic on existing residents before the completion of the new internal road network) *are* addressed in the EIS.

255

- Discussion of the main alternative sites considered is also not seen as relevant, given that the fundamental objectives of the project relate to the needs of Bilsthorpe.

Some effects not specifically identified in the EIA regulations, but considered to be relevant to the EIS, *are* included. These include employment, community and social effects, and the extent to which the proposals conform with, or contribute to the achievement of, statutory development plan policies and objectives. The effects considered in the EIS are listed below:

- community and social effects, including effects on commuting flows, employment opportunities and community and recreational facilities;
- effects on highways and traffic, and on rights of way;
- effects on existing water courses and on other infrastructure (e.g. sewers);
- employment effects;
- effects on landscape and visual amenity, including the impact on the general landscape setting and character of the area, the impact on views into the site from roads, public rights of way and residential properties, and the impact on views within and out of the site;
- effects on flora and fauna, including the loss, modification, reduction or extension of existing habitats on the site; the creation of new habitats on site; disturbance to or loss of species or animals; indirect or cumulative effects on habitats outside the site boundaries;
- changes in land-use, including agricultural land loss;
- effects on the cultural heritage;
- effects on buildings and other material assets;
- the extent to which the proposals conform with statutory development plan policies and objectives and/or contribute to the achievement of certain policies;
- the interaction between the development and the British Coal proposal to construct a 150 MW coal-fired power station in Bilsthorpe.

Consultations

During the preparation of the EIS, all statutory consultees and a number of other bodies were contacted, to obtain information about the site and its environs and to elicit initial comments on the proposals. The responses received from consultees were included in an appendix to the EIS. This allows the main issues raised at this stage to be easily identified, and shows how these comments were incorporated into later versions of the proposal and into the EIS. Examples of the type of issue arising out of the consultation process are summarized below.

The *Nature Conservancy Council* requested that the EIS should address "not only the impacts of the proposal on existing nature conservation resources and the measures to be taken to *minimize* these impacts, but also the steps to be taken to *create* further areas for nature conservation". In particular, they were keen to see proposals in the EIS for the creation of patches of the characteristic habitat of the Sherwood

Forest, such as heathland, acid grassland and open oak and birch woodland. The establishment of 50 ha of such habitat was a key element of the proposal and was discussed in some detail in the EIS.

The *Department of Transport* indicated that the proposal to create a new road junction on the A614 would probably meet with a "direction of refusal" on highway safety grounds. The Department asked that this view should be included in the EIS. However, the proposal was retained by the developers in the outline planning application. They argued in the EIS that the new junction would replace the existing one, rather than being additional, and that visibility for road users would be improved.

A watercourse flows through the proposed site, and this had existing flooding problems. The *National Rivers Authority* indicated that it would object if the development exacerbated these problems or caused new ones. The EIS mentions the possible need to provide a surface-water storage facility, to prevent such impacts. The need for such a facility would be determined by the outcome of further, more detailed studies. However, such studies were not carried out as part of the EIA process and are not reported in the EIS.

Severn Trent Water Ltd pointed out that the existing public sewers and Bilsthorpe sewage works would be inadequate to cope with the development. They would look to the developer to finance the provision of all off-site works required to service the development. Major off-site mains reinforcement would also be required to facilitate a water supply. The EIS states that a new trunk sewer would be laid and the existing sewage treatment works extended. It suggests that any adverse effects of these measures would be mitigated by conditions attached to any discharge consent granted by the NRA.

Records held at the *Nottingham Natural History Museum* were consulted during the preparation of the EIS. This source of information revealed two possible impacts not previously anticipated during the drawing up of the project proposals. The first of these was the possible effect on the Southwell Trail, a disused railway line passing through the site. The trail was found to be a Biological Grade 2 alert site, of district-wide importance. A section of the trail would be close to residential and amenity elements of the proposal, and was therefore at risk of damage by increased public use. The second potential impact emerged from an examination of records of protected species of birds, animals and plants on the site and in its environs. This revealed that part of the site may be contained within the breeding territory of the barn owl. As a result of these findings, additional mitigation measures were incorporated into the EIS. These included: (a) preventing construction works within 50 m of the Southwell Trail during the breeding season; (b) ensuring that the layout and design of residential areas minimized the potential abuse of the trail; (c) monitoring the effects of the development on the habitats of the trail; (d) support for survey and protection measures for the barn owl.

Local residents were also consulted. In October 1990, before the submission of the planning application, a public exhibition outlining the proposals was held at Bilsthorpe village hall. The developers also publicized the proposals in the local press at this stage. A total of 240 people visited the exhibition, and 130 completed

257

questionnaires about the proposals. A majority (62 per cent) were in favour of the scheme, a quarter were against, the remainder (13 per cent) undecided. The developers claimed that "this level of public support for a large-scale greenfield development is entirely unprecedented, but is based on a common perception of Bilsthorpe's problems". As a result of the comments received, alterations were made to the project proposals. The most notable of these were the provision of additional leisure and community facilities and the allocation of a larger area for the new village centre, to allow further facilities at a later date.

Prediction and assessment of significance

The EIS adopts a novel approach to impact prediction. For each predicted impact, an indication is given of the confidence in and probability of the prediction. This is followed by a qualitative assessment of the significance of the impact. Where appropriate, there is also an indication of the quantitative scale or magnitude of the impact (e.g. the number of dwellings affected, the percentage increase in traffic flows). Although this basic approach was followed in other statements, the Bilsthorpe EIS develops it further by using a series of standard terms relating to the probability of predictions and the qualitative assessment of effects. Each term is identified on a numerical scale of 1–7 (Table 9.5). The use of this scale was intended to provide a

Table 9.5 Bilsthorpe EIS – approaches to confidence/probability of predictions, and qualitative assessment.

Confidence/probability of predictions:
7   Absolute certainty
6   Near certainty/very high probability
5   High probability – to be expected
4   Likelihood/normal anticipation – to be anticipated
3   Seriously anticipated possibility
2   Possibility
1   Remote possibility

Qualitative assessments of effects:
7   Total/consuming/eliminating
6   Profound/considerable/substantial
5   Material/important
4   Discernible/noticeable/significant
3   Marginal/slight/minor
2   Unimportant/inconsequential/indiscernible
1   Irrelevant/no effect
Positive effects are followed by a + sign and negative effects by a – sign

A number of examples illustrate the technique:
(a)   "The habitat would certainly be eliminated (7, 7–)" (i.e. total confidence of total impact).
(b)   "There is a possibility that the habitat could be substantially damaged (2, 6–)" (i.e. relatively low probability, but a relatively high level of impact if it occurred).
(c)   "It would be expected that slight damage may occur to the habitat (5, 3–)" (i.e. relatively high probability, but a relatively low level of impact).

(Adapted from David Tyldesley & Associates 1990)

consistent meaning to each term, relative to other terms, every time it appeared in the text of the EIS.

Clearly, this approach has advantages and disadvantages. The main weakness is the large amount of subjective interpretation potentially involved. The classification of certain impacts as "unimportant", "minor" or "significant" may be affected by value judgements about the relative importance of different types of impact (e.g. economic and social versus landscape or ecological impacts). There may also have been an understandable tendency for those preparing the EIS to view the beneficial effects of the development as of more significance than adverse effects. The approach would seem to be less relevant for impacts that involved detailed technical calculations or models (e.g. the effects on air or water quality of specific emissions or discharges). Nevertheless, the approach has distinct advantages. It represents a useful means of describing the significance and likelihood of potential impacts in a consistent way throughout each section of an EIS, for a wide variety of different impacts.

Predicted impacts regarded as "at least significant" (i.e. 4–7 on the numerical scale) are easily identified in the EIS. They include a total of 12 adverse effects and 29 beneficial effects. The significant adverse effects identified consist mainly of the effects of increased traffic on existing residents and the adverse visual impacts of the development. Other significant effects include the irreversible loss of agricultural land, a potential increase in the flooding problems of an existing watercourse and a likely conflict with two policies in the approved structure plan. Half of the significant beneficial effects identified consist of community, social, employment and recreational benefits; these include the creation of about 600 jobs, a reduced dependency on the colliery, a reduction in the proportion of out-commuters, improved shops and community facilities, and greater recreational opportunities, including the use of open space and rights of way. The other significant benefits of the development mainly consist of the landscape and ecological benefits of the establishment of the new woodland/landscaping areas.

Modifications to the development introduced during the EIA process

The developers state that "environmental objectives have been a fundamental element of the scheme from its commencement. Environmental assessment has been a continuous process throughout the preparation of the project . . . The proposed layout of the development and landscaped areas has therefore been modified in minor ways, many times". The EIS lists the most important changes introduced on environmental grounds during the preparation of a succession of nine different proposals plans. Examples of such changes, along with their main justification, are given below:

- an extension of the landscaped area in the west of the site (undertaken three times), to refine the landscape and visual effects and to enhance the Sherwood Forest regeneration and other LPA objectives;
- a reduction in the amount of residential development proposed, from 49 to 40 ha, to achieve a better balance between the numbers of economically active

residents and job opportunities, and to minimize the residential and visual amenity effects of the new development on the existing village;

- the relocation of one of the two industrial sites (undertaken twice), to improve its relationship with the proposed new road access and landscaped areas;
- the introduction of a wedge of public open space between the historic part of the existing village and the new residential development, to minimize the residential and visual amenity effects of the new development on the existing village;
- the introduction of improved leisure and community facilities in the proposal, and an extension of the area for the new village centre, concern about which was widely expressed by the local community during the public exhibition of the proposals;
- a modification of the western boundary of the development, to avoid potential effects on archaeological resources identified during the preparation of the EIS.

Approach to mitigation measures and to environmental monitoring

Many *mitigation* measures are proposed in the EIS, most of which were adopted during the preparation of the various plans for the proposals, rather than after the undertaking of the EIA. In other words, mitigation measures were incorporated into the design and layout of the project, rather than added on after the EIA. For those mitigation measures that *did* result from the findings of the EIA, the EIS suggests that mitigation should generally be considered "for any negative effects which are at least anticipated and at least significant (i.e. 4–7)". A possible danger with this approach is that mitigation measures may not be considered for impacts with a low level of probability, but a high level of significance.

The EIS suggests that the proposed mitigation measures could be included either in conditions imposed on any planning permission or in an agreement between the LPA and the developer under Section 106 of the Town and Country Planning Act 1990 and Section 33 of the Local Government (Miscellaneous Provisions) Act 1982. The EIS provides a suggested list of conditions to control the development, and a list of matters that could be subject to agreements with the LPA.

The EIS draws attention to the need for continued *monitoring* of baseline environmental information, in order that (a) the proposals could be refined to take account of changing circumstances and (b) the predictions on environmental effects could be confirmed. Specific monitoring proposals are put forward in the EIS, and it is suggested that these could be incorporated in conditions or agreements associated with any planning permission. The responsibility for and funding of each aspect of monitoring are discussed in the EIS. Two examples of the monitoring measures proposed are given below:

- The establishment of the proposed woodland and heathland areas would need careful monitoring. Different techniques for the re-establishment of these habitats are recommended and the monitoring of their outcomes is seen as essential. This should include (a) their landscape/visual effects and (b) changes in and development of habitats. The latter would necessitate the regular surveying of plant and animal communities in all new areas. Relevant organizations would be invited to contribute their expertise to re-establish these habitats and develop an appropriate monitoring methodology.

- The EIS recommends that an annual monitoring report should be produced on behalf of the developers, identifying the amount and type of new development and reporting on relevant planning considerations. The monitoring report would examine the following specific issues: demographic changes, the origin of new residents, the workplaces of residents, house type and tenure, the creation and availability of jobs, the number of economically active residents, transport and communication networks, and the progress made towards the fulfilment of objectives and targets. The report would be made available to the LPAs, the parish council, the village trust (if established), British Coal, the applicant and any other bodies suggested by the local planning authorities.

A case study of EIA for a new settlement: Great Common Farm, Cambridgeshire

Introduction

The second case study concerns the proposal to construct a new settlement west of Cambridge, on land at Great Common Farm and Bourn Airfield. The scheme was originated by the University of Manchester, with backing from Stanhope Properties plc. An outline planning application was submitted in April 1989. The scheme was one of eight new settlement proposals along the A45 corridor, considered at a joint public inquiry held between February and July 1990. All eight schemes were rejected by the Secretary of State in March 1992, although the inquiry inspector had recommended the granting of outline planning permission in the case of the Great Common Farm application (DOE 1992c).

The application site is located about six miles west of Cambridge, immediately to the south of the A45 (Fig. 9.2). The existing small villages of Caldecote and Highfields lie to the east of the site, and open farmland forms the western and southern boundaries of the site. An unclassified road bisects the site, linking the village of Bourn with the A45. The application site, which totals just over 400 ha, consists of two principal landholdings, Great Common Farm (owned by the University of Manchester) and the disused Bourn Airfield.

The proposed scheme, application and planning context

The proposal for the 400 ha site envisaged the construction of 3,000 houses, a town centre, a business park and industrial area, recreational space (including a golf course) and large new areas of woodland. A business park covering 35 ha would be constructed on the disused airfield, providing just over 1 million ft^2 of office and research space. The existing industrial uses on the site would be retained, and an adjoining 10 ha of land would be allocated for new industrial development. About

261

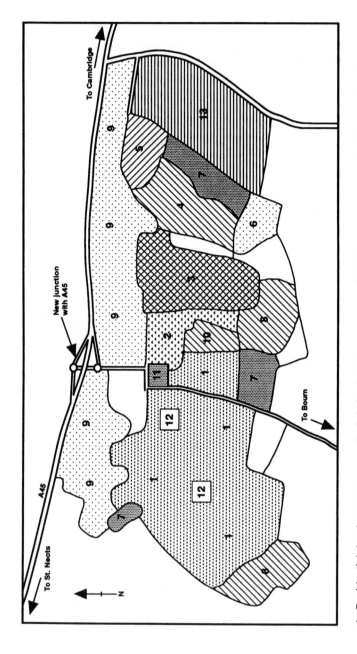

1. Residential development
2. Town centre
3. Business park
4. Industrial area
5. Existing industry
6. Existing woodland
7. New woodland
8. Playing fields
9. Golf course
10. Town park
11. Entry 'green'
12. Primary school
13. Existing residential area

Figure 9.2 Proposed new settlement at Great Common Farm, west of Cambridge. (Adapted from Land Use Consultants 1989)

50 ha of new woodland would be planted on the periphery of the site, mainly between the development and the village of Highfields. The existing road junction with the A45 would be substantially modified, with the dualling of a 1.25 mile stretch of the A45 fronting the site and the construction of a new grade-separated interchange. Access to the new settlement from this interchange would be via a new dual link road. Development would be phased over a 15-year period, resulting in an annual average building rate of about 200 houses.

The impetus for the scheme was provided by the review in the late 1980s of the Cambridgeshire Structure Plan. The review had identified the need for two new settlements in the county and had identified broad areas of search, along the A10 corridor to the north of Cambridge and along the A45 corridor to either the west or the east of Cambridge (Breheny et al. 1993). Policy 20/2 of the approved structure plan outlined requirements that would need to be satisfied by any new settlement proposals along the A45 corridor:

Provision would be made for a new settlement on the A45 corridor west or east of Cambridge which would:

- be close enough to Cambridge to make a significant contribution to its development needs, but located outside the green belt;
- complement the existing settlement pattern and not prejudice the extension planned for Papworth Everard;
- make use as far as possible of land which was under-used or of little environmental value and minimize the loss of high-quality agricultural land;
- minimize infrastructure costs and flood risks;
- provide the opportunity for business park development;
- be capable of accommodating about 3,000 dwellings with some reserve capacity for future expansion, 2,000 of which should be available before 2001;
- provide safe and easy access to the A45 trunk road.

The Secretary of State approved the structure plan review with modifications on 21 March 1989. Two weeks later, an outline planning application was submitted to South Cambridgeshire District Council to create a new settlement at Great Common Farm. Land Use Consultants was commissioned in May 1989 to undertake an EIA of the proposals, and a preliminary EIS was submitted to the district council in July 1989. The final version of the EIS (Land Use Consultants 1989) was submitted in December 1989, eight months after the submission of the outline planning application. Meanwhile, Ove Arup and Partners (consulting engineers) were commissioned to prepare a detailed report on the traffic and highways impacts of the development. This report was also submitted in December 1989, but its findings were not included in the EIS.

The scope of the EIS

All the applications for new settlement proposals along the A45 corridor were called in by the Secretary of State for consideration at a joint public inquiry. Given this context, the scope of the EIS appears to have been largely determined by reference to the issues thought likely to be relevant to the Secretary of State's consideration of the various proposals. These issues were determined by advice from the DOE and DOT, and later by the Secretary of State's statement under Rule 6 of the Town and Country Planning (Inquiries Procedures) Rules 1988, i.e. the call-in letter. The Rule 6 statement identified the following issues as relevant to the consideration of the applications:

1 *The relationship of the developments proposed to development plan policies, namely*:
   (a) the policies in the Approved Replacement Cambridgeshire Structure Plan (see above);
   (b) the green belt local plan, and any subsequent modifications proposed by the County Council;
   (c) the policies proposed in the South Cambridgeshire Draft Local Plan.

2 *The scale and nature of each development, including its expected final size beyond the plan period.*

3 *The appropriateness of each site, in terms of*:
   (a) its physical capacity;
   (b) the landscape;
   (c) nature conservation interests, particularly SSSIs;
   (d) architectural and historic interest, including listed buildings and conservation areas;
   (e) the loss of agricultural land and the re-use of derelict land;
   (f) susceptibility to flooding and the adequacy of any proposed flood-prevention works.

4 *The effects of each development on*:
   (a) the highway network and other transport facilities;
   (b) the character of the existing towns and villages in the area;
   (c) the housing and labour markets of the area, particularly the provision of low-cost homes;
   (d) the amenities of the area, including other services and facilities proposed or affected by the development.

The scope of the EIS was also influenced by a consideration of the "specified information" in Schedule 3 of the EIA regulations, assessment of the environmental issues identified at earlier stages in the project proposals, and an examination of the characteristics of each of the other Cambridge new settlement proposals. The resulting list of impacts considered in the EIS is outlined below:

- effects on microclimate and air pollution;
- effects on geology and soils;

- effects on surface water and groundwater; effects of foul drainage;
- ecological impacts, including effects on existing habitats and the creation of new habitats;
- effects on land-use, including effects on agriculture, existing industry and infra-structure on the site, existing users of Bourn Airfield and recreation;
- landscape and visual impacts;
- effects on human beings, including (a) effects on the existing population (on local properties, the coalescence of settlements, the generation of additional traffic and noise, and socio-economic effects) and (b) the characteristics of the new community (population and community structure, employment creation, physical form, architecture, community facilities and accessibility);
- effects on the cultural heritage;
- effects on material assets.

Methods used to identify, predict and assess impacts
The EIS is arranged so that the treatment of all environmental issues is structured in a similar way. The examination of each topic is divided into five distinct sections, outlined below. This type of EIS structure has since been incorporated into the DOE's good practice guidance on the preparation of environmental statements (DOE 1995). The five sections are as follows:

(a) A factual description of the *existing situation*, based on the findings of surveys conducted on site as well as on desk studies.
(b) A description of the *potential impacts* of the development and the size of the area potentially affected. The relative importance of these impacts is defined as follows:

- Impacts are of national or regional importance if the effects would be sufficiently important to be relevant at the level of national policies or regional strategies. For example, direct impacts on an SSSI or National Nature Reserve would be of national importance, since such sites are part of a national register of protected nature conservation sites. The employ-ment impact of the business park element of the new settlement might be regarded as of regional importance, if it is likely to affect strategies to direct employment development to particular parts of the region.
- Impacts are of county-wide importance if they would affect strategic deci-sions at a county level, i.e. the scale of impact would be such that it could undermine (or support) structure plan policies, such as those outlined in Policy 20/2 above.
- Impacts are of district-wide importance if they would be relevant within the context of a local plan. For example, this might include the long-term impact of increased traffic on local roads or impacts on existing local industry.

265

- Impacts are of local importance if they would be largely contained within the site itself. Examples might include the impact of construction traffic, effects on local properties, or increased noise levels.

The assessment of "relative importance" is based on the professional judgement of the relevant specialists involved in the preparation of the EIS. The inclusion of potential impacts in the EIS does not mean that they are necessarily expected to occur. Rather, they are included to indicate the issues that have been considered during the design process and investigated in the subsequent prediction of impacts.

(c)   A description of the *predicted impacts* of the development, for both the construction and operational stages. In most cases, the predicted impacts cover a smaller range of issues and are of less significance than the potential impacts. This is because of the incorporation of mitigation measures during the design process to eliminate or minimize those features likely to cause significant adverse effects. The predictions are informed views based on the professional judgement of the relevant specialists, rather than statements of fact.

(d)   The scope for *mitigation* or amelioration of the predicted impacts.

(e)   A summary of the *residual predicted impacts* and a judgement about their *significance*. Impacts are classified as having "major", "some", "minor" or "no" significance. The assessment of the level of significance of particular impacts is again based on professional judgement. As with the Bilsthorpe EIS, the temptation to regard beneficial impacts as more significant than adverse impacts is a potential weakness of such an approach.

Although the EIS clearly sets out the method of determining the relative importance of potential impacts, there is obviously scope for subjective interpretation. Indeed, inconsistencies are apparent in the assignment of levels of importance to particular impacts in the EIS. For example, the need for safe and easy access to the A45 trunk road is specifically identified as a requirement of the new settlement in Structure Plan Policy 20/2. For this reason, this issue is regarded as of county-wide importance. However, certain other matters clearly identified in Policy 20/2 are classified as of only district-wide importance. These include the possible coalescence of communities, the loss of agricultural land and the re-use of derelict land.

One reason for these apparent inconsistencies stems from the specific circumstances of the Cambridge new settlement proposals – the explicit support in the approved structure plan for a new settlement along the A45 corridor and the holding of a joint public inquiry to consider all eight applications together. The principal issue at the inquiry was therefore *where* to locate the new settlement, rather than *whether* a new settlement was needed. In these circumstances, impacts that would arise wherever the new settlement was located, or that did not differ significantly from those that would arise at alternative sites, might be seen as less critical in the final weighting of the competing proposals. Such impacts may therefore be regarded as less important than would otherwise be the case. This reasoning certainly appears to have been used in deciding the relative importance of certain impacts in the EIS.

9.4 EIA and Waste-Disposal Projects

The extent of EIA activity

Assessment of the environmental impacts of waste-disposal projects represents a very significant part of EIA activity in the UK. According to Frost & Wenham (1996), such projects accounted for almost 20 per cent of all environmental statements submitted in the UK between July 1988 and September 1995. They identify over 470 EISs for waste-disposal projects, over half (260) involving landfill/landraise, the remainder mainly comprising incineration (80) and waste-water and sewage treatment projects (85) (Frost & Wenham 1996) (see Table 9.6). These figures are much higher than those quoted by Petts & Eduljee (1994a), who note that just over 200 waste-disposal EISs had been submitted up to December 1992. At that time, this represented approximately 15 per cent of all EISs in the UK.

There is some evidence that EIA activity has slackened. For example, the numbers of EISs submitted for waste incineration and waste-water and sewage treatment projects appear to have peaked in the early 1990s, rapidly declining in more recent years (see Frost & Wenham 1996). This peak in activity partly reflects legislative changes, including the raising of emission standards for clinical-waste incinerators in 1991 (see Petts & Eduljee 1994a). Nevertheless, despite the more recent decline in the number of proposals, waste-disposal developments continue to represent a significant proportion of EIA activity in the UK.

Table 9.6 Number of EISs submitted for waste disposal projects in the UK, July 1988–September 1995.

Project type	Schedule no.	No. of EISs	% of total
Landfill/landraise	1.9(2), 2.11c	262	55.4
Waste-water/sewage treatment	2.11d	86	18.2
Incineration	1.9, 2.11c,d	79	16.7
Waste-recycling schemes	2.11c	6	1.3
Waste depots	1.9, 2.11c	20	4.2
Other waste treatment	2.11c	17	3.6
Other waste disposal	1.9, 2.11c	3	0.6
TOTAL: All waste-disposal EISs		473	100.0

(*Source:* Derived from Frost & Wenham 1996)

Determining the need for EIA

For the purposes of EIA, waste-disposal projects in the UK fall within the scope of the Town and Country Planning (Assessment of Environmental Effects) Regulations

267

1988. Such developments can be either Schedule 1 or Schedule 2 projects according to the type of wastes involved. The following are classified as Schedule 1 projects, for which EIA is mandatory:

(a) waste-disposal installations for the incineration or chemical treatment of "special wastes";
(b) landfill operations involving the disposal of "special wastes".

Special wastes are those wastes which are particularly hazardous to human health; they are defined under the Control of Pollution (Special Wastes) Regulations 1980 (Barron 1994).

In addition, the following developments are classified as Schedule 2 projects, for which EIA may be required:

(c) installations for the disposal of "controlled wastes", or waste from mines or quarries, not falling within Schedule 1 above;
(d) sites for depositing sludge.

Controlled wastes are defined under the Environmental Protection Act 1990; they consist of household, commercial, industrial, demolition and construction wastes, as well as sewage sludge when landfilled or incinerated. Agricultural, mining and quarrying wastes – although accounting for the bulk of wastes produced in the UK – are not defined as controlled wastes (Barron 1994).

The Department of the Environment has provided guidance on the circumstances in which EIA is likely to be required for Schedule 2 projects (DOE/WO 1989). This notes that:

... installations, including landfill sites, for the transfer, treatment or disposal of household, industrial and commercial wastes ... with a capacity of more than 75,000 tonnes a year may well be candidates for EA ... Except in the most sensitive locations, sites taking smaller tonnages of these wastes, Civic Amenity sites, and sites seeking only to accept inert wastes (demolition rubble, etc.) are unlikely to be candidates for EA.

It is likely that under the implementation of the amended Directive this guidance will be refined, including a reduction in size criteria (e.g. annual capacity). Petts (1993) notes that, in practice, most waste-disposal EISs – possibly in excess of 80 per cent – appear to have been for Schedule 2 projects, with a high level of voluntary submission by developers. Unlike some other project types (e.g. roads), project-specific guidance on EIA for waste-disposal developments is limited (see Barron 1994, Petts & Eduljee 1994a). This lack of guidance might be expected to be reflected in EIS quality, a topic which is the concern of the next section.

The quality of environmental statements and the EIA process for waste disposal projects

Studies of EIS quality

A number of recent studies have examined the quality of environmental statements for both incineration and landfill projects in the UK (see, for example, Jones 1991, McMahon 1996, Radcliff & Edwards-Jones 1995, Rowan 1993). Most of these studies have employed the widely used review methodology developed by Lee & Colley (1992), thereby allowing comparisons to be made with the quality of EISs for other project types. The consensus from these studies appears to be that waste-disposal EISs, although of highly variable quality, are on the whole slightly worse than EISs for other project types. In practice, given the generally low standard of EISs as a whole, this means that a high proportion of waste-disposal EISs are of poor quality. The quality of landfill EISs appears to be poorer than that for incinerators (see Petts 1993).

One of the most recent reviews of the quality of EISs for waste-disposal projects is that undertaken by McMahon (1996). This involved a review of ten EISs for landfill projects in Northern Ireland, the statements reviewed having been submitted between 1991 and early 1995. Using the review criteria developed by Lee & Colley, the majority of the sample of EISs (60 per cent) were found to be of "satisfactory" quality (grades A, B or C). However, only 20 per cent were classified as "good" (grades A or B), and 20 per cent were classified as "poor" (grades E or F). These findings appear to be broadly in line with those from studies of EISs for a range of project types (see Ch. 8).

Other studies have reached rather more critical conclusions. For example, Jones (1991) compared the quality of landfill EISs in England with the quality of those in the Netherlands, and found that only three out of the 16 English landfill EISs (fewer than 20 per cent) could be considered "satisfactory", compared with all nine of the Dutch EISs reviewed. Only one English EIS was assessed as "good", and no fewer than six (38 per cent) were considered "poor" (Jones 1991). A study by Rowan (1993) reviewed the quality of landfill EISs published in the UK before mid-1992. This revealed that fewer than 40 per cent of statements were "satisfactory", again using the Lee & Colley review criteria.

The main areas of weakness in landfill EISs appear to be in the identification and evaluation of impacts, the treatment of alternatives and mitigation measures, and the involvement of the public and interest groups before the submission of the EIS (Jones 1991, McMahon 1996). For example, 60 per cent of the EISs reviewed by McMahon were assessed as poor in their treatment of alternatives, and none were assessed as good. In half the statements reviewed, earlier – rejected – alternatives were not re-appraised if severe impacts were predicted. As McMahon notes in relation to the consideration of alternative sites: "In most cases the considerations of ownership and economic interests will have dictated the location of a site . . . *long before the planning application or EA procedures are set in motion* (McMahon 1996; emphasis added).

This suggests, if true, that EIA for such projects is not truly iterative, site selection being a "given" rather than arrived at through the EIA process (see also Petts 1993, Petts & Eduljee 1994a). However, subsequent government guidance recommends the discussion of main alternatives as a matter of normal practice, and the implementation of the amended Directive will further strengthen the consideration of alternatives (see Section 4.5).

To date there have been fewer published studies of the quality of EISs for waste-incineration projects, although Petts (1993) notes that, in contrast to the generally poor quality of landfill EISs: "some incinerator EAs have been seen to be leading the development of assessment techniques in relation to certain impacts, [such as] public health risk assessment". The study by Radcliff & Edwards-Jones (1995) examined the quality of EISs for clinical-waste incinerators in the UK. Although these authors did not assign comprehensive gradings to the statements reviewed, the study revealed a number of weaknesses, both in the statements and in the wider EIA process for these developments. These weaknesses can be summarized below:

- The range of impacts assessed showed a wide variation, with three of the 13 EISs reviewed considering fewer than half of the potential impacts associated with waste incineration.
- Certain impacts were rarely addressed. For example, although air quality, noise, traffic and visual impacts were assessed in almost all the EISs reviewed, nuisance and impacts on ecology were less frequently addressed, and soil quality and socio-economic impacts were assessed in only a small minority of cases. This suggests poor scoping of impacts (see also Petts 1994, Petts & Eduljee 1994a).
- Although most impacts were quantified, their significance was very rarely addressed, other than for air quality and – to a lesser extent – for noise impacts. Only one EIS considered the significance of all the predicted impacts identified.
- Baseline data tended to be limited, were not collected for all the impacts considered and, in almost half the EISs reviewed, did not appear to be used to determine the significance of predicted impacts.
- The treatment of alternative sites and technologies was patchy. Only five of the 13 EISs identified alternative sites, and even fewer (two) discussed alternative technologies for the incineration process.
- The mitigation measures listed in the EISs tended to be confined to a limited range of impacts, mainly air pollution and noise.
- The commitment to monitoring was weak; no provision was made for the monitoring of impacts other than those on air quality. Although eight of the 13 EISs discussed proposals for the monitoring of air quality, several of these consisted only of tentative statements of intent rather than firm commitments (see also Petts & Eduljee 1994b).

Similar weaknesses in waste-disposal EISs in general have been identified by other commentators. For example, Petts & Eduljee (1994a), in their wide-ranging review of waste-disposal EIA, identified the following common weaknesses in UK waste-disposal EISs:

- A lack of rigour in the scoping of potential impacts. For example, transport impacts are often poorly scoped, with little consideration of public perceptions of traffic nuisance and risk. Poor scoping also results in a limited consideration of certain impacts in EISs, not always arising from a judgement that such effects will not be significant. Examples include ecological impacts, landscape and visual impacts (especially in cases where proposals are located in existing industrial areas) and the effects of vibration (as distinct from noise).
- The limited treatment of indirect impacts arising from off-site ancillary developments, such as the landfilling of residues from a new waste-incineration facility.
- A tendency to – wrongly – equate compliance with licensing and IPC authorization requirements with a proof of the lack of significant environmental impacts (see also Sections 9.3, 9.4).
- Inadequate information in EISs on the methodologies used and assumptions made. This can result in delays, since LPAs are likely to request further information on such matters.

Concerns about the wider EIA process

The case study by Radcliff & Edwards-Jones (1995) raises a number of concerns about the adequacy of the wider EIA process, including the roles of some of the main participants in the process. Although in most cases the local planning authority had been involved in scoping discussions prior to the submission of the EIS, a number of the planning authorities involved identified areas in which the EIS could have been improved. These included the need for improved predictive methods for air quality impacts and – to a lesser extent – the inclusion of more baseline data and the consideration of alternatives. The EIS authors also felt that improvements could have been made, including the better presentation of information, a more detailed description of the methodologies used and greater consultation with external bodies before the submission of the EIS. A majority of the authors felt that the quality of the EIS had been constrained by the lack of financial resources allocated to the EIA.

A disturbing finding was the apparent divergence between the assessments of quality of those involved in the EIA process and those derived from more formal review criteria, such as those of Lee & Colley. This divergence accounts for the fact that a number of LPAs appeared willing to accept and make use of poor quality EISs:

> For example in [one EIS] in which virtually no baseline data were presented and no quantification of impact magnitude was made, the [EIS] author declared that there were "no major omissions" in the [EIS]. Similarly in [another EIS], the planning authority declared itself to be satisfied with the quality of an environmental statement which contained baseline data for only two potential impacts, noise and nuisance, neither of which was expressed in a quantitative manner. (Radcliff & Edwards-Jones 1995)

271

Other commentators have identified a number of further weaknesses in the EIA process for waste-disposal projects (see, for example, Petts 1993, Petts 1994, Petts & Eduljee 1994a, Therivel et al. 1992, Weston 1994). Many of these are related to the lack of a strategic dimension to EIA. This results in the following deficiencies in project-level EIA:

(a) The poor treatment of the combined and cumulative impacts of a proposal where it is one of several in an area. (Some studies of the cumulative impacts of multiple development proposals have been undertaken in the UK, but not generally by project developers. For example, Her Majesty's Inspectorate of Pollution (HMIP) undertook an assessment of the air pollution implications of a number of incinerator and power station schemes proposed in the early 1990s in the East Thames corridor of Greater London (see also Street 1997)).
(b) A lack of attention to global impacts on air quality and climate.
(c) A limited consideration of the need for schemes, alternative processes or technologies, or alternative sites.

Further problems are created by the separation of the planning and pollution authorization systems for certain waste-disposal projects. This separation means that EIA at the planning stage is often based on incomplete information about the design and layout of the project, with project definitions frequently subject to change after the submission of the EIS (Bird & Therivel 1996, Petts & Eduljee 1994a). We discuss these problems in greater detail in the next section.

The problem of overlaps between planning and pollution controls

Planning and pollution controls

Waste-disposal and treatment projects are unusual in that they require two forms of statutory consent – planning permission under the Town and Country Planning Act 1990, and licences or authorizations under separate pollution control legislation. Waste-incineration projects require integrated pollution control (IPC) authorizations; these are issued by the Environment Agency under Part I of the Environmental Protection Act 1990. Landfills and waste-treatment plant require waste-management licences, issued by the Environment Agency under Part II of the same legislation. This dual consent procedure – involving both planning and pollution control – can create overlap and conflict, and has important implications for the EIA of waste-disposal and treatment projects. In order to understand these implications further, it is necessary to consider the nature of pollution control in the UK and its interaction with the planning system (see also Brock 1993a, b, DOE 1992b, 1994, Petts & Eduljee 1994a, Sheate 1994, UKELA/IEA 1993).

Waste-incineration projects require both planning permission and integrated pollution control (IPC) authorization. The concept of IPC was introduced in Part I of the Environmental Protection Act 1990. This legislation created a single system of pollution control for major industrial processes, covering all emissions – whether to air, water or land. IPC covers so-called "prescribed processes", including waste incineration but not landfill or waste treatment. IPC requires all operators of pre-scribed processes to apply for authorization to operate the processes. Processes subject to IPC are divided into Part A and Part B processes. For Part A processes, which include waste incineration, the authorization is issued by the Environment Agency (by HMIP before April 1996) and deals with releases of substances to air, water and land. For a Part B process, the authorization is issued by the relevant local authority and deals only with releases to the air (Brock 1993a). An application for IPC authorization is normally submitted after the separate application for plan-ning permission, and in most cases it will not be considered until after the deter-mination of the planning application.

Conditions are imposed on all IPC authorizations, specifying the limits for re-leases, the monitoring requirements and the operational controls (DOE 1994). All authorizations are subject to the general condition that the "best available tech-niques not entailing excessive cost" (BATNEEC) will be used to prevent, minimize or render harmless releases of prescribed substances into the environment. BATNEEC focuses on the use of appropriate techniques to control discharges to the environ-ment. The definition of techniques in BATNEEC includes process and plant design, including site layout, hardware and management systems. In addition, for Part A processes, regard must be had to the "best practicable environmental option" (BPEO) in determining the most appropriate environmental route for the release or disposal of wastes. An examination of the BPEO involves an analysis of alternative options to determine the one which results in the least damage to the environment as a whole, balanced against costs (Petts 1994).

Landfill and waste-treatment schemes are not subject to IPC. Instead, pollution from such schemes is controlled by a system of waste-management licences, under Part II of the Environmental Protection Act 1990. As with the system of IPC, condi-tions are attached to all such licences, including the use of BATNEEC to control the release of substances into the environment. Planning permission must be granted for a scheme before a waste-management licence is issued.

Overlaps and conflicts: implications for EIA
The dual consent procedure for waste-disposal and treatment projects has a number of implications for EIA, of which the following are of particular importance:

(a) the modification of project design and layout during the pollution control authorization process, from that earlier agreed or considered at the planning stage;

(b) the duplication of information requirements, with project promoters asked to submit much of the same information twice to serve the requirements of both planning and pollution control;

(c) the differences between planning and pollution control authorities in the defini-tion of their respective roles and in the interpretation of BATNEEC.

Project modifications: the definition of BATNEEC used in both IPC and the waste-licensing system embraces, among other considerations, "the design, construction, layout and maintenance of the buildings in which [the process] is carried on" (Environmental Protection Act 1990). This means that the pollution control authorization process, which involves the application of BATNEEC to control releases of substances, can often result in modifications to the design and layout of a project. Since design and layout are also matters subject to planning control, conflicts between pollution control and the planning system can and do arise (see Brock 1993a, UKELA/IEA 1993). Conflicts will occur in cases in which the pollution control authorization involves modifications to the design or layout earlier agreed in the planning permission.

Examples of the matters over which conflicts can occur include site access arrangements, the layout of buildings and chimney stack heights. For example, the height of the chimney stack for a new waste incinerator will be examined by the LPA at the planning stage largely as a landscape or visual matter. However, at the later IPC authorization stage, the Environment Agency may insist on a taller stack than envisaged at the planning stage, to ensure the better dispersion of emissions (Brock 1993a, b, UKELA/IEA 1993). Such an increase in stack height may have important implications for the visual impact and acceptability of the incinerator proposal. Such impacts may not have been anticipated in the EIS for the scheme, unless some form of worst-case analysis were included in the EIS; in addition, and more seriously, the LPA may have been unable to take into account the more severe visual impact when deciding the planning application.

Difficulties of this kind are important for EIA, because of the uncertainty they create about the eventual design and layout of a project. Such uncertainty makes it difficult for developers to satisfactorily address certain impacts in their environmental statements, since a project design may be subject to change during the later pollution control authorization process. Local planning authorities may also have doubts about the reliability of EIS predictions based upon a provisional design that may be subject to later modification. Indeed, a recent study into these matters concluded that many EIS predictions appear to be invalidated by project design changes arising from the pollution control authorization process (see Bird & Therivel 1996). This is an important finding, throwing doubt upon the usefulness of EIS predictions and, by extension, upon the integrity of the EIA process as a whole.

The duplication of information: much of the information provided in the EIS at the planning stage is also relevant to the application for pollution control authorization. This has led to concerns among project promoters about the possible duplication of information and to uncertainty about the different requirements of the two processes (UKELA/IEA 1993). For example, for projects covered by IPC, certain information must be supplied by operators with their applications for IPC authorization. This includes details of the techniques to be used to prevent releases to air, water or land, as well as information on any proposed releases of prescribed substances. An assessment of the environmental consequences of such releases must also be carried out (Brock 1993b). Of course, the EIS submitted at the earlier planning stage will also have examined, among other matters, the potential impacts of the project on air, water and land, and outlined proposed mitigation measures. There is, therefore, a potential for overlap and duplication in the information required at the planning and pollution authorization stages, with developers being asked to submit much of the same information twice. This has led to concerns among developers that consent procedures can become "unduly slow and cumbersome" (UKELA/IEA 1993).

The differences of interpretation between LPAS and pollution control authorities: the use of BATNEEC as the means of controlling pollution in the IPC and waste-licensing systems involves an inherent element of uncertainty, not only about the eventual design and layout of a project, but also about the extent to which pollution itself will be controlled. As Weston (1994) has argued, to gain authorization under the Environmental Protection Act 1990 does not mean that a plant will not cause pollution. Rather, it means that the pollution released will be at levels which have been reduced by the most cost-effective and technically efficient means available, i.e. by the application of BATNEEC:

> Compliance with BATNEEC means that the operator cannot be prosecuted or be guilty of a statutory nuisance. It does [not] mean that there is no risk to health and the environment, or that the operation will not cause a loss of amenity to local residents. In other words, [the local planning authority is faced with] . . . residual uncertainty over impacts when the [planning] application or appeal is to be determined. (Weston 1994).

A number of LPAS appear to be doubtful about the ability of the pollution control authorization process to adequately protect their interests in waste-disposal and treatment projects (Mylrea 1994, Petts & Eduljee 1994a, Sheate 1994). Essentially, these doubts concern the pollution control authorities' interpretation of BATNEEC, as well as concerns about the effectiveness of monitoring and enforcement activities. Since BATNEEC involves both a techniques element (BAT) and a cost element (NEEC), there is always a possibility that the pollution control authority may place greater emphasis than the planning authority would wish on the cost element:

> NEEC could mean that [the Environment Agency] might not impose the maximum levels of emission control on the basis that, in their view, the cost of so doing would be excessive. The result could be a disagreement between what [the Environment Agency] would consider acceptable and what the planning authority would consider acceptable in terms of emission levels. (Mylrea 1994)

This uncertainty about pollution-related impacts is one of the main reasons for the attempts by LPAS to impose planning conditions and legal agreements covering such matters. This is despite strong government and judicial guidance to the contrary (see DOE 1994, Kitson & Harris 1994, Mylrea 1994, Petts & Eduljee 1994a). Government guidance, as outlined in *PPG23: planning and pollution control* (DOE 1994), makes it clear that, although LPAS should take account of the impact of potential emissions when considering planning applications, the control of these emissions is the responsibility of the relevant pollution control authority. The government has stressed that this does *not* mean that applicants can ignore the pollution implications of their proposals at the planning stage and in their EISs.

The uncertainty created by the use of BATNEEC appears to be rarely acknowledged in environmental statements. Indeed, compliance with BATNEEC is often cited in EISs as proof that the proposals will not cause a nuisance and will therefore be environmentally acceptable:

> A review of [waste disposal EISs] reveals a tendency to place much faith in the authorisation and licensing systems, to such an extent that evidence of no significant environmental impact is often inferred because the facility will comply with all the license or authorisation requirements, rather than presenting a full identification, prediction and assessment of potentially significant effects. (Petts & Eduljee 1994b)

Possible solutions

Various solutions to the problems of overlap and conflict between planning and pollution controls have been suggested (see Brock 1993a, Petts & Eduljee 1994a, Sheate 1994, UKELA/IEA 1993). Most commentators have focused on the conflict between the IPC system and the planning system. Suggested measures have included the following:

- Improved arrangements for formal consultation between LPAs and the Environment Agency, which are currently seen as weak.
- An optional procedure in which the applications for planning permission and IPC authorization would be considered together, with a single EIS meeting the requirements of both. Such a procedure would allow for solutions on matters such as chimney stack heights to be agreed between the planning authority and the Environment Agency, and to be common to the conditions on the planning permission and IPC authorization (UKELA/IEA 1993).
- Arrangements which would allow a developer to submit a single EIS, serving the requirements of both planning and pollution control, even if the two applications were not considered simultaneously.

Brock (1993a, b) urges caution on the last of these recommendations, arguing that the two forms of environmental assessment required for planning and IPC, although involving overlapping elements, are in fact quite distinct, with a number of important differences. First, the assessment required for IPC is concerned with a much narrower range of impacts than the EIA for planning purposes, focusing only on emissions to air, water and land. Other impacts, such as those on traffic, noise or landscape, are not within the remit of IPC. Secondly, the assessment for IPC tends to be more detailed, dealing with emissions and mitigation measures in much greater technical detail than the EIA for planning purposes. The IPC assessment must also strike a balance between the mitigation of effects and cost, as required by BATNEEC. By contrast, EIA at the planning stage does not need to specify the precise techniques which will be used to mitigate environmental effects, nor does it need to consider the cost of proposed mitigation measures (Brock 1993b, UKELA/IEA 1993).

> Bearing in mind ... the differences in the breadth and subject matter of the environmental assessment for planning, and the examination for IPC, it is unlikely that one document can serve both purposes. They are different processes and need to be recognised as such. (Brock 1993a)

Recent planning policy guidance from the UK Department of the Environment, in *PPG23: planning and pollution control* (DOE 1994), urges LPAs to discuss potentially polluting developments with the Environment Agency at an early stage, in order to reduce the possibility that conflicting requirements might be imposed on developers. The department also recommends, as suggested above, that applications for planning permission and IPC authorization should be submitted in parallel wherever possible: "This will help minimise delays and enable conditions that are likely to be imposed under pollution controls, such as minimum chimney heights, to be taken into account in the planning decision" (DOE 1994).

The DOE stops short of recommending the submission of a single EIS serving both planning and pollution control purposes, but notes that much of the information included in the planning EIS is likely to be similar to that provided with the application for pollution control authorization.

9.5 Summary

This chapter has examined the application of the Town and Country Planning (Assessment of Environmental Effects) Regulations to proposals for new settlements and waste-treatment and disposal facilities. For new settlements, there is ambiguity over the need for EIA, with little guidance provided by the DOE's indicative criteria and thresholds. In practice, a significant minority of schemes have escaped the need for formal EIA. Predictably, the size of a new settlement is an important influence on the need for EIA, but other factors are also at work.

Most new settlement proposals are eventually considered at a public inquiry, raising the issue of the extent to which EIA and the consideration of the EIS are incorporated into such inquiries. There is a need for further research in this important area (see, for example, Blackmore et al. 1997; Jones & Wood 1995; Weston 1997). The promotion of new settlement schemes through the local plan process, as recommended in government planning policy guidance, raises questions about the appropriate timing of the submission of the EIS. It could be argued that, for such projects, formal EIA should be reported on at an earlier stage than the submission of the planning application. Indeed, some developers appear to be adopting this approach voluntarily, although there are concerns about the extent of publicity and consultation in such cases. An examination of two case studies of specific new settlement proposals reveals some features of good EIA practice. Of particular interest are the scoping process, the role of consultation, the treatment of mitigation and monitoring, the approach to prediction and the assessment of impacts.

The examination of waste-disposal projects reveals great variability in the quality of EISs, possibly reflecting the lack of project-specific EIA guidance for such developments. A particular problem for the EIA of waste-disposal projects is the potential for overlap and conflict between planning and pollution controls. This can result in modifications to projects after the EIS stage, the duplication of information requirements and uncertainty about pollution-related impacts in EISs.

References

Barron, J. 1994. Waste management. In *Environmental assessment: a guide to the identification and mitigation of environmental issues in construction schemes*, Construction Industry Research and Information Association (ed.), CIRIA Special Publication 96. London: CIRIA.

Bird, A. & R. Therivel 1996. Post-auditing of environmental impact statements using data held in public registers of environmental information. *Project Appraisal* 11(2), 105–16.

Blackmore, R., C. M. Wood, C. E. Jones 1997. The effect of environmental assessment on UK infrastructure project planning decisions. *Planning Practice and Research* 12(3), 223–38.

Breheny, M. J., T. Gent, D. Lock 1993. *Alternative development patterns – new settlements*. London: HMSO.

Brock, D. 1993a. IPC authorisations vs. planning permission. *Environmental Policy and Practice* 2(4), 347–54.

Brock, D. 1993b. *Environmental impact assessment and integrated pollution control: overlaps in the planning and IPC systems*. Paper presented to Institute of Environmental Assessment Conference on Effective Environmental Assessment, October, Birmingham.

Bulleid, P. 1997. Assessing the need for EIA. In *Planning and environmental impact assessment in practice*, J. Weston (ed.), 26–41. Harlow: Longman.

David Tyldesley & Associates 1990. *Bilsthorpe 2000 +: environmental impact statement*. Hucknall, Nottinghamshire: DTA.

David Tyldesley & Associates 1991. *Bilsthorpe 2000 +: a sustainable future*. Hucknall, Nottinghamshire: DTA.

DOE 1991. *Monitoring environmental assessment and planning*. London: HMSO.

DOE (Department of the Environment) 1992a. *Planning policy guidance note 3 (revised): housing*. London: HMSO.

DOE 1992b. *Planning, pollution and waste management*. London: HMSO.

DOE 1992c. *Secretary of State's decision letter and Inspector's report: Cambridgeshire new settlements, A45 corridor*. London: DOE.

DOE 1994. *Planning policy guidance note 23 – planning and pollution control*. London: HMSO.

DOE 1995. *Preparation of environmental statements for planning projects that require environmental assessment: a good practice guide*. London: HMSO.

DOE 1996. *Revision of planning policy guidance note 23 – "planning and pollution control" on waste issues*. Consultation draft document. London: HMSO.

DOE/Welsh Office 1989. *Environmental assessment: a guide to the procedures*. London: HMSO.

Frost, R. & D. Wenham 1996. *Directory of environmental impact statements July 1988–September 1995*. Oxford: Oxford Brookes University, School of Planning.

Harris, R. 1992. Integrated pollution control in practice. *Journal of Planning & Environment Law*, 611–23.

Jones, C. E. & C. M. Wood 1995. The impact of environmental assessment on public inquiry decisions. *Journal of Planning and Environment Law*, 890–904.

Jones, J. M. 1991. *Environmental impact assessment for the disposal of waste on land: the experience of England and the Netherlands*. (MSc dissertation, Faculty of Biological Sciences, University of Wales Aberystwyth, Aberystwyth).

Kitson, T. & R. Harris 1994. A burning issue? Planning controls, pollution controls and waste incineration. *Journal of Planning and Environment Law*, 3–7.

Lambert, A. J. & C. M. Wood 1990. UK implementation of the European Directive on EIA – spirit or letter? *Town Planning Review* **61**(3), 247–62.

Land Use Consultants 1989. *Great Common Farm, proposed new settlement and business park: environmental statement*. London: LUC.

Lee, N. & D. Brown 1992. Quality control in environmental assessment. *Project Appraisal* **7**(1), 41–5.

Lee, N. & R. Colley 1992. *Reviewing the quality of environmental statements*, 2nd edn. Occasional Paper 24. Department of Planning and Landscape, University of Manchester.

McMahon, N. 1996. Quality of environmental statements submitted in Northern Ireland in relation to the disposal of waste on land. *Project Appraisal* **11**(2), 85–94.

Mylrea, K. P. 1994. Drawing the dividing line between planning control and pollution control: Gateshead MBC v Secretary of State for the Environment and Northumbria Water Group plc. *Journal of Environmental Law* **6**(1), 93–106.

Petts, J. 1993. *Environmental assessments for the waste industry*. Paper presented to IBC/IEA Conference on Advances in Environmental Impact Assessment, November, London.

Petts, J. 1994. Incineration as a waste management option. In *Waste incineration and the environment*, R. E. Hester & R. M. Harrison (eds). Cambridge: Royal Society of Chemistry.

Petts, J. & G. Eduljee 1994a. *Environmental impact assessment for waste treatment and disposal facilities*. Chichester: Wiley.

Petts, J. & G. Eduljee 1994b. Integration of monitoring, auditing and environmental assessment: waste facility issues. *Project Appraisal* **9**(4), 231–41.

Radcliff, A. & G. Edwards-Jones 1995. The quality of the environmental assessment process: a case study on clinical waste incineration in the UK. *Project Appraisal* **10**(1), 31–8.

Rowan, S. P. 1993. *The current standards and practice of the environmental assessment of landfill projects in the UK* (MSc thesis, University of Manchester).

Royal Commission on Environmental Pollution 1993. *Seventeenth report: incineration of waste*. London: HMSO.

Sheate, W. R. 1994. *Making an impact: a guide to EIA law and policy*. London: Cameron May.

Slater, D. 1992. *Integrated pollution control – the role of environmental assessments and environmental management systems*. Paper presented to First Membership Conference of the Institute of Environmental Assessment, July, Birmingham.

279

Street, E. 1997. EIA and pollution control. In *Planning and environmental impact assessment in practice*, J. Weston (ed), 165–79. Harlow: Longman.

Therivel, R. 1991. *Directory of environmental statements 1989–1990*. Oxford: Oxford Polytechnic, Impacts Assessment Unit, School of Planning.

Therivel, R., E. Wilson, S. Thompson, D. Heaney, D. Pritchard 1992. *Strategic environmental assessment*. London: Earthscan/RSPB.

Tyldesley, D. 1991. *Best practice for urban infrastructure environmental assessments*. Paper presented to IBC/IEA Conference on Advances in Environmental Assessment, October, London.

UKELA/IEA 1993. *Overlaps in the requirement for environmental assessment*. A report of the United Kingdom Environmental Law Association and Institute of Environmental Assessment Working Party on Environmental Assessment and Integrated Pollution Control. Lincoln: IEA.

Ward, S. (ed.) 1992. *The garden city: past, present and future*. London: E & FN Spon.

Weatherhead, P. 1994. Burning issues in a policy vacuum. *Planning Week* 2(14), 12–13.

Weston, J. 1994. Assessments at the appeal cutting edge. *Planning* 1075, 24–5, 28.

Weston, J. (ed.) 1997. *Planning and environmental impact assessment in practice*. Harlow: Longman.

Weston, J. & M. Hudson 1995. Planning and risk assessment. *Environmental Policy and Practice* 4(4), 189–92.

Wood, C. M. & C. E. Jones 1991. *Monitoring environmental assessment and planning*. London: HMSO.

Note

1   The direction, dated 26 June 1989, is listed in the *Journal of Planning and Environment Law*, 1989, p. 856.

CHAPTER 10

Environmental impact assessment of projects not subject to planning control

10.1 Introduction

This chapter examines the application of EIA to projects not covered by town and country planning legislation. Two case-study sectors are used to illustrate some of the principal issues involved in the EIA of such projects. Sections 10.2 and 10.3 consider the role of EIA in the assessment of major road schemes. The planning of new trunk road and motorway schemes in the UK has been the sole responsibility of the Department of Transport (DOT) and latterly of the Highways Agency, at least until the recent advent of privately financed toll-road proposals. This has aided the development of a consistent approach to the assessment of such proposals, including the consideration of alternatives. A critique of existing practice in road scheme EIA is outlined in Section 10.2, and we examine a case study of a recent motorway proposal in Section 10.3, in an attempt to uncover some of the strengths and weaknesses of the current approach to EIA. Section 10.4 then discusses the application of EIA to projects in the electricity supply industry. As with trunk roads, this sector had an established record in the evaluation of environmental impacts long before the implementation of EC Directive 85/337. However, the privatization of the industry means that many developers have become involved in the planning and EIA process in the sector, for a wide range of projects. This raises questions about the consistency of approach towards EIA within the industry. Other developments, such as the EC Directive on Large Combustion Plants and the introduction of integrated pollution control, also raise important issues of relevance to EIA.

10.2 EIA of trunk road and motorway proposals

The planning of trunk road and motorway proposals

The planning and decision-making process for new trunk road and motorway schemes in England and Wales involves well-established and often lengthy procedures. Some

281

knowledge of the process is necessary if we are to understand how and at what point environmental impacts are assessed and taken into account in the planning of such schemes. The planning process for new trunk roads and motorways can be divided into six main stages, each of which consists of a number of elements (Sheate & Sullivan 1993):

1   the early stages, including entry to the Roads Programme;
2   a public consultation;
3   an announcement of the preferred route;
4   a statutory publication, and submission of the environmental statement;
5   the public inquiry;
6   events after the inquiry.

Further details on each of these stages are provided below (see also DOT 1992a, Hopkinson et al. 1990, NAO 1994, Sheate & Sullivan 1993, Tromans 1991). The entire planning process, from the identification of the need for a scheme to the eventual start of construction, can take many years, periods of 10–15 years being not untypical. This initial discussion describes the situation prevailing before the publication of new guidance on road scheme EIA in 1993. More recent changes in the procedures are outlined later in this section.

Early stages
The first step in the emergence of a new trunk road or motorway scheme is the *formal recognition by the Department of Transport (DOT) of a need or problem in a particular area*, often prompted by lobbying from local authorities, members of parliament, the business community or local residents. The need for a new road scheme may reflect traffic congestion or a high accident rate on existing routes, or may spring from environmental considerations. For example, village bypass schemes are often motivated by a desire to reduce the effects of heavy through-traffic on residents living alongside the existing route.

Once a problem has been formally identified, so-called "route identification studies" and "scheme identification studies" are undertaken by the DOT. A *route identification study* is carried out in cases where the problem area involves a major route or transport corridor (e.g. the Trans-Pennine corridor). Such studies may consider alternative modes, including public transport solutions. Studies typically involve only a broad-brush consideration of environmental matters, focusing on existing problems and the identification of sensitive areas rather than on estimates of the impacts of new roads (DOT 1992a). Any detailed assessment of impacts is difficult at this early stage, since the precise line of the new or upgraded routes may not be known. Once a route has been identified, it is divided into smaller schemes for further study.

Scheme identification studies are carried out for particular parts of a route or for stand-alone proposals that do not form part of a larger route. Such studies, usually undertaken by consulting engineers, are essentially feasibility studies to investigate whether the problems identified can be solved and whether the solutions are likely to be economically and environmentally acceptable (DOT 1992a). They examine the options available for solving the problems identified, including do-nothing and do-minimum options. The latter typically involve on-line improvements to the existing route. The assessment of environmental impacts is again limited at this stage, although sensitive areas are likely to be identified, and potential effects noted. Initial findings will also be presented on the nature of the existing environment, including landscape quality. The likely visual impacts of the alternatives under consideration may also be considered (DOT 1992a).

After consultation within the DOT, a decision is made by the Secretary of State for Transport whether to place the scheme within the government's National Trunk Roads Programme. The announcement that a scheme has been admitted into the Programme includes a statement of the significant environmental effects identified at that stage.

Public consultation

Following a further assessment of the alternative scheme options, *public consultation* takes place on the options under consideration. The public consultation stage includes a local exhibition presenting the different scheme options. These will typically include alternative lines for the same route as well as do-nothing or do-minimum options. The need for the scheme or alternative modes are rarely presented (Sheate & Sullivan 1993). The first detailed environmental assessment of a scheme is undertaken during the period leading up to its public consultation. Consultations are also held at this stage with the relevant local authorities and statutory environmental organizations. The environmental effects of the alternatives are summarized and presented in a formal tabular framework, designed to allow the effects of each to be compared. A period is allowed for comments on the various options presented.

Announcement of the preferred route

After public consultation, *the Secretary of State makes a decision on the preferred route* for the scheme. This route is then announced, its design is developed, and a detailed environmental appraisal of the preferred route design takes place. The period between public consultation and the announcement of the preferred route can be a lengthy one, especially for controversial or problematic schemes.

> [Publication of the preferred route] is an important watershed in the progress of the scheme: from then on the design and appraisal process will concentrate on the preferred route ... Departmental commitment to the preferred route will [also] increase commensurately, although discussions may continue to be held with objectors to try to resolve objections informally. (Tromans 1991)

Statutory publication

The next main stage in the process is *the publication of draft statutory orders* for the line of the new road (line orders), as well as for any necessary modifications to side roads (side road orders). Orders for the compulsory purchase of land are usually published at a later stage. Most new trunk roads or motorways are divided into a number of sections for planning, consent and contractual purposes. This means that a set of draft orders is published for each of these sections (Sheate & Sullivan 1993). The line of the route indicated in the draft orders may differ from the preferred route previously announced by the Secretary of State. The publication of draft orders is an important stage in the process, since it is at this point that *the environmental statement required by the EC Directive is published*. A period is allowed for objections to the draft orders, and negotiations between the DOT and objectors may then take place, in an attempt to resolve objections.

The public inquiry

Unless all objections can be resolved by negotiation, *the draft orders are considered at a public inquiry*. The environmental statement and any comments on it by statutory consultees will be considered at the inquiry, along with a wide range of other evidence. For further details on the inquiry stage of the process, see Sheate & Sullivan (1993) and Tromans (1991).

After the inquiry

The inquiry inspector prepares a report outlining recommendations, including modifications to the draft orders (e.g. changes in the line of the route, improved mitigation measures). The report is submitted for joint consideration by the Secretaries of State for Transport and the Environment.

The Secretaries of State consider the inspector's report and the objections raised at the inquiry. They must also consider the environmental statement for the scheme and any opinions expressed by statutory environmental bodies or members of the public. They then publish the inspector's report, and at the same time announce their decision on the scheme, either confirming or rejecting the orders, with or without modifications. If the Secretaries of State do not accept all the inspector's recommendations, they must notify those involved in the inquiry and allow them to make written representations. In some cases, the inquiry may be reopened (OECD 1994).

The necessary land for the scheme is then acquired, often involving compulsory purchase orders, which are considered at a separate inquiry. Contracts are then prepared and let, incorporating mitigation measures. Finally, the construction of the road takes place, and the road is opened to traffic.

The environmental appraisal of trunk road and motorway schemes prior to the implementation of the EC directive

The formal appraisal of proposed road schemes in the UK dates back to the early 1960s. Early methods focused on the traffic and economic implications of schemes (Bruton 1985, Simpson 1992). The consideration of environmental matters in road appraisal was not formalized until the late 1970s. Currently, road appraisal by the

DOT comprises three main elements: (a) a traffic appraisal; (b) an economic appraisal (using cost–benefit analysis); (c) an environmental appraisal. Each of these elements is described below (see also DOT 1992a, Macpherson 1993, Sheate & Sullivan 1993, Simpson 1992, Tromans 1991). The description of environmental appraisal methods relates to the position before the implementation of EC Directive 85/337 in July 1988. The implementation of the Directive is discussed later in the section.

Traffic appraisal

The traffic appraisal for a scheme involves an analysis of current traffic flows on the existing road network, and forecasts of future traffic growth on the network in the absence of the proposed scheme. Forecasts are then made of likely traffic flows along the proposed new network, by re-assigning those movements likely to be transferred from the old to the new networks. "High growth" and "low growth" forecasts are prepared, using the DOT's National Road Traffic Forecasts (DOT 1992a). Before 1994, no allowance was made in these forecasts for the possibility that a new road scheme might generate additional traffic movements over and above those that would otherwise have occurred (see DOT 1994a, Sheate & Sullivan 1993).

Economic (cost–benefit) appraisal

An economic or cost-benefit appraisal is also carried out for all proposed schemes. This involves calculating the economic costs and benefits likely to result from the proposed scheme. These are then compared with the costs and benefits associated with the existing network (the "do-nothing" option), or of the existing network modified by those changes that are likely to occur regardless of whether or not the proposed scheme goes ahead (the "do-minimum" option) (DOT 1992a). The DOT uses a computer program known as COBA to calculate the costs and benefits of these various options. The COBA program was first introduced in 1973, although it has undergone periodic updating since then (see Simpson 1992). The COBA calculations take account of the following rather narrow range of costs and benefits:

- the value of travelling time;
- the value of accidents;
- vehicle operating costs;
- the construction and preparation costs of the scheme, including land and other compensation costs;
- the cost of maintenance.

In other words, COBA is essentially concerned with the benefits to road users (re-duced journey times, fewer accidents, lower vehicle operating costs), set against the construction and maintenance costs of the scheme (Tromans 1991). Costs and ben-efits to the wider community or to the environment are not generally expressed in monetary terms and are not considered in COBA. Costs and benefits are calculated for the construction period and a 30-year life, and then discounted to determine their present value. The difference between the "net user cost savings" of a scheme (compared with the do-nothing or do-minimum options) and its net capital costs is

285

known as the scheme's "net present value" (NPV). If the NPV is positive, then the scheme is justified in economic terms. It is normal for the NPVs for all feasible scheme options to be compared; the option with the highest NPV represents the best "value for money". However, this option does not receive automatic support, since account will also be taken of the environmental impacts of each of the options under consideration (DOT 1992a, Tromans 1991). Nevertheless, it has been argued that the weight given to the COBA result in the early stages of the planning and selection of schemes is considerable (see Hopkinson et al. 1990, Sheate & Sullivan 1993).

> The real problem with COBA . . . is that it dominates . . . the choice of options which are put to the public. Only those which the Department is prepared to build are offered at public consultation. Other solutions which might be preferable on environmental grounds, but are "uneconomic" in COBA terms, are either not revealed or shown as "discarded". . . Major mitigation, such as a tunnel rather than a cutting, is likely to increase construction costs so substantially as to make COBA negative and therefore a non-viable option from the DOT's point of view. This makes it very difficult to present alternative schemes that will be taken seriously by the DOT (Sheate & Sullivan 1993).

Environmental appraisal

Prior to the publication of new guidance in 1993, the appraisal of the environmental impacts of road schemes was carried out using the so-called "framework method" developed during the late 1970s. After initial investigations, the DOT had concluded in the mid-1970s that the inclusion of environmental and social factors in the cost–benefit appraisal of schemes was impracticable – because it was felt that such impacts could not satisfactorily be valued in monetary terms. However, it was accepted that more formal arrangements should be devised for the consideration of such non-economic impacts in the appraisal of schemes (DOT 1992a). Consequently, in 1976 the DOT established an independent Advisory Committee on Trunk Road Assessment (ACTRA) to investigate these matters and to make recommendations. ACTRA's report, and the later report of the new Standing Advisory Committee on Trunk Road Assessment (SACTRA), established in 1978, developed a possible approach to the appraisal of environmental and other non-economic impacts (DOT 1978, 1979). ACTRA and SACTRA recommended that techniques should be developed to describe and evaluate the importance of the following types of impact:

- accidents;
- effects on pedestrians;
- the loss of buildings;
- noise;
- visual intrusion;
- air pollution;
- disruption during construction;

- effects on employment opportunities;
- agricultural land-take and severance;
- community severance;
- effects on the intrinsic value of important environmental assets and landscape.

It was recommended that all these effects, plus the economic costs and benefits of a scheme, should be presented in a comprehensive framework. This was essentially a large table or matrix summarizing all the impacts of the proposed scheme, whether economic or environmental and whether expressed in monetary terms or not. It was argued that the framework should be structured in such a way that the effects of the scheme on different incidence or "user groups" could be readily identified. Five groups were identified by ACTRA and SACTRA:

- road users directly affected by the proposed scheme;
- non-road users directly affected by the scheme;
- those concerned with the intrinsic value of the area affected by the scheme;
- those indirectly affected by the scheme;
- the financing authority.

Frameworks were to be prepared at various stages in the planning of schemes, the most important being those prepared for the public consultation stage and the public inquiry (DOT 1979). The framework would contain separate columns for the various scheme options, as well as for the do-nothing and do-minimum options, allowing all alternatives to be compared on a common basis. This would make decisions between options easier, with account taken of all factors, including those not valued in monetary terms (DOT 1979, 1992a). The use of the framework approach, as recommended by ACTRA and SACTRA, subsequently became standard practice – albeit in a slightly modified form – in the appraisal of road schemes by the DOT.

The framework approach was set out formally in the DOT's *Manual of environmental appraisal* (MEA), published in 1983 (DOT 1983). The MEA, which was essentially a "how to do it" guide to environmental appraisal, consisted of three main parts: (a) the structure of the framework and general DOT advice on environmental appraisal; (b) advice on techniques for assessing a range of specific environmental impacts; (c) a list of the chief published sources for further reference.

The MEA specified a number of "appraisal groups" around which the framework was to be structured, although these were slightly different from those recommended by ACTRA and SACTRA. The relevant groups were as follows (see also Macpherson 1993, Simpson 1992, for further details):

Group 1 – the effects on travellers;
Group 2 – the effects on the occupiers of property;
Group 3 – the effects on the users of facilities;
Group 4 – the effects on policies for conserving and enhancing the area;
Group 5 – the effects on policies for development and transport;
Group 6 – financial effects.

287

Table 10.1 Part of a UK Department of Transport trunk road proposal appraisal framework.

GROUP 2: OCCUPIERS. Sub-Group: Residential

Effect	Units	Proposed scheme	Do nothing	Comments
Properties demolished	Number of properties	1	0	Cost of acquisition and demolition is included in Group 6 (financial effects)
Noise increase	Number of houses experiencing an increase of:			Noise changes take into account proposed mitigation measures. The changes are the difference between the forecast for 2012 and the existing levels in 1997 prior to the opening of the road. The units are dB(A) L10 for 18 hours, 6am to midnight.
	More than 16 dB(A) L10	9	0	
	11–15 dB(A)	22	0	
	6–10 dB(A)	8	0	
Noise decrease	Number of houses experiencing a decrease of:			Properties are along the existing A556 trunk road. The changes are as described above for noise increases.
	More than 16 dB(A) L10	59	0	
	11–15 dB(A)	0	0	
	6–10 dB(A)	7	0	
	3–5 dB(A)	7	0	
Visual obstruction	Number of properties within 300 m of centreline subject to:			
	High	6		
	Moderate	3		
	Slight	7		
Visual intrusion	High	8 (15)	0	Numbers take account of proposed landscaping measures. Figures in brackets are without landscaping
	Medium	18 (30)	0	
	Low	31 (12)	0	
Reduction of existing severance		Substantial relief to properties in Mere and Bucklow Hill fronting A556	None	

(*Source*: DOT/Allott & Lomax (DOT 1992))

Frameworks were to be prepared for the range of options considered at the public consultation stage, and – at the later public inquiry stage – for the Secretary of State's preferred route.

The bulk of the MEA consisted of technical guidance on the assessment of various types of environmental impacts. Advice was provided on the measurement and description of 11 impact categories, namely:

(a) traffic noise;
(b) visual impact;
(c) air pollution;
(d) community severance;
(e) effects on agriculture;
(f) heritage and conservation areas;
(g) ecological impact;
(h) disruption due to construction;
(i) pedestrians and cyclists;
(j) the view from the road;
(k) drivers' stress.

The MEA recommended that each of these impact categories should be examined in the environmental appraisal of schemes, although certain impacts (such as air pollution) were not expected to be relevant in all cases (see Macpherson 1993). The main findings emerging from the assessment of these impacts were to be summarized in the framework, having been allocated to the relevant appraisal group (or groups) (DOT 1992a). For example, community severance might affect Group 1 (e.g. drivers and cyclists), Group 2 (e.g. residents) and Group 3 (e.g. the users of bridleways and footpaths). The framework presented as many impacts as possible in monetary terms, such as the savings in journey times or vehicle operating costs derived from COBA. As many as possible of the remaining impacts were presented in quantitative form, for instance the number of houses affected by specified noise increases or levels of visual intrusion, or the number of ramblers affected by the severance of an existing footpath. Any remaining issues that could not be quantified, such as the effects on policies for conserving the area, were simply described in summary form in the framework. Table 10.1 shows part of an appraisal framework for a proposed motorway scheme. The MEA and the framework approach were subject to much criticism before their replacement by new guidance in July 1993 (see, for example, CPRE 1991a, DOT 1992a, Hopkinson et al. 1990, Macpherson 1993, Sheate & Sullivan 1993).

The implementation of EC Directive 85/337

For trunk road and motorway schemes, EC Directive 85/337 was implemented by means of the Highways (Assessment of Environmental Effects) Regulations 1988. These regulations amended the existing legislation, under which consent is given for such roads, the Highways Act 1980, inserting a new section (Section 105A). The DOT subsequently issued guidance, in the form of departmental standard HD 18/88, indicating how the Directive and the provisions of the amended Act were to be followed in practice (DOT 1989). This guidance was about the following issues: (a) Which schemes should be subject to formal EIA under the terms of the Directive? (b) At what stage in the planning and consent process for new schemes should the environmental statement required by the Directive be published? (c) What should be the scope and format of these environmental statements?

The need for EIA

Departmental standard HD 18/88 listed the types of road scheme that require mandatory EIA (Annex I schemes) and provided thresholds and criteria to be used to decide whether EIA was required for other (Annex II) schemes. The guidance stated that the following types of scheme would require formal EIA:

- All new motorways and "special roads" (i.e. express roads restricted to certain types of traffic, accessible from controlled junctions and on which stopping and parking is prohibited). These schemes fall within Annex I of the EC Directive, for which EIA is mandatory.
- All new trunk roads over 10 km in length.
- Other new trunk roads over 1 km in length which pass:

    (a) through or within 100 m of certain designated areas, i.e. national parks, sites of special scientific interest, conservation areas or nature reserves; or

    (b) through an urban area where at least 1,500 dwellings lie within 100 m of the centre line of the proposed road.

- Motorway and other trunk road improvements likely to have a significant effect on the environment.
- Exceptionally, other new trunk roads which do not fall within the criteria above, but which are judged to have a significant impact on the environment.

In addition, the New Roads and Street Works Act 1991 introduced a requirement for mandatory EIA for all private-sector road schemes, including toll motorways, bridges and tunnels. Such schemes fall within the definition of "special roads" and are therefore classified as Annex I projects. An early example was the toll bridge from the Kyle of Lochalsh to the Isle of Skye in Scotland (Sheate & Sullivan 1993).

It should be noted that environmental appraisal is carried out by the DOT on all trunk road schemes, including those deemed not to meet the criteria set out in departmental standard HD 18/88. Although no environmental statement is published for such a scheme, details of its environmental appraisal will be presented at the public inquiry into the scheme (Sheate & Sullivan 1993).

Publication, scope and format of the Environmental Statement

The environmental statement for a new trunk road or motorway scheme is published at the same time as the draft line orders for the scheme. The statement is considered at the public inquiry into the draft orders, along with other supporting information. Until July 1993, guidance on the appropriate content of the EIS was provided by the DOT's departmental standard HD 18/88 (DOT 1989). This advised that an environmental statement for such a scheme should include the following elements:

(a)   a description of the proposed scheme and its site;
(b)   an outline description of the measures proposed to mitigate adverse environmental effects;
(c)   sufficient data to identify and assess the main effects that the scheme is likely to have on the environment;
(d)   a non-technical summary.

In cases where the alternative options presented at the public consultation stage had environmental effects significantly different from those of the published scheme, the environmental statement was also to include the following two elements:

(e)   a summary description of the main alternatives presented at public consultation (although details of the environmental effects of these alternatives need not be included);
(f)   the reasons for the choice of the published scheme.

The DOT indicated that the tabular framework derived from the MEA (DOT 1983) would be sufficient to allow the main environmental effects of a scheme to be identified and assessed. It was therefore recommended that the framework should be used as the means of presenting information on environmental effects in the EIS. Certain types of impact were specifically excluded from consideration in the EIS, including those listed in Annex III (Clause 4) of the EC Directive. These include a range of indirect and secondary impacts, as well as matters such as the description of forecasting methods used. The guidance also implied that certain types of impact listed in Annex III (Clause 3) would rarely need to be addressed in a road scheme environmental statement. These included impacts on climate, soil and water (CPRE 1991a, DOT 1989, Sheate 1994). New guidance on the preparation of environmental statements for trunk road and motorway schemes was published by the DOT in July 1993 (DOT 1993); we discuss it later in this section. However, it should be noted that the new arrangements apply only to newly emerging schemes entering the Roads Programme and not to schemes already in progress (DOT 1993, Sheate 1994).

A critique of the practice of EIA for trunk road and motorway schemes

The treatment of environmental matters in trunk road appraisal and decision-making and, more specifically, the way in which the EC Directive on EIA was implemented by the DOT have been the subjects of much criticism in the UK. Perhaps the most influential critique of existing practice was that made in the early 1990s by the DOT's own Standing Advisory Committee on Trunk Road Assessment (SACTRA)

(DOT 1992a). However, other important contributions to the debate have been made by the Royal Commission on Environmental Pollution (RCEP 1994) and the National Audit Office (NAO 1994), as well as by conservation bodies such as the Council for the Protection of Rural England (CPRE 1991a, b, 1992) and a range of independent commentators and academics.

The main criticisms levelled at the DOT and its procedures for assessing the environmental impacts of new schemes can be grouped into the following broad categories:

(a) A number of trunk road schemes have escaped the need for formal EIA, including those transitional schemes which were in the planning pipeline before the implementation of the EC Directive in July 1988.

(b) Formal environmental assessment and the publication of the environmental statement take place at too late a stage in the planning, design and appraisal of new road schemes. This results in a limited treatment of alternatives, and means that environmental assessment is not truly iterative.

(c) The division of road schemes into small sections, each of which is planned, designed and assessed separately, limits the assessment of impacts at a more strategic level. As a result, combined and cumulative impacts are poorly treated.

(d) The quality of DOT environmental statements is generally poor and compares unfavourably with that for other project types. In particular, the use of the framework approach to present environmental information is inappropriate.

(e) Indirect and secondary impacts are not given sufficient emphasis.

(f) The treatment of mitigation is weak. There is a lack of clear commitment to specific measures and an emphasis on aesthetic and landscape matters; there are few attempts to assess the likely effectiveness of mitigation measures.

(g) The monetary valuation of environmental impacts is not well developed.

(h) Until the appearance of new guidance in July 1993, the DOT's guidance on environmental assessment, contained in the MEA, was inadequate and in need of updating.

The need for EIA and the special case of transitional schemes

For trunk road and motorway schemes, the need for EIA is determined by the DOT itself. In other words, it acts as both developer and competent authority. This is rather different from the arrangements for planning projects under the Town and Country Planning Regulations, in which the competent authority – usually the LPA – is in most cases distinct from the developer and has the discretion to require an EIA (CPRE 1991b). Guidance on the need for EIA for trunk road and motorway schemes was, until July 1993, provided by the DOT's departmental standard HD 18/ 88 (DOT 1989). This provides thresholds and criteria for use within the DOT when deciding the need for EIA for an individual scheme (see above for further details). The thresholds and criteria contained in the guidance have been criticized for being arbitrary and inflexible. Concerns have been expressed that the thresholds relating to sensitive areas such as sssIs may be set at too high a level. Combined with a too rigid interpretation of such thresholds, this may mean that schemes with potentially significant impacts avoid the need for EIA (see Box & Forbes 1992, CPRE 1991b, RCEP 1994). In practice, it appears that a large number of trunk road schemes, including major bypasses, have not been subject to formal EIA (CPRE 1991b).

Section 105A(7) of the amended Highways Act 1980 excluded the need for EIA in cases where the draft line orders for a road scheme had been published prior to July 1988. Such schemes were considered by the DoT to have already entered the planning pipeline before the EC Directive took effect, and were therefore deemed not to be subject to the requirements of the Directive. This interpretation has been challenged by a number of commentators and has been the subject of complaints to the European Commission (see Kunzlik 1996). The basis of these complaints is Article 2.1 of the EC Directive. This states that: "Member States shall adopt all measures necessary to ensure that, *before consent is given*, projects likely to have significant effects on the environment . . . are made subject to an assessment with regard to their effects." (Emphasis added.)

One interpretation of this requirement is that, once the Directive came into force, in July 1988, projects in certain categories – including trunk road and motorway schemes – could not lawfully be granted consent until they had been subject to EIA. This would be the case irrespective of whether the scheme had entered the planning pipeline before July 1988. This view was reinforced by the fact that consent for such pipeline projects might not be granted until several months or even years after the Directive came into force (Kunzlik 1996). This interpretation was supported by the European Commission. In October 1991, the Commission issued a well-publicized letter of formal notice to the UK government concerning non-compliance with the EC Directive. Among other points, the Commission argued that the exemption of pipeline projects from the requirements of the Directive was unlawful (Salter 1992; see also Ch. 6).

The pipeline issue affected a number of major road schemes in the UK, including the controversial M3 motorway extension through Twyford Down in Hampshire (Bryant 1996). However, lengthy delays in the Commission's enforcement procedures meant that its judgement on the pipeline issue had little practical effect on the progress of such schemes (Kunzlik 1996). For example, construction work at Twyford Down was started shortly after the Commission had initiated enforcement proceedings against the UK Government over the pipeline issue in general and the Twyford Down scheme – amongst others – in particular.

The timing of EIA and the treatment of alternatives

As noted earlier, the detailed environmental appraisal of a scheme tends to take place only after the announcement of the preferred route, during the period leading up to the publication of draft line orders. Prior to the entry of schemes into the Roads Programme and the selection of route options for public consultation, the DoT's assessment of environmental impacts tends to be limited (NAO 1994). The environmental statement for a proposed scheme is published with the draft line orders for the preferred route option. By this late stage in the planning and design process for the new road, many important decisions will already have been taken, not least the crucial decision that a new road is preferable to alternative options, such as public transport solutions (CPRE 1991a, b). The DOT's commitment to the detailed alignment of a route will also have become fixed by this stage, and any variations will be difficult and will involve delays and extra costs (Tromans 1991). Although the appraisal process may address alternative routes, usually only the impacts of the preferred route are outlined in the environmental statement. This has led to the claim that the emphasis in DOT road appraisal is on: "mitigating [the] damaging effects of the preferred option . . . rather than modifying the proposals or option to eliminate or minimise the likely damaging effects on the environment" (Sheate & Sullivan 1993).

In other words, it is claimed that EIA for road schemes is not iterative, in that it tends to take place *after* the process of route selection, and is used to assess the impacts of the preferred route. It is argued that EIA must also have a role at an earlier stage in the planning of schemes, before or in parallel with the route selection process. EIA at this earlier stage would be used to assess the impacts of the alternative route options (and possibly of other solutions involving alternative modes), and would influence the choice of the preferred route.

The DOT's own Standing Advisory Committee on Trunk Road Assessment (SACTRA), in its influential review of road scheme EIA, called for the formal presentation of the results of EIA studies at a much earlier stage in the planning of schemes (DOT 1992a). It recommended that, for all schemes, Stage 1 and Stage 2 reports should be submitted, as outlined in the DOT's *Design Manual for Roads and Bridges* (DOT 1993).

The need for a more strategic level of assessment
Because of the way in which a trunk road and motorway scheme tends to be planned, assessed and given consent in a series of small sections, even when forming part of a larger route, EIA at the scheme level inevitably involves a rather piecemeal approach to assessment (DOT 1992a). For example, proposals for the east–west Folkestone to Honiton trunk route along the south coast of England were divided into more than 30 small sections, including 16 separate bypass schemes (CPRE 1991a). There is no opportunity to carry out a strategic assessment of the whole route in such cases. This is important, since decisions taken on a scheme by scheme basis can tend to foreclose later options (NAO 1994).

In addition, EIA on individual schemes finds it difficult to address the wider, cumulative and global environmental issues to which road schemes *in aggregate* can give rise (see Lee & Walsh 1992). Examples of such impacts include emissions of greenhouse gases or accumulated losses from the national stock of sensitive ecological sites (DOT 1996). To capture such impacts adequately, an earlier, more strategic level of assessment is needed. SACTRA, in its report on road scheme EIA, argued that such strategic appraisal should be guided by explicitly stated environmental policy objectives (DOT 1992a). These objectives would be set by central government and might relate either to *global* issues (such as the problem of carbon dioxide emissions) or to *national* or *strategic* concerns (such as the conservation of endangered species or habitats, or the preservation of nationally important heritage sites). SACTRA urged that the performance of road schemes in meeting such agreed objectives should be formally assessed at an early stage in the planning of new roads: "No [road] scheme should be admitted into the Roads Programme until its performance against these strategic objectives and constraints has been evaluated and reported in outline" (DOT 1992a).

SACTRA recommended that the effects of schemes on *regional* and *local* environmental policies and objectives adopted by LPAS should also be assessed in a similar way, again before the schemes entered the Roads Programme. A number of other commentators have suggested that similar, objectives-led systems of strategic environmental assessment should be applied to the transport sector as a whole (see, for example, CPRE 1992, Ferrary 1994, Sheate 1992).

The quality of environmental statements for road schemes

A number of independent reviews of the quality of environmental statements for trunk road and motorway schemes have been published in recent years. The most important are perhaps those carried out by SACTRA (DOT 1992a) and the National Audit Office (NAO 1994). Other reviews have focused on specific aspects of the environment statement, such as the treatment of ecological impacts (Treweek et al. 1993). Other, more general studies of the quality of EISs prepared for a range of project types allow comparisons to be made between road scheme and other EISs (Lee & Brown 1992, RSPB 1995).

Perhaps the most influential critique of the quality of environmental statements for road schemes was that provided by the DOT's Standing Advisory Committee on Trunk Road Assessment (SACTRA) (DOT 1992a). SACTRA carried out a review of a small sample of EISs, all of which had been prepared in the first few years after the implementation of the EC Directive. SACTRA was surprised by the brevity of most road scheme EISs and was particularly critical of the way information on environmental effects was presented in such statements. According to SACTRA, the format and content of EISs appeared to have been strongly influenced by the longer-established MEA framework, which had been developed by the DOT to summarize the various effects of different scheme options in a common tabular format.

> It seems that a view has been taken [by the DOT] that the Framework already contains the core of the information required by the [EC] Directive. . . . Thus, in every case which we have examined, the Framework had been reproduced without any adaptation, including all the economic data produced by COBA, *as the centrepiece of the Environmental Statement.* (DOT 1992a, emphasis added.)

SACTRA argued that the use of the MEA framework in EISs was inappropriate. It weakened and over simplified the treatment of environmental impacts, by limiting the presentation of such effects to a tabular summary format only. If environmental effects were to be summarized, then the non-technical summary was the appropriate place for this to be done. The remainder of the EIS should be much expanded, with a full description of the relevant environmental data, more attention to the importance and significance of impacts, and details provided of the methodologies used in the assessment (DOT 1992a). Other commentators, while accepting some of the limitations of the framework approach, have argued that its use is not without its benefits. For example, in its review of road scheme EIA, the National Audit Office concluded that the universal use of the MEA framework to structure DOT environmental statements ensured a useful consistency of approach. The common format provided by the framework helped identify the principal impacts of schemes (NAO 1994).

Other studies have widely criticized road scheme environmental statements, focusing less on the presentation of information and more on the substantive content of the statements. The following main weaknesses have been identified:

- The description of a proposed scheme and its site can be poor. For example, in a review of the treatment of ecological impacts in road scheme EISs, Treweek et al. (1993) discovered that almost half the statements reviewed failed even to indicate the length of the proposed schemes. Even fewer EISs provided figures for land-take, and none gave detailed breakdowns of the areas of wildlife habitat likely to be affected by the scheme.
- The description of baseline conditions can be a weak area. For example, the description of different types of land-use and habitat is inconsistent, making it difficult to summarize those habitats potentially affected in a coherent way (Treweek et al. 1993).
- Potential impacts are not always quantified and in some cases are not discussed at all. For example, in the study by Treweek et al., the ecological impacts of the scheme were quantified in only three of the 37 statements reviewed, and were not considered at all in five statements. Other studies have noted that, in some cases, certain adverse impacts are not quantified (NAO 1994).
- There is a tendency for EISs to refer only to the more obvious and direct impacts, such as habitat loss. More complex and indirect impacts, such as habitat fragmentation, are considered in only a small minority of statements (Treweek et al. 1993).
- Certain types of impact are rarely addressed in EISs. These include construction stage impacts, including the impacts of wastes generated during construction, and the cumulative effects generated by other road building in the vicinity of a scheme (NAO 1994).
- Assessment of the importance or significance of impacts is generally poor (NAO 1994).
- The proposed mitigation measures are not always fully described, and their likely effectiveness is rarely discussed. Where mitigation measures *are* described, most statements are vague about whether and within what period the measures

will actually be implemented (NAO 1994, Treweek et al. 1993). Treweek et al. found that the emphasis in most EISs was almost entirely on landscaping and tree planting, designed to minimize visual or aesthetic impacts. Major realignments of routes were not considered in any of the statements reviewed.

- Assessment methodologies and the quality of environmental data upon which the EIS is based are varied. Most EISs appear to have been prepared without the benefit of new survey work, and some ecological surveys have been carried out at obviously inappropriate times (Treweek et al. 1993).

Although such studies appear highly critical of road scheme EISs, it should not be assumed that the deficiencies identified are peculiar to such schemes. Indeed, one recent study of the treatment of ecological and nature conservation issues in EIA has highlighted the existence of similar weaknesses in environmental statements for a wide range of project types (RSPB 1995). Some studies, however, allow a direct comparison between the quality of EISs prepared for road schemes and that of those prepared for other types of project. Lee & Brown's (1992) review of a sample of early EISs, published between July 1988 and early 1991, found that EISs for trunk road and motorway schemes tended to be below average quality. Using the review criteria developed by Lee & Colley, only 25 per cent of such statements were judged to be satisfactory (Lee & Brown 1992, Lee & Colley 1992). Some commentators have argued that the apparently poor quality of EISs for such schemes may be associated with the dual role of the DOT as both developer and competent authority. This means, it is claimed, that EISs lack rigorous independent scrutiny, and that it may weaken the pressure on developers to be sufficiently thorough and objective when assessing the negative environmental impacts of their projects (Lee 1992, Sheate & Sullivan 1993).

Most of the critical studies referred to above were concerned with early examples of road scheme EISs, in most cases published before 1993. More recent research, examining statements prepared during 1993 and 1994, points to a general improvement in environmental statements (DOE 1996), and in those for road schemes in particular. For example, a recent study into the treatment of nature conservation in EIA, although not using formal review criteria, concluded that EISs for road schemes were of "above average" quality (RSPB 1995). Statements for road schemes were characterized by thorough baseline data, explicit consideration of alternatives and the description of extensive mitigation measures. This apparent improvement may well be linked to the publication of revised guidance on road scheme EIA in July 1993.

The treatment of indirect and secondary impacts

Guidance on the appropriate content of road scheme EISs was, until July 1993, provided by the DOT's departmental standard HD 18/88 (DOT 1989). This guidance was rather more narrowly defined than that for planning projects contained in the Town and Country Planning Regulations. In particular, much of the information listed in Annex III of the EC Directive was explicitly excluded from consideration in road scheme EISs (CPRE 1991a, Sheate 1994). The departmental standard stated that none of the issues listed in Annex III (Clause 4) of the Directive need be addressed in a road scheme EIS. These include a range of indirect and secondary impacts, including impacts arising from (a) the existence of the project; (b) the use of natural resources; (c) the emission of pollutants, the creation of nuisances and the elimination of waste.

Such impacts can be wide-ranging and potentially significant. Effects arising from the existence of a project might include consequential development pressures and associated land-use changes, which have been shown to be highly significant for certain road schemes (see, for example, Headicar & Bixby 1992, RCEP 1994). A new road scheme may also generate additional or induced traffic, which may in turn give rise to indirect impacts on energy use, air, soil and water (DOT 1994a, b, RCEP 1994). Impacts arising from the use of natural resources might include the environmental effects of increased extraction of aggregates for road construction (CPRE 1993). Finally, impacts resulting from the emission of pollutants and the elimination of wastes might include the long-term effects of pollutants on adjacent vegetation, the contribution to greenhouse gases and acid rain and the need for disposal sites for unwanted spoil (Box & Forbes 1992, CPRE 1991a).

In addition to excluding all of the above types of impact, the DOT's departmental standard implied that certain types of more direct impact, listed in Annex III (Clause 3) of the EC Directive, would not normally need to be addressed in a road scheme EIS (DOT 1989). These included impacts on climate, soil and water. One commentator has argued that this could be seen as pre-empting the EIA process, the purpose of which is to assess the likely magnitude and significance of potential impacts (CPRE 1991a). It also highlights the difficulties of EIA at the individual scheme level in giving adequate consideration to wider and cumulative impacts.

The treatment of mitigation

For road schemes, three main types of mitigation measure can be distinguished (Box & Markham 1994):

(a) *avoidance* of the impact (e.g. choosing another route to avoid a valued ecological site);
(b) *reduction* of the impact (e.g. modifying the route's alignment so that the road skirts a valued site rather than passes through it, or so that the land-take required by the road is reduced);
(c) *compensation* for impacts which are unavoidable (e.g. the creation of new habitats to replace those lost or the relocation of environmental assets).

Within each of these broad categories, a wide variety of mitigation measures is available (see Box & Forbes 1992, English Nature 1994). A number of commentators have argued that the DOT's approach to mitigation focuses primarily on the *reduction* of adverse impacts, too little on the *avoidance* and *offsetting* of impacts (CPRE 1991a, DOT 1992a, Sheate & Sullivan 1993, Treweek et al. 1993). This is partly a reflection of the limited powers available to the DOT to compensate for the loss of environmental assets (Box & Markham 1994). Another concern is the lack of clarity in environmental statements about the commitment to mitigation measures. For example, the National Audit Office study of road scheme EIA concluded that, for most of the EISs examined in its review:

> The reader could not determine whether the [mitigation] measures proposed were sufficient or justified, or whether other measures would be more effective. Where mitigation measures were described, there often appeared to be uncertainty whether and to what timescale the work would be carried out. (NAO 1994)

New arrangements for the disclosure of mitigation measures were introduced with the new guidance on road scheme EIA published by the DOT in July 1993 (DOT 1993).

The monetary valuation of environmental impacts

As previously discussed, road appraisal by the DOT does not generally assign monetary values to the environmental assets affected by the particular scheme. This partly reflects the discredited attempts to value such assets in early cost–benefit studies, such as the Roskill Commission's study into the site of a third London Airport (Bruton 1985, DOT 1992a). The debate about whether environmental assets can and should be valued in monetary terms is a long-standing one (see, for example, Hopkinson et al. 1990, Pearce et al. 1989). Nevertheless, some commentators have argued that in the case of road schemes, a limited extension of monetary valuation to certain environmental effects is both desirable and feasible (see, for example, DOT 1992a, NAO 1994).

SACTRA, in its deliberations on road scheme EIA, recommended an extension of monetary valuation to as many environmental effects as practicable, and the inclusion of such money values in the cost–benefit analysis of road schemes (DOT 1992a). It was felt that certain environmental costs could already be valued satisfactorily, and that these values could be readily incorporated within existing cost–benefit calculations. These included the direct costs of mitigation or environmental protection measures associated with a road scheme, and the costs imposed directly upon polluters by the government, in the form of taxation, as part of the "polluter pays" principle. As regards the valuation of other environmental effects, SACTRA recommended a more cautious approach, with the experimental application of various monetary valuation techniques in a series of pilot schemes.

The need for revised guidance on environmental assessment

Until July 1993, guidance on EIA for trunk road and motorway schemes was provided by the DOT's *Manual of environmental appraisal*, first published in 1983, and the Departmental Standard HD 18/88 (DOT 1983, 1989). SACTRA, in its review of road scheme EIA, called for the replacement of the MEA by a new and expanded manual covering a range of issues not previously addressed, including the impacts on land-take, the loss of open space, water pollution, vibration and "personal stress". It was also recommended that the new manual should provide guidance on the assessment of road schemes against national, regional and local environmental policies and objectives. Many of SACTRA's recommendations were reflected in the new guidance published by the DOT in 1993. This new guidance, and other recent changes in assessment practice, are outlined in the next section.

Recent developments in road scheme planning and EIA

New guidance on road scheme EIA

In July 1993 the DOT published a new *Design manual for roads and bridges.* Volume 11 of the new manual provides guidance on the environmental assessment of new road schemes (DOT 1993), and replaced both the *Manual of environmental appraisal* (DOT 1983) and the guidance contained in departmental standard HD 18/88 (DOT 1989). The new design manual introduced a number of important changes to the environmental appraisal of road schemes and was widely welcomed as offering the prospect of a significant improvement in EIA practice (see, for example, Lewis 1994, NAO 1994, Sheate 1994). The manual incorporates many of the changes recommended by SACTRA (DOT 1992a), most of which were accepted by the DOT (DOT 1992b, Wood 1992). The principal features of the new guidance are outlined below.

The new design manual provides advice on the general principles of EIA and on the techniques for assessing specific types of impact, and makes recommendations for reporting the findings of EIA. The bulk of the manual comprises detailed guidance on assessment techniques. This is essentially an updating of the advice in the *Manual of environmental appraisal* (DOT 1983), although a number of issues not previously considered are now included. These include impacts on water quality, drainage, geology and soils, as well as guidance on assessing the impacts of schemes on policies and plans. The guidance stresses that all the impacts that are likely to be significant should be assessed, including indirect, secondary and cumulative impacts.

The new guidance provides for the formal assessment and reporting of environmental impacts at an earlier stage in the planning of new schemes. Three main stages in the assessment of schemes are identified:

1   *Prior to the entry of a scheme into the Trunk Roads Programme.* The new manual advises that the role of assessment at this stage is to "identify the main environmental advantages, disadvantages and constraints associated with broadly defined routes or corridors". Any relevant constraints (such as population centres, historic buildings and ecological sites) should be mapped, allowing an initial assessment of the potential impacts of a variety of possible route corridors. The guidance advises that assessment at this stage will involve desk studies; only in exceptional circumstances will site visits be necessary, other than for landscape assessment. Findings are to be presented in a new Stage 1 report,

before the entry of the scheme to the Roads Programme. This will replace the report on the scheme identification study prepared under the old arrangements (NAO 1994).

2 *Prior to public consultation on alternative route options.* Assessment at this stage should "identify the factors and effects to be taken into account in the selection of the route options to be presented at public consultation, and should identify the environmental advantages, disadvantages and constraints associated with each of these options". Findings should be presented in a Stage 2 report, which will be more detailed in its treatment of impacts than the Stage 1 report, drawing additionally upon information obtained from site visits.

3 *Prior to the publication of the EIS, but after the selection of the preferred route.* Assessment at this stage should be of the preferred route design, culminating in the publication of the environmental statement. It is recommended that the EIS should comprise three parts:

- volume 1: a comprehensive document drawing together all the relevant information about the proposed scheme, baseline conditions, mitigation measures and environmental effects;
- volume 2: containing detailed assessments of the environmental effects of the scheme, by type of impact;
- a non-technical summary: summarizing the main points in volume 1.

The guidance stresses that all mitigation measures described in the EIS should be firm commitments. Arrangements are outlined for the description of those measures not yet fully agreed. The EIS should include details of the proposed arrangements for the monitoring and maintenance of mitigation measures.

The treatment of alternatives in the EIS is little changed from that recommended in the earlier guidance, with only a brief description required of the alternative route options studied and the reasons for their rejection.

The new guidance advises that, in certain cases, a more strategic level of assessment, covering the combined impacts of several related schemes, may be necessary.

Consideration of longer routes or a number of related schemes together can give a clearer sense of the impacts of the proposal seen as a whole and may allow a better choice of alignment and design in both environmental and traffic terms. It will also help to ensure that schemes which should be assessed together at later stages, because of the interaction of their environmental effects, are not considered in isolation (DOT 1993).

Such strategic assessments would take place during the Stage 1 assessment – i.e. before the entry of the scheme into the Roads Programme. The guidance (DOT 1993), however, strikes a note of caution: "Since schemes . . . have been initiated and progressed with different timescales the adoption of such [a strategic] approach may not be possible in practice." This suggests that the number of such strategic assessments may be limited in practice (NAO 1994, Sheate 1994).

The tabular framework document outlined in the *Manual of environmental appraisal* is replaced in the new guidance by "environmental impacts tables" (EITs). These, like the MEA framework, are designed to summarize the main environmental impacts of a proposed scheme compared with the do-nothing or do-minimum options. It is suggested that an EIT will have the following structure:

(a) Effects of a scheme on the following appraisal groups:
   1 Local people and their communities
   2 Travellers
   3 The cultural and natural environment
   4 Policies and plans
(b) Land-use table
   This should list and quantify the existing land-uses required for the scheme, including the land required only during construction.
(c) Mitigation table
   This should identify agreed mitigation measures, describing their location, purpose, anticipated benefit and, where possible, estimated cost.

Cost–benefit and financial effects – which were included in the MEA framework – are excluded from the new EITs (Sheate 1994). The use of EITs is recommended at each stage in the assessment process, with more detailed information included as a project develops. However, it is stressed that the EIT is only one means of presenting information; other forms of presentation, including written text and maps, are also recommended.

Other recent changes in assessment procedures and practice

A number of important changes in trunk road scheme assessment procedures and practice, and in the planning of such schemes, have taken place since the publication of the new design manual in 1993. The main changes are summarized below:

(a) The responsibility for trunk road planning, construction and maintenance was transferred to a new independent Highways Agency in April 1994. The DOT (now incorporated in the new Department of the Environment, Transport and the Regions) retains responsibility for all transport policy.
(b) The DOT has accepted recommendations by the Royal Commission on Environmental Pollution and others for the closer integration of trunk road planning with both land-use planning and the planning of other transport infrastructure (see CPRE 1992, RCEP 1994). In 1996, the DOT invited comments on a proposal to integrate formal consultation on the trunk road programme within the existing system of regional planning guidance (DOT 1996).
(c) The procedures for the internal review of EISs within the Highways Agency have been strengthened (NAO 1994).
(d) The DOT has commissioned further research into the monetary valuation of a range of environmental impacts, and it published a literature review on the subject in 1996 (DOT 1996, NAO 1994, Tinch, 1996);

(e)   In 1994, SACTRA published a report on the extent to which new trunk roads generate additional or "induced" traffic, rather than merely redistribute the same amount across the road network (DOT 1994a). SACTRA's report identified the circumstances in which such induced traffic might be generated, and recommended a number of changes to scheme appraisal methods. The DOT accepted the main conclusions of the report and issued new guidance on the assessment of schemes (DOT 1994b). Before this, the DoT's cost–benefit calculations had failed to allow for the effects of induced traffic. The likely significance of such induced traffic is now assessed for all new trunk road schemes, although the techniques for carrying out such assessments are still being developed (DOT 1994b, 1996);

(f)   SACTRA's recommendations for a more strategic level of assessment have been reflected in the emergence of a number of studies assessing alternative options within transport corridors, as well as the more formal consideration in certain road scheme EISs of the cumulative impacts associated with adjacent or linked developments (see, for example, ERM 1994a, b, Scottish Office 1994, Wilson 1994);

(g)   At a more general level, the DOT has examined the scope for an assessment of the "total and cumulative environmental effects of the trunk roads programme as a whole", and has commissioned a feasibility study into the subject. It has accepted that such an assessment would allow a better judgement to be made between the roads programme and other transport options (DOT 1996). All this must now be set in the context of a new government approach to transport policy, with a shift away from road building towards a more integrated approach, with more support for traffic management and public transport (DETR, 1997).

10.3   A case study: the M6–M56 Link Road Scheme

Introduction

This section presents a case study of the EIA for a new motorway scheme in northwest England (DOT 1992c). Its environmental statement was submitted towards the end of 1992, shortly after the publication of the critical report into road scheme EIA by the DOT's Standing Advisory Committee on Trunk Road Assessment (SACTRA) (DOT 1992a), but prior to the emergence of the new guidance contained in the DOT's design manual (DOT 1993). It is therefore of interest to examine how this transitional example of road scheme EIA dealt with some of the issues raised by SACTRA and other commentators. Of particular interest are the scope of the EIA, the way environmental information was presented in the EIS and the treatment of mitigation, alternatives and indirect or consequential impacts.

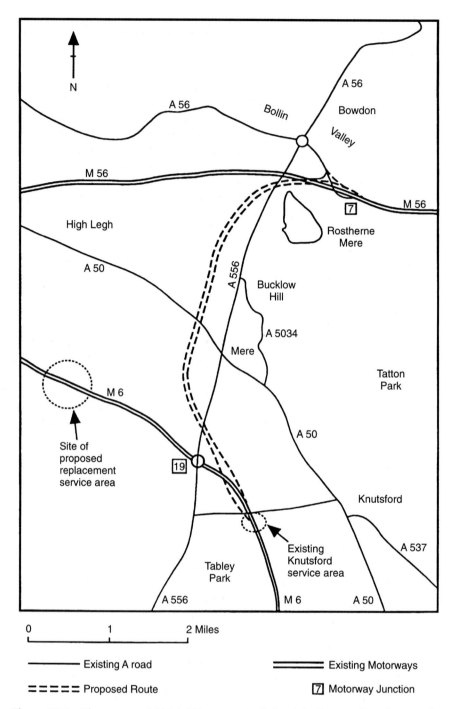

Figure 10.1 The proposed M6–M56 motorway link road. (*Source:* based on DOT/ Allott & Lomax (DOT 1992))

The proposal

The proposed scheme, known as the A556(M), involved the construction of a new motorway link between the existing M6 and M56 motorways, between Knutsford and Altrincham in Cheshire (see Fig. 10.1). The existing link was provided by the A556 trunk road, which was mainly a four-lane single carriageway. This route, which served much of the motorway traffic between the cities of Manchester and Birmingham, was seen as problematic. It was characterized by a high volume of traffic, much of it consisting of heavy goods vehicles, peak hour congestion and a poor accident record. The do-nothing option and on-line improvements to the existing road were rejected in favour of a replacement link road. This was to be a three-lane dual carriageway for most of its 6.5 mile (10 km) length. The existing A556 route would lose its trunk road status when the new scheme was completed.

The land surrounding the existing A556 was rural and predominantly agricultural. The relevant local authority, Cheshire County Council, had identified three areas of special landscape value in the vicinity, at Tatton Park to the east, the Bollin Valley to the north and Tabley Park to the south-west. Ecologically designated sites included the important site of Rostherne Mere, near the existing junction with the M56 motorway. This was designated a site of special scientific interest (SSSI), a national nature reserve and a wetland of international importance under the Ramsar Convention on account of the waterfowl which nested at the site (DOE/DOT 1995). There were also six designated sites of biological importance, as well as many patches of undesignated woodland, unimproved grassland, lane-side hedges and ponds of local importance; wildlife corridors connected these features (DOT 1992c).

The background to the proposal and the key planning stages

The proposed scheme originally entered the National Trunk Roads Programme in 1987 as a stand-alone scheme. However, it later became part of a much longer route, known as the Greater Manchester Western and Northern Relief Road (GMWNRR). This route, divided into three main parts, involved the construction of new relief roads around the northern and western fringes of the Greater Manchester conurbation. The A556(M) scheme represented Stage I. Six alternative routes for the A556 scheme, including an on-line improvement of the existing road, were considered by the DOT before the public consultation, although only one route option was presented at the consultation stage in November 1989.

The announcement of the preferred route took place in December 1990 and, following the detailed design and assessment of this route, draft line orders for the scheme and the environmental statement were published in October 1992. Modifications to the draft orders and an addendum to the environmental statement were published in February 1993. A public inquiry into the draft orders was held between October and December 1993. Supporters of the scheme at the inquiry included the local authorities in the area, Cheshire County Council and Macclesfield Borough

Council, as well as the Council for the Protection of Rural England and the Mere Residents Association (representing residents along the existing A556 route). Objectors included adjacent local authorities (such as Trafford Metropolitan Borough Council), environmental bodies such as English Nature and Friends of the Earth, and many local residents and members of the public. Approval for the scheme, confirming the draft line orders in modified form, was given in July 1995 (DOE/DOT 1995). A separate inquiry on the compulsory purchase orders necessitated by the scheme was held in the spring of 1997.

The scope and format of the environmental statement

The environmental statement for the scheme in fact comprised four separately bound volumes rather than a single document (DOT 1992c). The main document described the proposed scheme and summarized its main impacts, while three further technical volumes dealt with a range of specific impacts in greater depth. The impacts addressed in the EIS appear to have been based largely on those listed in the DOT's *Manual of environmental appraisal* (DOT 1983). However, certain additional effects not identified in the MEA, including impacts on water and drainage, were also examined as part of the EIA for the scheme and were included in the EIS.

The presentation of the information on environmental impacts in the EIS differed significantly from that found in earlier examples of road scheme EISs and criticized by SACTRA and others (DOT 1992a) (see Section 10.2). The tabular MEA framework was not used as the main means of presenting information. Instead, the framework was relegated to the status of an appendix to the EIS, and the environmental impacts of the scheme were discussed in detail in the main body of the text of the statement. There was no attempt to structure this discussion around the six appraisal groups in the MEA framework. More detailed assessments, prepared by various consultants, were incorporated into three additional, separately bound documents, which formed part of the EIS. These included an agricultural assessment, an ecological survey and assessment, an archaeological assessment, an air quality report, and a road traffic and construction noise report. Detailed technical reports in support of the conclusions in the EIS were not provided for certain types of impact, including effects on water and drainage, landscape, severance and construction stage impacts. However, supporting documents on water and drainage impacts were provided by the DOT at the subsequent public inquiry.

The treatment of mitigation

The mitigation measures proposed for the scheme were praised by the inspector at the public inquiry. Referring to the proposed landscaping plans for the scheme, the inspector concluded that: "in my opinion, the mitigation measures proposed by the Department for this scheme by way of the depression of the greater part of the route, and the mounding and planting, are exceptional. The praise from the CPRE ... and others is well merited" (DOE/DOT 1995).

It was noted in Section 10.2 that in the mitigation measures for road schemes the reduction of adverse effects, rather than the prevention or offsetting of impacts, tends to be emphasized. The mitigation measures proposed in the EIS for the A556(M) scheme appeared to be mainly designed to prevent or reduce adverse impacts, especially those on the landscape, visual intrusion and noise. However, some of the measures suggested could be regarded as means of offsetting or compensating for adverse effects. These mainly concerned the ecological impacts of the scheme proposals. The EIS conceded that sites for habitat creation would largely be limited to the motorway verges and the areas enclosed by the new motorway access roads. The creation of scrub and woodland, species-rich grassland and shallow pools or ponds with extensive, dense cover alongside, and the improvement of existing ponds, were recommended. The EIS stated that "the creation of [such habitats] will enhance the wildlife-carrying capacity of the area, and may encourage birds such as the barn owl and hobby to breed". It contained a range of suggested mitigation measures, rather than a detailed outline of specific commitments: more specific measures were to be devised during the detailed design stage (DOT 1992c).

The treatment of alternatives

Six alternative schemes were considered by the DOT prior to the public consultation stage in November 1989. These included a do-minimum option (involving an on-line improvement of the existing A556 route) and five alternative off-line routes, including the preferred scheme. Compared with the preferred scheme, these alternative routes were characterized by a more westerly alignment and/or a more westerly location for the new interchange with the M56 motorway. Those alternatives not considered included the do-nothing option and alternative modes, such as public transport or park-and-ride. The EIS stated that, as the major strategic route for motorway traffic between the Midlands and Manchester: "only a new road was judged to be appropriate or effective in coping with the forecast growth in demand for traffic movement" (DOT 1992c).

Five of the six alternatives under consideration were rejected by the DOT before the public consultation, including the do-minimum option and all the more westerly route options. Consequently, only one option was presented at the public consultation stage in late 1989. Such single-option consultation, although not the norm, is not unusual (NAO 1994). The route – in somewhat modified form – was confirmed as the Secretary of State's preferred route at the end of 1990. Further modifications were made during the process of detailed route design prior to the submission of draft line orders for the scheme and the environmental statement in October 1992. The EIS for the scheme was therefore submitted almost two years after the announcement of the preferred route and three years after the public consultation stage. This sequence of events encourages the belief that the crucial decisions about the general alignment of the route had been taken long before the appearance of the EIS, and indeed before the public consultation stage. It is therefore not surprising that the treatment of alternatives in the EIS was far from satisfactory.

The relevant DOT guidance at the time, contained in Departmental Standard HD 18/88 (DOT 1989), indicated that a road scheme EIS should include a brief description of the alternatives considered at the public consultation stage and the reasons for the choice of the preferred route. Therefore, since only one option for the A556(M) scheme was presented for public consultation, it would appear that the EIS did not need to address the issue of alternatives at all. Notwithstanding this, a brief description of all six of the original options considered before the public consultation was included in the EIS. The reasons for the choice of the preferred scheme were also outlined, although this did not amount to a detailed comparison of the various options. For example, the comparison of the environmental impacts of the different options occupied only one page of the EIS, most of the discussion focusing on the relative economic, traffic and safety implications of the schemes.

The EIS stated that "an assessment of the five Do Something options revealed that there was little to choose between them in environmental terms", except that a more serious visual impact arose out of the routes involving a more westerly location for the interchange with the M56, which would be sited in a conspicuous location. This conclusion may or may not be true, but the EIS did not contain the detailed information on the environmental effects of each option to support such a statement. Indeed, the more serious impact of the westerly M56 interchange locations was questioned by English Nature in its comments on a draft version of the EIS:

> The choice of the most easterly option [for the interchange] has been based on traffic, operational, safety and economic grounds. There appears to have been *no* consideration of the considerably greater impact of the chosen location on Rostherne Mere [a Ramsar site, national nature reserve and SSSI]. (English Nature, in DOT 1992c)

English Nature argued that the impacts of the scheme on the important site of Rostherne Mere, including visual intrusion and increased air pollution and noise levels, had not been adequately addressed in the EIS. Other objectors made similar comments at the subsequent public inquiry into the proposals (see DOE/DOT 1995). Subsequent to the public consultation and the announcement of the preferred route, significant alterations were made to the proposed route, which, the DOT argued, reinforced its selection. However, rather unhelpfully, the EIS did not clearly identify these alterations, nor did it justify them, either in environmental or any other terms. Further modifications to the proposals were made shortly after the submission of the EIS; these were described in an addendum to the EIS, published in February 1993.

Although the EIS contained only a very limited treatment of alternatives, the issue of alternatives was one of the main preoccupations of the subsequent public inquiry held during 1993. No fewer than 12 main alternatives proposed by objectors were considered at the inquiry (DOE/DOT 1995). Most of them (seven) involved route realignments, ranging from minor adjustments to the proposed scheme to entirely different route corridors. The other alternatives involved minor modifications to the side road orders or other design changes, such as the placing of part of the route in a cutting rather than on an embankment. Four of these alternatives were subsequently accepted by the Secretaries of State in their decision on the scheme, following recommendations by the inquiry inspector. These included: a slight westerly realignment of the route near its junction with the M6, to avoid the Mere Estate; putting the northern section of the route in a cutting rather than on a high embankment, to reduce the visual impact to and from Rostherne Mere and the Bollin Valley; bringing together the northern and southern carriageways along part of the route; and providing a replacement bridge, to retain access along an important side-road severed by the scheme (DOE/DOT 1995).

A strategic level of assessment

As we noted above, at the time of the publication of draft orders the proposed scheme was part of a much longer route, known as the Greater Manchester Western and Northern Relief Road (GMWNRR). The GMWNRR was divided into three main stages, each of which was to be subject to separate planning, EIA and consent procedures. The A556(M) scheme represented Stage I of the GMWNRR. Stage II involved the construction of a motorway link between the M56 and the M62 to the west of the Manchester conurbation, and Stage III continued the route along the M62 corridor around the northern perimeter of the conurbation. Public consultation on Stage II of the route took place in October 1992, at the time of the publication of draft orders for the A556 scheme. Despite this background, the EIS for the scheme made no explicit reference to the existence of the proposed GMWNRR, or to the relationship of the proposed scheme to Stages II and III of the route. No strategic assessment of the environmental consequences of the whole route appears to have taken place in this case. Although Stage II of the GMWNRR was subsequently abandoned by the DOT, following overwhelming public opposition to the proposals, the failure of the EIS even to mention the existence of the GMWNRR concept was unfortunate.

Indirect and consequential effects

The proposed scheme, as described in the draft orders and the EIS, was expected to have important implications for the existing motorway service area (MSA) alongside the M6 at Knutsford. The scheme proposals involved closing the existing north-facing slip-roads onto the M6, with the result that the service area would no longer be open to either northbound or southbound M6 traffic. This would therefore have left a gap of almost 40 miles between the nearest existing service areas at Sandbach and Charnock Richard, and might have been expected to result in a demand for a replacement MSA site in the vicinity of the Knutsford area. The need for such a replacement MSA, its possible site and the environmental effects of its development and operation were not addressed in the EIS for the A556(M) scheme.

However, at an earlier stage in the development of the scheme, the DOT identified a replacement site for such a service area and submitted a notice of proposed development on the site. This site was east of Arley Hall, some three miles north-west of the existing Knutsford service area (see Fig. 10.1). The DOT asked its consultants to include the site in their ecological survey and impact assessment carried out during the design and assessment of the A556 scheme. However, before the publication of draft orders for the scheme, the Department abandoned its plans to develop the Arley Hall site. This was because of changes to the planning regime for MSA provision introduced in August 1992. These transferred responsibility for identifying new MSA sites, seeking planning permission and acquiring the necessary land, from the DOT to the private sector (see Sheate & Sullivan 1993). As a result, the EIS for the A556(M) scheme did not identify the Arley Hall site or any other site for a replacement MSA. This means that a major form of consequential development resulting directly from the scheme, and with potentially significant environmental effects, was not addressed in the EIS.

A somewhat ironic postscript is that shortly after the submission of the EIS the scheme proposals were subject to further modifications, which involved the retention of access to the existing Knutsford MSA for M6 traffic. These changes therefore removed the need for a replacement MSA site. Whether EISs for motorway schemes should or could discuss the need for the provision or replacement of MSAs – and the environmental impacts of such provision – is open to debate. What is clear is that the present arrangements do not require any consideration of such consequential development. The removal of responsibility for MSA provision from the DOT has reinforced the separation between the planning and environmental assessments of motorway proposals and their associated service areas.

Other matters

The ready availability of environmental statements to interested members of the public is an important part of the proper functioning of the EIA process. Arrangements for copies of EISs to be made available in specified locations for perusal by the public are included in the various EIA regulations, including those for trunk road and motorway schemes. Government guidance also indicates, in relation to projects covered by the Town and Country Planning Regulations, that "the developer should make a reasonable number of copies of the statement available for sale to the public. A reasonable charge reflecting printing and distribution costs may be made" (DOE/WO 1989). It may also be expected that this guidance would be applicable to projects subject to other regulations, including road schemes.

The environmental statement for the A556(M) scheme was priced at £120 for Volume 1 (the main body of the EIS) and at £277 for all four volumes, including the detailed technical reports. Such charges appear far from reasonable, and are likely to have restricted the ability of members of the public to purchase copies of the EIS in this case. Although copies of the non-technical summary were made available free of charge, by its very nature this document contained only a brief description of the scheme and its consequences.

Conclusions

The case study indicates that some of SACTRA's recommendations appear to have been incorporated into the DOT's more recent EISs. The most notable changes are the consideration of impacts not included in the MEA (e.g. water and drainage), the abandonment of the framework as the means of presenting information on environmental effects and the inclusion of detailed technical reports and assessments in the EIS. The quality of road scheme EISs has undoubtedly improved substantially compared with the earlier examples criticized by SACTRA. However, there are continuing concerns about the quality of the wider EIA process for major road schemes. The arrival of the EIS at a time when many of the key decisions about the scheme had already been made and the limited treatment of alternatives, and of indirect and consequential impacts, are well illustrated by the case study.

10.4   EIA and the electricity supply industry

Introduction

The consideration of the environmental impact of project proposals in the UK electricity supply industry pre-dated the introduction of EC Directive 85/337 by several years. The evaluation of certain environmental impacts, especially those on air quality, can be traced back to the early years of the Central Electricity Generating Board (CEGB) in the late 1950s and early 1960s. At this time, a series of large 2000 MW coal-fired power stations were being planned. Baseline studies of sulphur dioxide (SO_2) concentrations and dust-fall in the vicinity of the proposed sites were carried out prior to operation, and monitoring was continued throughout the operational life of these new stations. Such information was used to test the calculations of stack-plume rise and the dispersion of pollutants, and it assisted in the development of present-day dispersion models (Manning 1991). Since its formation in 1957, the CEGB had also been obliged under the Electricity Act 1957 to take account of the effects of its projects on "local amenity", i.e. the natural beauty of the

311

countryside, the flora, fauna and geological and physiographical features of special interest, and buildings and other objects of architectural or historical interest.

Over the years, a considerable body of expertise was built up within the CEGB on the environmental implications of its projects. Environmental research was undertaken within the Board into a variety of issues, such as air pollutants, the projects' effects on the ecology of rivers and estuaries and the restoration of ash dumps to agricultural use (CEGB 1979, Howells & Gammon 1980, Sheail 1991). The CEGB also commissioned research from outside bodies, an example being studies of the socio-economic impacts of its proposals carried out by Oxford Polytechnic during the 1980s (Glasson 1984, Glasson et al. 1987).

By the early 1980s, after initial misgivings, the CEGB had become a firm supporter of the draft EC Directive on EIA. Sheail suggests that a pivotal event in helping to change attitudes may have been the Sizewell B public inquiry, the length of which made "a more formal environmental assessment procedure [begin] to appear more attractive" (Sheail 1991). Indeed, the Board was one of several major utilities, including the coal, oil and gas industries, to submit EISs with its proposals before the implementation of the Directive. Statements were prepared in 1987 and early 1988 for a PWR power station at Hinkley Point, three coal-fired stations at Fawley, Kingsnorth and West Burton, and a flue gas desulphurization plant at the existing Drax station. By the time the Directive came into force, substantial experience had been gained within the industry in producing EISs for a range of different projects. Sheail (1991) provides an excellent historical review of the treatment of environmental matters within the CEGB.

Projects subject to EIA under EC Directive 85/337

For projects within the electricity supply industry (ESI), EC Directive 85/337 was implemented by means of the Electricity and Pipeline Works (Assessment of Environmental Effects) Regulations 1989. Revised regulations were issued in early 1990 as a result of the privatization of the ESI under the Electricity Act 1989. In addition, certain projects in the ESI are covered by the Town and Country Planning (Assessment of Environmental Effects) Regulations. The need for EIA for different types of project is summarized below (see also Robson et al. 1994).

Projects subject to the secretary of state's consent[1] (and deemed planning consent): Electricity and Pipeline Works (AEE) Regulations:

- the construction or extension of a nuclear power station (Schedule 1);
- the construction or extension of a non-nuclear generating station with a heat output of at least 50 MW (Schedule 1 if 300 MW or more, Schedule 2 otherwise);
- the construction or diversion of an oil or gas pipeline at least 10 miles long (Schedule 2);

- the placement of an overhead transmission line (other than a service line) at least 10 miles long (Schedule 2).

For Schedule 2 projects, EIA is required "where the Secretary of State takes the view that the project would be likely to have significant environmental effects".

Projects not subject to the Secretary of State's consent: Town and Country Planning (AEE) Regulations:

- the construction or extension of a non-nuclear thermal power station, an installation for the production of electricity, steam and hot water (e.g. combined heat and power) or an installation for hydroelectric energy production, with a heat output of less than 50 MW (Schedule 2);
- the construction or diversion of an oil or gas pipeline of shorter than 10 miles (Schedule 2);
- the placement of an overhead transmission line shorter than 10 miles (Schedule 2).

Before 1994, there was some ambiguity in the regulations about the need for EIA for wind-power proposals. Such projects will almost always have a heat output below the 50 MW threshold and therefore do not require the Secretary of State's consent. They are therefore covered by the Town and Country Planning (AEE) Regulations, but were not specifically identified in either Schedule 1 or Schedule 2. Before 1994, a number of EISs for wind-turbine developments were requested by LPAS and the DOE under Schedule 2–3(a), i.e. "an installation for the production of electricity, steam and hot water". However, the CPRE correctly argued that this category would appear to refer to combined heat and power schemes rather than wind turbines, which do not produce steam or hot water. The CPRE expressed concern that, given the lack of any explicit inclusion of wind-power developments in Schedule 2, a developer challenging the need for EIA might well be successful (CPRE 1991b). This ambiguity was resolved by statutory instrument SI 677, published in 1994, which extended the range of projects covered by the Town and Country Planning (AEE) Regulations. Wind-farms were one of the three project categories added to Schedule 2 of the Regulations at this time (Bond 1997). Guidance on wind-farm EIA has also been published in recent years (DOE/WO 1993, FOE 1995; see also Coles & Taylor 1993, Hinson 1994), and wind-farms are now explicitly recognized in Annex II of the amended EC Directive.

Overhead transmission lines under 10 miles in length do not require the Secretary of State's consent and therefore again come under the Town and Country Planning (AEE) Regulations. However, such projects were defined as permitted development under the General Development Order 1988 and do not therefore require planning permission. Consequently, there is "no effective consent procedure to which an EIA requirement can be tied . . . EIAS are therefore not generally required for [such developments]" (CPRE 1991b). Statutory instrument SI 417, published in 1995, allows a developer to ask the local planning authority for an opinion on whether EIA should be carried out in cases of permitted development (Bond 1997).

Current issues in EIA in the electricity supply industry

The quality of environmental statements

A number of studies have suggested that the quality of environmental statements for projects in the electricity supply industry is above average. Lee & Brown (1992), in their study of EISs submitted up to early 1991, examined the quality of a large number of EISs using the familiar Lee and Colley review method (Lee & Colley 1992). They concluded that statements prepared under the Electricity and Pipeline Works Regulations were of well above average quality; 100 per cent were graded as "satisfactory" (grades A, B or C), and 50 per cent were graded as "good" (grades A or B). For EISs prepared under the Town and Country Planning Regulations, the proportion of satisfactory statements was much lower, at only 41 per cent, with less than a quarter (24 per cent) rated as good. Other studies have reached broadly similar conclusions. For example, Therivel et al. (1992) reviewed 59 EISs submitted before early 1992, for projects involving energy production, including power stations, wind-farms, transmission lines and pipelines. They concluded that: "Overall, the EISs were very good: using standard EIS review criteria (e.g. Lee & Colley [1992]), they would on average rank among the top one-quarter of EISs prepared to date."

Divided consent and EIA procedures for individual project components

Almost all the new power station proposals submitted since the implementation of the EC Directive on EIA have been gas-fired developments (Manners 1997). Although existing power station sites have been selected in a number of cases, these projects have usually necessitated the construction of new gas pipelines to the sites, as well as additional transmission connections off site. Typically each of these project components is carried out by a different company and is the subject of separate consent and EIA procedures. A good example is provided by the proposal to construct a 1,725 MW gas-fired combined heat and power (CHP) station at the existing ICI chemical complex at Wilton, Teesside, whose EIS states:

The overall project involves five major components:

1. A new natural gas pipeline from the North Sea ... to a landfall in the Teesside area to be built and run by others.
2. A gas reception and processing facility in the Teesside area to be built and run by others.
3. [The] combined heat and power (CHP) plant [at ICI Wilton] ...
4. A [gas] pipeline from the processing facility to the CHP facility to be built and run by others.
5. National Grid system upgrades by the National Grid Company will be necessary [i.e. transmission lines and substation improvements].

Each component will require a separate environmental assessment, with individual planning applications being submitted to the relevant authority as appropriate. (Cremer & Warner 1990)

The application for the CHP generating station was approved by the Secretary of State for Energy in November 1990. It later transpired that the project would necessitate the construction of extensive 400 kV overhead lines through open countryside (CPRE 1991b, Sheate 1994, 1995). This led the CPRE to lodge a formal complaint with the European Commission about the Secretary of State's decision. The CPRE argued that consent had been granted "without the implications of overhead transmission lines being properly addressed as part of the project EIA . . . The need for transmission lines from a power station is entirely dependent on the existence of the project and therefore should be considered as part of the EIA". The implication was that the Secretary of State had granted consent "without having [all] the necessary information available on which to base his decision" (CPRE 1991b).

This example is by no means unique and clearly highlights a problem with the current procedures. The separate consent procedures for different components of the same project divide the responsibility for EIA among many developers. EIA may well be carried out and reported on at different times for each component, depending on the timing of each consent application. The result is that all the environmental effects of the whole project are not assessed and presented together, for consideration by the Secretary of State (Sheate 1995).

The consideration of adjacent concurrent developments
A related issue concerns the extent to which the EIA of a project in the ESI takes account of adjacent development proposals. A review by Therivel et al. (1992) suggests that such cumulative impacts are rarely addressed. For example, only one out of eight EISs for pipelines and transmission lines considered cumulative impacts. EISs for power stations and wind-farms were worse still; none of the 17 EISs studied dealt with such impacts. An example is provided by two proposals for adjacent gas-fired power stations at Killingholme in Humberside. The CEGB's successor companies, National Power and PowerGen, submitted applications within a few months of each other for almost identical developments on adjacent sites. However, neither EIS examined the interaction between the two proposals or their likely cumulative impacts. Other examples are provided by applications for wind-farms on adjacent sites, often involving the same developer. The consideration of cumulative impacts in such cases is usually inadequate, if not non-existent.

Manning suggests that such examples raise a number of issues. Although the combined impacts of the different proposals "might be acceptable even if all the schemes went ahead . . . this would have to be demonstrated by an integrated assessment. Otherwise, a form of 'first come, first served' approach would arise by default" (Manning 1991). But who should carry out such an integrated assessment? If developers are required to take account of other related projects in the locality, problems can be anticipated. Access to the necessary information about competing companies' proposals may be problematic, and developers will surely wish to present the impacts of their own projects in a light more favourable than those of their competitors. An alternative would be for the local authority concerned to carry out itself or commission an integrated assessment of the various proposals. Given that

315

projects emerge at different times and may be subject to modification, when should such an assessment be carried out? In addition, if the assessment were to draw on information from each developer, "there [could] be scope for chaos if each interested party uses different assessment methods" (Manning 1991). An example would be the use of a different pollutant dispersion model by each developer to assess the impact of atmospheric emissions. Street (1997) describes the innovative approach to the assessment of cumulative air pollution impacts adopted by one LPA, Kent County Council.

It should be pointed out that for certain projects integrated assessments have been undertaken. An example concerns two applications for wind-farms on adjacent sites in Powys. The applications were submitted within a month of each other by the same developer. In this case, the local planning authority and the developer agreed that a single EIS should be submitted covering both sites.[2]

The treatment of alternatives

Ideally, EIA of individual projects in the ESI should include an assessment of alternative sites, fuels and technologies, or refer the reader to a higher tier of EIA where such an assessment has taken place. The final choice between these various alternatives should result from this EIA process. This does not mean that the least environmentally damaging options will necessarily be selected. However, it does mean that environmental considerations will be weighed with the relevant technical, commercial and other factors in arriving at the final decision. In practice, the consideration of alternatives in EISs for projects in the ESI tends to be rather limited. Therivel et al. (1992) report the results of a review of almost sixty energy-sector EISs submitted since the implementation of the EC Directive. They found that project-level EISs generally dealt poorly with issues of need, site selection and alternative locations and processes. However, some differences were apparent between different project types in the treatment of these issues. For example, EISs for pipelines and transmission lines tended to deal well with the issues of alternative routes, since the main way of mitigating impacts was to consider more sensitive route alignments. By contrast, EISs for power stations and wind-farms displayed a much more limited approach, with little or no reference to alternative locations or processes. For example, EISs for wind-farms, although often including a discussion of general siting criteria, made few references to specific alternative sites.

The need for an EIA of plans and programmes in the electricity supply industry

The limited treatment of cumulative impacts and alternatives in EIA at the project level suggests the desirability of an earlier, more strategic level of assessment. The need for an EIA of plans and programmes in the electricity supply industry has been recognized by a number of bodies in recent years and would appear to have become more pressing in the current privatized regime. For example, it has been argued that an EIA should be conducted into the national plan to implement the 1988 EC Large Combustion Plants (LCP) Directive (CPRE 1990, Sheate 1994).

The LCP Directive requires Member States to draw up programmes for the progressive reduction of emissions of SO_2 and NO_X (nitrous oxides) from large combustion plants. The National Plan drawn up by the UK government to comply with the Directive required phased reductions in SO_2 emissions of 20, 40 and 60 per cent by 1993, 1998 and 2003 compared with the 1980 level. Emissions of NO_X were to be reduced by 15 and 30 per cent by 1993 and 1998. As part of the National Plan, Her Majesty's Inspectorate of Pollution (HMIP) set limits for total annual emissions by National Power and PowerGen. For example, National Power was required to reduce its total SO_2 emissions from 1,600 kilotonnes in 1991 to 660 kilotonnes in 2003. Within this broad framework, the generators were left to decide how they would achieve these reductions. Possible responses by the ESI to meet HMIP's targets include the following:

- the use of low-sulphur coal in existing coal-fired power stations;
- the installation of flue gas desulphurization (FGD) equipment in existing larger coal-fired power stations;
- a move away from coal towards other fuels, especially gas, which involves minimal SO_2 emissions – in practice, this has been the main response of the generators and has entailed the construction of new gas-fired power stations.

Other possible responses might have included demand-side measures, such as improved energy efficiency. However, such measures are not the responsibility of the privatized electricity generators. The CPRE (1990) has argued that an EIA of all the alternative ways of reducing emissions, including demand-side measures, should be conducted by the government. Such an assessment would have identified the options likely to cause the least environmental damage before decisions about the use of different fuels and abatement technologies were made by the generators. Project-level EISs cannot perform such a role, since they are unable to influence the priority to be given to each option in the total programme to reduce emissions.

An issue of particular concern to the CPRE (1990) was the priority to be accorded to the installation of FGD plant as a way to reduce emissions and the choice between the alternative FGD methods available. It comments: "CPRE is . . . concerned about the potential impact of retro-fitting FGD equipment to power stations when the wider environmental effects of such a programme have not been adequately assessed beforehand". Two main FGD methods are available: the limestone-gypsum method and the regenerative process. These methods have different environmental implications, and one method may be more suitable at certain power station sites than at others. The limestone–gypsum method: (a) requires large quantities of quarried limestone; (b) produces large quantities of gypsum (some of which will find a commercial market as plasterboard, while the remainder will have to be disposed of); (c) involves the need to transport these materials to and from a site. The regenerative method requires different materials and produces different waste products (with a potential commercial use in the chemicals industry). To date, the generators have consistently favoured the limestone–gypsum method (Sheail 1991). CPRE noted with concern that the EISs for FGD plants at Ratcliffe (Nottinghamshire) and Ferrybridge

317

(North Yorkshire) did not even mention the alternative regenerative process. It claimed that these project EISs were prepared without a comparative assessment of the environmental impacts of the two methods, either overall or at the specific power station sites concerned; or, if such an assessment was carried out, it was not reported in the EIS (CPRE 1990).

The potential for the application of EIA to other policies, plans and programmes in the ESI is discussed in more detail in Therivel et al. (1992) and Byron & Sheate (1997). Chapter 13 includes further discussion of strategic environmental assessment.

The types of impact to be addressed in the EIS

Local versus regional and global impacts. Most EISs for power station projects concentrate largely on the local impacts of the developments. For example, the consideration of atmospheric emissions will focus on the implications for SO_2 and NO_X concentrations near the site, much less so on the far-afield effects of these emissions. As Manning states, "it is unrealistic to expect something as complex as a global warming analysis to be applied in the context of individual planning applications for combustion sources" (Manning 1991). He goes on to argue for a clearer distinction to be made in the EIA process between local and wider regional and global issues. EIA at the project level should continue to focus on an assessment of local impacts, issues such as visual intrusion, noise and dust probably deserving more attention than at present. Manning notes that the wider regional and global impacts of projects are increasingly being controlled by national and international regulation, such as the EC LCP Directive. He suggests that EIA at the level of policies, plans and programmes is the appropriate context in which to deal with such concerns. However, in the period before the implementation of such a strategic level of EIA in the ESI, "the extent to which [project] EIAs should address wider issues is a matter for interpretation" (Manning 1991).

Socio-economic versus physical environmental impacts. EISs for projects in the electricity supply industry generally include considerations of socio-economic impacts as well as of the more conventional physical environmental impacts. This aspect of current practice is to be welcomed (Glasson & Heaney 1993). The formal prediction and monitoring of such impacts dates back to the late 1970s, when the CEGB commissioned research into the local social and economic effects of its construction projects and its existing operational stations. These studies have continued to the present day. Although socio-economic impacts can be negative as well as positive, developers have clearly welcomed the opportunity to include the employment and wider economic benefits of their schemes within the format of the EIS.

The treatment of uncertainty

The issue of uncertainty raises problems in EIA. Uncertainty caused by the continual refinement and modification of a project proposal during and after the preparation of the EIS is not unique to projects in the ESI (see Frost 1997). However, several factors suggest the likelihood of particular problems with such projects. First, power station design and layout may be dependent on the choice of the main

318

contractor to construct the station, a choice that is unlikely to have been made at the time of the EIS submission. For example, National Power's EIS for a gas-fired station at Didcot (Oxfordshire) makes the following statement: "contractors' plant and station designs are known to differ significantly . . . and the final plant configuration and layout will depend on the choice of main contractor following competitive tendering" (National Power 1990). The result is that the project as described in the EIS is specified in very provisional terms (Bird 1996). For example, the EIS identifies three different cooling options for the new station: (a) natural draught cooling towers, 114 m high; (b) mechanical draught cooling towers, 18–22 m high; (c) air-cooled condensers, about 30 m high. A fourth option was presented after the sub-mission of the EIS. Clearly, each of these options would have very different visual impacts (including the visibility of plumes as well as of the towers).

A second reason for uncertainty is that certain design details will be subject to approval or modification by the Environment Agency as part of the pollution control authorization process (Bird 1996; Bird & Therivel 1996; see also Ch. 9 for a more detailed discussion). Again, the Didcot EIS states that: "exhaust gases from [each] boiler would be discharged to the atmosphere via a chimney, the height of which . . . is subject to approval by HMIP, but is expected to be about 65 metres above ground level". The EIS states that there would be six of these chimneys, although again this would be subject to approval by HMIP (now incorporated into the Environment Agency). Any modifications to stack height or the number of stacks would have implications for the assessment of both air quality and visual impacts. Bird notes that the final approved design of the Didcot B station, after the pollution control authorization process, showed a number of major modifications to the design described in the EIS. For example, instead of six chimneys 65 metres high, the final design involved two 85-metre chimneys. Similarly, the final choice of cooling system comprised banks of air-cooled condensers less than 17 metres high; the design assumed in the EIS, and used in the landscape and visual assessment, had comprised two cooling towers 114 metres high. As Bird comments:

> Such major design changes must make [many of] the EIS predictions irrelevant, especially [those contained in] the landscape impact assessment. The EIS for [this development] is more of a discussion document than a finalised statement of the environmental impacts [of] the development. (Bird 1996)

Several approaches are available to developers when dealing with such uncertainties in their EISs. The first would be simply to assess the effects of the most likely option; a second would be to assess the option likely to give rise to the most significant impacts (the worst case); a third would be to assess the effects of all realistic options in the EIS. The second and third approaches would appear to be the most satisfactory ways of acknowledging uncertainty in the EIS.

Local authority review

Although most power station proposals are subject to the Secretary of State's con-sent, local authorities will of course wish to examine the physical environmental and socio-economic consequences of such projects. Any objection by a local authority will generate a public inquiry into the proposals. However, the complexity of certain impacts arising out of power station projects is likely to give rise to difficulties for local authorities in their consideration of such schemes. Local authorities may not possess the in-house expertise to conduct a thorough review of all aspects of an EIS. They may therefore decide to use outside consultants to review project proposals and EISs in such cases.

An example is provided by the decision of Oxfordshire County Council to commission environmental consultants to review National Power's EIS for its proposed gas-fired power station at Didcot (Environmental Resources Ltd 1991). The consultants were asked to review specific sections of the EIS, namely those dealing with atmospheric emissions, noise and vibration, water use and (more briefly) socio-economic issues. More detailed studies on noise and air-quality impacts, involving new survey work and the use of different pollutant dispersion models, were also commissioned. Impacts not included in the external review were those on flora, fauna and transport (which were reviewed in-house by the county council) and landscape and visual effects (which were reviewed by consultants commissioned by the district council). Other issues, strictly outside the remit of the EIS, were also addressed in the review. These included the need for the project, possible alternative fuels, the effects of a FGD installation at the existing coal-fired station at Didcot, and the scope for a combined heat and power scheme on the site. The review was carried out within five weeks, and its results were used by the county council during the public inquiry into the scheme proposals.

10.5 Summary

The formal environmental appraisal of trunk road schemes had been in operation well before the implementation of EC Directive 85/337. Indeed, the existing methods and procedures of environmental appraisal heavily influenced the way in which the Directive was interpreted by the DOT. The influential report by the Standing Advisory Committee on Trunk Road Assessment (SACTRA), published in 1992, raised a number of specific concerns about the implementation of the Directive. A number of SACTRA's recommendations, particularly those concerning the presentation of information and the range of impacts to be addressed, appear to have been incorporated into the most recent road scheme EISs. The quality of such EISs has undoubtedly improved substantially compared with the earlier examples criticized by SACTRA. Despite this, there are continuing concerns about the adequacy of the wider EIA process for major road proposals, including the arrival of the EIS late in the decision-making process, the limited treatment of alternatives, and indirect and cumulative impacts.

As with trunk road proposals, the assessment of environmental impacts in the electricity-supply industry pre-dated the introduction of EC Directive 85/337 by several years. Over the years, a considerable body of expertise was built up within the former CEGB on the environmental implications of its projects. Much of this experience has been inherited by the new privatized generating companies. Current weaknesses in the EIA of projects in the sector include the separate consent and EIA procedures for linked project components and the limited treatment of adjacent developments and alternatives. Flowing from these concerns, it has been argued that scope exists for a more strategic level of assessment in the electricity supply sector. This echoes the views expressed by SACTRA about trunk road schemes.

References

Bird, A. 1996. *Auditing environmental impact statements using information held in public registers of environmental information*. Working Paper no. 165. Oxford: Oxford Brookes University, School of Planning.

Bird, A. & R. Therivel 1996. Post-auditing of environmental impact statements using data held in public registers of environmental information. *Project Appraisal* **11**(2), 105–16.

Bond, A. J. 1997. Environmental assessment and planning: a chronology of development in England and Wales. *Journal of Environmental Planning and Management* **40**(2), 261–71.

Box, J. D. & D. Markham 1994. Roads, routes and resources. *Landscape Design*, June, 38–41.

Box, J. D. & J. E. Forbes 1992. Ecological considerations in the environmental assessment of road proposals. *Highways and Transportation*, April, 16–22.

Byron, H. & W. R. Sheate 1997. Strategic environmental assessment: current status in the water and electricity sectors in England and Wales. *Environmental Policy and Practice* **6**(4), 155–65.

Bruton, M. J. 1985. *Introduction to transportation planning*. London: UCL Press.

Bryant, B. 1996. *Twyford Down: roads, campaigning and environmental law*. London: E & FN Spon.

CEGB (Central Electricity Generating Board) 1979. *Electricity supply and the environment*. London: HMSO.

Chris Blandford Associates 1994. *Wind turbine power station construction monitoring study*. Bangor, Gwynedd: Countryside Commission for Wales.

Coles, R. W. & J. Taylor 1993. Wind power and planning – the environmental impact of windfarms in the United Kingdom. *Land Use Policy* **10**(3), 205–26.

CPRE (Council for the Protection of Rural England) 1990. *The proposed statutory plan for reductions of emissions of sulphur dioxide and oxides of nitrogen from existing large combustion plants in the United Kingdom*. Response to the government consultation paper on Implementation of the Large Combustion Plants Directive. London: CPRE.

CPRE 1991a. *UK implementation of environmental assessment for roads: a complaint to the European Commission*. London: CPRE.

CPRE 1991b. *The environmental assessment directive – five years on*. Submission to the European Commission's five-year review of Directive 85/337/EEC on environmental assessment. London: CPRE.

CPRE 1992. *Rules of the road*. London: CPRE.

CPRE 1993. *Driven to dig – road-building and aggregates demand*. London: CPRE.

Cremer & Warner Consulting Engineers & Scientists 1990. *Environmental assessment of proposed combined heat and power generating facility: technical report*. Cremer & Warner.

DETR (Department of the Environment, Transport and the Regions) 1997. *Consultation paper: developing an integrated transport policy*. London: DETR.

DOE (Department of the Environment) 1996. *Changes in the quality of environmental statements for planning projects*. London: HMSO.

DOE/DOT 1995. *Highways Act 1980, A556(M) Improvement (M6–M56 Link)*. Inspector's Report and Secretaries of State's Decision Letter. CNW 218/8/1/08, 7 July. Manchester: Government Office for the North West.

DOE/WO (Department of the Environment/Welsh Office) 1989. *Environmental assessment: a guide to the procedures*. London: HMSO.

DOE/WO 1993. *PPG 22 – renewable energy*. London: HMSO.

DOT (Department of Transport) 1978. *The report of the Advisory Committee on Trunk Road Assessment*. London: HMSO.

DOT 1979. *Trunk road proposals – a comprehensive framework for appraisal*. Report of the Standing Advisory Committee on Trunk Road Assessment. London: HMSO.

DOT 1983. *Manual of environmental appraisal*. London: HMSO.

DOT 1989. *Departmental standard HD 18/88: environmental assessment under EC Directive 85/337*. London: HMSO.

DOT 1992a. *Assessing the environmental impact of road schemes*. Report of the Standing Advisory Committee on Trunk Road Assessment. London: HMSO.

DOT 1992b. *Assessing the environmental impact of road schemes*. Response by the Department of Transport to the report by the Standing Advisory Committee on Trunk Road Assessment. London: HMSO.

DOT 1992c. *A556(M) improvement (M6–M56) environmental statement*, vols I and II. Prepared by Allott & Lomax Consulting Engineers. Sale, Greater Manchester: Allott & Lomax.

DOT 1993. *Design manual for roads and bridges*, vol. 11: *Environmental assessment*. London: HMSO.

DOT 1994a. *Trunk roads and the generation of traffic*. Report of the Standing Advisory Committee on Trunk Road Assessment. London: HMSO.

DOT 1994b. *Trunk roads and the generation of traffic*. Response by the Department of Transport to the report by the Standing Advisory Committee on Trunk Road Assessment. London: HMSO.

DOT 1996. *Transport – the way forward: the government's response to the transport debate*. London: HMSO.

DTI (Department of Trade and Industry) 1992. *Guidelines for the environmental assessment of cross-country pipelines*. London: HMSO.

English Nature 1994. *Roads and nature conservation: guidance on impacts, mitigation and enhancement*. Peterborough: English Nature.

Environmental Resources Ltd 1991. *Environmental review of the proposed combined-cycle gas turbine power station at Didcot: final report*. London: ERL.

ERM (Environmental Resources Management) 1994a. *Channel Tunnel Rail Link and M2 junction 1 to 4 widening: combined effects annex*. Annex to environmental statements for CTRL and M2 widening schemes. London: ERM.

ERM 1994b. *Setting forth: environmental appraisal of alternative strategies*. Edinburgh: ERM.

Ferrary, C. 1994. *A methodology for integrating environmental concerns into transport planning*. Paper presented to the 22nd European Transport Forum (the PTRC Summer Annual Meeting), 12–16 September, University of Warwick.

FOE (Friends of the Earth) 1995. *Planning for wind power – guidelines for project developers and local planners*, 2nd edn. London: FOE.

Frost, R. 1997. EIA monitoring and audit. In *Planning and environmental impact assessment in practice*, J. Weston (ed.), 141–64. Harlow: Longman.

Glasson, J. 1984. Local impacts of power station developments. In *Energy policy and land use planning*, D. Cope, P. Hills, P. James (eds), 123–46. Oxford: Pergamon.

Glasson, J., M. J. Elson, D. Van der Wee, B. Barrett 1987. *Socio-economic impact assessment of the proposed Hinkley Point C power station: final report*. Oxford: Oxford Polytechnic, School of Planning.

Glasson, J. & D. Heaney 1993. Socio-economic impacts: the poor relations in British environmental impact statements. *Journal of Environmental Planning and Management* **6**(3), 335–43.

Headicar, P. & B. Bixby 1992. *Concrete and tyres – local development effects of major roads: a case study of the M40*. London: CPRE.

Hinson, P. 1994. *Environmental assessment for wind farms*. Paper presented to 5th Annual Conference on Advances in Environmental Impact Assessment, IBC Technical Services Ltd, November, London.

Hopkinson, P. G., J. Bowers, C. A. Nash 1990. *The treatment of nature conservation in the appraisal of trunk roads*. Submission to the Standing Advisory Committee on Trunk Road Assessment. Peterborough: Nature Conservancy Council.

Howells, G. D. & K. M. Gammon 1980. *Role of research in meeting environmental assessment needs for power station siting*. Paper presented to British Ecological Society Conference on Ecological Aspects of Environmental Impact Assessment, April, University of Essex.

Jones, T. 1992. *Environmental impact assessment for coal*. London: IEA Coal Research.

Kunzlik, P. 1996. The lawyer's assessment. In *Twyford Down: roads, campaigning and environmental law*, B. Bryant (ed.). London: E & FN Spon.

Lee, N. 1992. EA and trunk roads. *Project Appraisal* **7**(4), 261–2.

Lee, N. & D. Brown 1992. Quality control in environmental assessment. *Project Appraisal* **7**(1), 41–5.

Lee, N. & R. Colley 1992. *Reviewing the quality of environmental statements*. Occasional Paper no. 24, 2nd edn. Manchester: University of Manchester, Department of Planning and Landscape.

Lee, N. & F. Walsh 1992. Strategic environmental assessment: an overview. *Project Appraisal* **7**(3), 126–36.

Lewis, M. 1994. Long-awaited and comprehensive. *Project Appraisal* **9**(4), 290–1.

Macpherson, G. 1993. *Highway and transportation engineering and planning*. Harlow: Longman Scientific and Technical.

Manners, G. 1997. Gas market liberalization in Britain: some geographical observations. *Regional Studies* **31**(3), 295–309.

Manning, M. 1991. *The role of environmental assessment in power system planning*. Paper presented to IBC/IEA conference on Advances in Environmental Assessment, October, London.

NAO (National Audit Office) 1994. *Department of Transport: environmental factors in road planning and design*. London: HMSO.

National Power 1990. *Environmental statement: Didcot B proposed combined-cycle gas turbine power station*. London: National Power.

OECD (Organization for Economic Co-operation and Development) 1994. *Environmental impact assessment of roads*. Paris: OECD.

Pearce, D. W., A. Markandya, E. B. Barbier 1989. *Blueprint for a green economy*. London: Earthscan.

RCEP (Royal Commission on Environmental Pollution) 1994. *Eighteenth report: transport and the environment*. London: HMSO.

Robson, A., T. L. Shaw, C. J. L. Taylor 1994. Electricity generation. In *Environmental assessment: a guide to the identification and mitigation of environmental issues in construction schemes*. Construction Industry Research and Information Association, CIRIA Special Publication no. 96. London: CIRIA.

RSPB (Royal Society for the Protection of Birds) 1995. *Wildlife impact – the treatment of nature conservation in environmental assessment*. Sandy: RSPB.

Salter, J. R. 1992. Environmental assessment: the challenge from Brussels. *Journal of Planning and Environment Law*, January, 14–20.

Scottish Office 1994. *Setting forth: strategic assessment summary report*. Edinburgh: Scottish Office Industry Department, Roads Directorate.

Sheail, J. 1991. *Power in trust: the environmental history of the Central Electricity Generating Board*. Oxford: Oxford University Press.

Sheate, W. R. 1992. Strategic environmental assessment in the transport sector. *Project Appraisal* 7(3), 170–4.

Sheate, W. R. 1994. *Making an impact: a guide to EIA law and policy*. London: Cameron May.

Sheate, W. R. 1995. Electricity generation and transmission: a case study of problematic EIA implementation in the UK. *Environmental Policy and Practice* 5(1), 17–25.

Sheate, W. R. & M. Sullivan 1993. *Campaigners' guide to road proposals*. London: Council for the Protection of Rural England / Transport 2000.

Simpson, M. 1992. Transport evaluation of highway schemes. *Project Appraisal* 7(4), 197–203.

Street, E. 1997. EIA and pollution control. In *Planning and environmental impact assessment in practice*, J. Weston (ed.), 165–79. Harlow: Longman.

Therivel, R., E. Wilson, S. Thompson, D. Heaney, D. Pritchard 1992. *Strategic environmental assessment*. London: RSPB/Earthscan.

Tinch, R. 1996. *Valuation of environmental externalities*. London: HMSO.

Treweek, J. R., S. Thompson, N. Veitch, C. Japp 1993. Ecological assessment of proposed road developments: a review of environmental statements. *Journal of Environmental Planning and Management* 36(3), 295–307.

Tromans, S. 1991. Roads to prosperity or roads to ruin? Transport and the environment in England and Wales. *Journal of Environmental Law* 3(1), 1–37.

Wilson, E. 1994. *Progress with strategic environmental assessment*. Paper presented to 5th Annual Conference on Advances in Environmental Impact Assessment, IBC Technical Services Ltd, November, London.

Wood, D. A. 1992. *Assessing the environmental impact of road schemes: the SACTRA report*. Paper presented to the 20th European Transport Forum (the PTRC Summer Annual Meeting), September, UMIST, Manchester.

Notes

1 In England, the consent of the Secretary of State for Energy was required prior to the incorporation of energy matters into the Department of Trade and Industry. The consent of the Secretary of State for Trade and Industry is now required.
2 The applications were submitted in 1991 by a joint venture company led by Ecogen Ltd, for wind-farms on sites at Penrhyddlan and Llidiartywaun in Montgomeryshire District, Powys.

CHAPTER 11
Comparative practice

11.1 Introduction

This chapter takes different developed EIA systems from countries in Europe, North America, Australia and Asia to illustrate the range of existing EIA systems and to act as comparisons with the UK and EC systems discussed earlier. These systems include several elements of good practice, including "model" systems generally praised for their comprehensiveness and effectiveness (e.g. from the Netherlands, Canada). For each country, a case study is presented to highlight some of the successes and failures of the systems in practice.

The Netherlands was chosen as an example because the country is well known for its progressive and well-developed environmental policies, which are based on the principle of sustainable development. The Dutch EIA system incorporates a particularly high level of public consultation, and uses an independent EIA Commission to scope each EIA and subsequently review its adequacy. The case study of a demonstration integrated gasification combined-cycle power station shows the broad range of alternatives addressed and the efficiency of the system.

Canada is also known for its progressive environmental policies. Its federal EIA system has good procedures for public participation and review, and a particular strength lies in its emphasis on the monitoring of a project's actual impacts after construction. Canada's provinces have separate EIA systems for projects under their jurisdiction. The case study of British Columbia reveals some of the strengths and weaknesses of one provincial EIS system, and proposals for and perspectives on reform.

In *Australia* the responsibility for EIA is also shared between the national and state governments. However, a high level of government discretion and the low level of public participation render this system probably less powerful than the systems of Canada and the Netherlands. The case study of the third runway at Sydney Airport shows that environmental considerations may well be marginalized in the economic debates on a project.

Japan has no national legal requirement for the preparation of EIAs, in contrast to the other systems discussed here, but instead it has EIA guidelines for the national ministries to follow. At the local level, some EIA regulations and (mostly) guidelines have also been established. Generally this system seems to ensure that the most environmentally harmful proposals are avoided. The case study of the Trans-Tokyo Bay Highway shows how the different national and local EIA procedures interact.

Table 11.1 gives an overview of the status of EIA systems worldwide, to the best of the authors' knowledge. Member States of the European Union are not listed, as they are discussed in greater depth in Appendix 2. For countries where an early EIA regulation has recently been updated, the table tries to give the more recent date. The table also lists selected post-1989 references regarding EIA in these countries, where they exist. Generally case studies of individual projects are only included where these address the general EIA system as well. For some countries, such as the USA and Australia, so much information exists that only a very few references are listed. Roe et al. (1995) and the Netherlands Commission for Environmental Impact Assessment (1996) have published good listings of EIA guidelines from around the world, which are not shown in the table. Regular updating of information on EIA systems in other countries can be found in the University of Manchester's *EIA Newsletter* and in *Environmental Impact Assessment Review*.

So far, this book has concentrated on EIA as it is carried out in industrialized Western countries. It is there that environmental problems first emerged, EIA systems were first developed, and the most experience with EIA has been amassed over the years. However, as can be seen from Table 11.1, many EIA systems have emerged elsewhere in the world, particularly in the past ten years. In Asia, EIA systems vary, from Hong Kong's sophisticated procedures, which include monitoring and application to strategic actions, to no EIA system in Laos. Most Asian EIA systems take the form of regulations established in the late 1980s and early 1990s. In South America, although Colombia established one of the world's first EIA systems in 1973, and Brazil established its EIA regulations in 1981, most other EIA systems are less than 10 years old, and many countries have guidelines rather than regulations. EIA in Africa is less advanced than in other parts of the world. About a dozen African countries have EIA systems, virtually all established since 1990, although many *ad hoc* EIAs have been carried out in response to donor agencies' requirements. Because many of these EIA systems have been established quite recently, generally only limited, if any, information is available about their effectiveness. Section 11.3 reviews some of the achievements and problems encountered in countries with emerging EIA systems. Despite institutional and technical hurdles, EIA is becoming increasingly accepted, codified and used as an integral part of project planning by increasingly well-trained indigenous staff. Finally, Section 11.4 discusses the important role of international funding institutions, such as the World Bank, the United Nations and the European Bank for Reconstruction and Development, in developing and spreading good EIA practice for the projects and programmes they fund.

Table 11.1 Existing EIA systems worldwide, and selected references.

Country	Guideline (G) or regulation (R)	Date of implementation	References
AFRICA, MIDDLE EAST			Kakonge 1994, 1995; Kakonge & Imevbore 1993
Botswana	R*		Okaru & Barannik 1996
Burkina Faso	R		Okaru & Barannik et al. 1996
Egypt			
Ethiopia	–	–	Okaru & Barannik et al. 1996
Ghana	R	1994	Allotey 1994, Allotey & Amoyaw-Osei 1996a, 1996b, Baryeh et al. 1996, Ofori 1991, Okaru & Barannik et al. 1995
Israel	R	1982	Amir 1992, Brachya 1993
Kenya	R*		Okaru & Barannik 1996, Hirji & Ortolano 1991
Madagascar	R	1995	Peters 1994
Mauritius	R	1991	Okaru & Barannik et al. 1996
Mozambique	R	1994	
Namibia	R	1994	Okaru & Barannik et al. 1996
Nigeria	R	1992	Dung-Gwom 1996, Okaru & Barannik et al. 1996
Oman	R?	1993	
South Africa	R*		Okaru & Barannik et al. 1996, Moshoeshoe & Malatsi 1996, Sowman et al. 1995, Wiseman 1996
Tanzania	–	–	IIED 1995, Kamukala 1992, Okaru & Barannik et al. 1996, UNEP 1994
Uganda	R	1995	Douthwaite 1996
Zambia	R	1990	Baryeh et al. 1996
Zimbabwe	R*	1994	Chaibva 1996, Mubvami 1996, Okaru & Barannik et al. 1996
AMERICAS (except USA)**			Chico 1995a, LaRovere & Baraton 1996, Moreira 1992
Bolivia	G	1993	
Brazil	R	1981	Brito & Moreira 1995, Fowler & Dias De Aguiar 1993, Hacon 1990
Canada	R	1992	CEAA 1996, Wood 1995
Caribbean	G	1991	
Chile	G	1993	
Colombia	R	1973	
Costa Rica	G	1994	
Mexico	R	1988	Gomez et al. 1996, Pisanty-Levy 1993, Sanchez-Silva & Cruz-Ulloa 1994, Tortajada-Quiroz 1996
Paraguay	R	1993	
Peru	R	1992	Iglesias 1996
Uruguay	R	1994	de Mello 1995
Venezuela	R	1983	Chico 1995b

Table 11.1 (cont'd)

Country	Guideline (G) or regulation (R)	Date of implementation	References
ASIA			Briffett 1995, 1996, Nay Htun 1992, Sadar & Si 1994, Welles 1995, Xie et al. 1996
Bangladesh	G*		
Bhutan	G	1993	
China	R	1990s	Fearnside 1994, Ortolano 1996, Sinkule & Ortolano 1995, Wenger & Huadong 1990
Hong Kong	R	1990	Au 1996, Coombs 1993, Reed 1994
India	G	1989	Modale 1994, Valappil et al. 1994, Vizayakumar & Mohapatra 1991
Indonesia	R	1986	Coles 1992, Smith & van der Wansem 1995
Japan	G	1984	Barrett & Therivel 1991, Kurasaka 1996, Takabe 1994
Korea	R	1979	Hahn 1993, Han et al. 1996, Kim & Murabayashi 1992
Malaysia	R	1987	Coles 1992, Ibrahim 1992, Nor 1991
Nepal	G	1992	Devkota 1996
Pakistan	G	1986	
Papua New Guinea	R	1978	
Philippines	R	1977	Cardenas 1995, Ross 1994, Smith & van der Wansem 1995
Singapore	–	–	Briffett 1994, 1995
South Pacific	T		Onorio & Morgan 1995, Schoeffel 1995, SPREP 1993
Sri Lanka	R	1984	Smith & van der Wansem 1995
Taiwan	R	1994	Leu et al. 1996
Thailand	G	1979	Tongcumpou & Harvey 1994
Vietnam	R	1994	Le Duc, Tran Van & Nierynck 1997
AUSTRALIA, NEW ZEALAND, ANTARCTICA			
Antarctica	–	–	Burgess et al. 1992, UK FCO 1995
Australia	R	1987	Harvey 1998, Harvey & Ferguson 1994, Harvey and McCarthy 1997, Thomas 1996, Wood 1993, 1995, Wood & Bailey 1994
New Zealand	R	1991	Dixon & Fookes 1995, Hughes 1996, Montz & Dixon 1993, Morgan 1995, Wood 1995
EUROPE (except EU*)**			
Armenia	R	1995	Ter-Nikoghosyan 1996
Baltic States	R*		
Belarus	R	1992	
Bulgaria	R	1991	Veleva 1996
Croatia	R	1984	
Czech Republic	R	1992	Branis 1994, Branis & Kruzikova 1994, Cizkova & Zenaty 1993, Romanillos et al. 1996
Estonia	R	1992	EBRD 1994, Peterson 1995

Table 11.1 (cont'd)

Country	Guideline (G) or regulation (R)	Date of implementation	References
Hungary	R	1993	EBRD 1994, Mondok 1995, Radnai 1993
Kazakhstan	R*	1991	
Latvia	R	1990	EBRD 1994
Lithuania	R*	1995	EBRD 1994, Grönlund 1996
Moldova	R	1996	
Norway	R	1990	
Poland	R	1990	Commission for EIA in Poland Bulletin, EBRD 1994, Jendroska & Sommer 1994, Kassenberg 1994, Rzsezot 1995
Romania	R	1990	Romanillos et al. 1996
Russian Federation	R	1991	Cherp 1995, 1996, Cherp & Khotuleva 1996
		1995	
Slovak Republic	R	1994	Bianchi & Rosova 1995, EBRD 1994, Husenicova 1996, Huskova 1994, Kozova & Drdos 1995, Pavlickova et al. 1995, Vrbensky & Kolocany 1995
Slovenia	R	1994	EBRD 1994
Switzerland	R	1985	Ruchti 1993
Ukraine	R	1991	

Note: G = guideline, R = regulation, T = training manual, – = no known EIA regulations or guidelines, * = partial, ** See Chapter 2 for USA and constituent states, *** = European Union (EU) Member States (discussed in greater depth in Appendix 2)

11.2   Developed EIA systems

The Netherlands

The Netherlands, with its small area of densely populated and highly industrialized land, has developed a worldwide reputation for powerful and progressive environmental legislation. The National Environmental Policy Plan (NEPP) of 1989 and an update (NEPP-plus) of 1990 (Ministry of Housing, Physical Planning and Environment 1989, 1990) established a national environmental strategy based on the concept of sustainable development. These plans in turn were based on an earlier report, *Concern for tomorrow*, published by the National Institute of Public Health and Environmental Protection, which specified targets (e.g. for emission controls) that would need to be met if the Netherlands were to achieve sustainable development.

330

The Netherlands, on the basis of NEPP and NEPP-plus, is implementing far-reaching changes in a wide range of areas. For instance, waste-disposal policies aim to reduce waste disposal on land from 55 to 10 per cent of all wastes produced by 2000; targets have been set for air pollution emissions for 2000, broken down by sector; and transport and land-use policies are being jointly developed to minimize the need for car travel. The cost of these measures is expected to be 0.9–2.6 per cent of Dutch GNP in 2010.

Legislative framework

EIA in the Netherlands, as in the UK, is required by EC Directive 85/337. This is implemented through the Environmental Protection (General Provisions) Act of June 1986, the EIA Decree of May 1987, which designates activities subject to EIA and the Notification of Intent EIA Decree of July 1987, which designates the contents and requirements of the notification of intent. EIA is required for a so-called "positive list" of projects that are considered to have a significant impact on the environment. This list is based on Annexes I and II of Directive 85/337, with further additions, and with the exemption of projects that are expected to have no serious harmful environmental consequences.[1] In addition, EIA is currently required for sectoral plans on waste management, the supply of drinking water, energy and electricity, and some land-use plans. EIA procedures are also being developed for other policies, plans and programmes; this is discussed further in Chapter 13.

Procedures

Figure 11.1 summarizes the Netherlands' EIA procedures. Once a (public or private) developer decides to carry out an activity included in the "positive list", the competent authority is informed, a "notification of intent" is published, and the Minister of Environment notifies the EIA Commission.[2] Once the Commission is notified, the chairman sets up a panel of five to eight experts, which carries out a scoping exercise to assess the range and magnitude of the impacts and the alternatives that the EIS should cover. Within two months, it must present an advisory note of project-specific guidelines to the competent authority. The competent authority in turn produces formal EIS guidelines for the action, specifying the alternatives and the main environmental impacts that the EIS must address. The first scoping stage must take place within three months.

The developer is responsible for preparing the EIS, which must include:

- a statement of the purpose and reason for the activity;
- a description of the activity and "reasonable" alternatives (including that least harmful to the environment and the do-nothing option);
- an overview of the specific decision(s) for which the EIS is being prepared and of decisions already taken regarding that activity;
- a description of the existing environment, and the expected future state of the environment if the activity is not carried out;

331

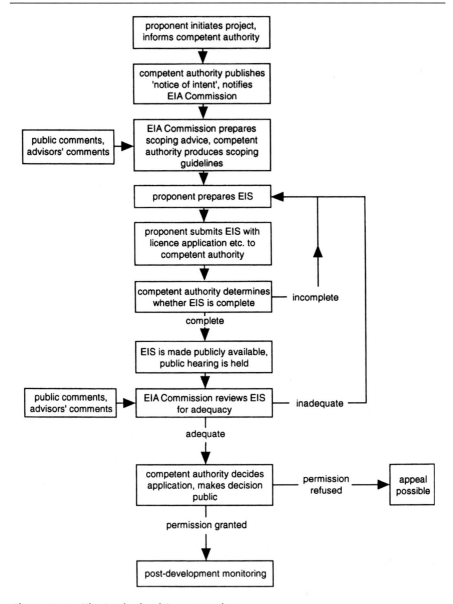

Figure 11.1 The Netherlands' EIA procedures.

- a description of the environmental impacts of the proposed activity and alternatives, and the methods used to determine these impacts;
- a comparison of the activity with each alternative;
- gaps in knowledge;
- a non-technical summary.

The developer submits the EIS to the competent authority, which has six weeks to decide whether it meets the criteria of the guidelines or whether any corrections or amendments are required. The findings of this inspection are made public. Once it has been accepted, the EIS is made publicly available for one month. During this time, a public hearing must be held, and the public may comment on the EIS. At this time bodies such as the Regional Inspector for Environmental Protection also provide advice to the competent authority concerning the contents of the EIS. A record of the public review and other advice is then passed to the EIA Commission.

The Commission receives a copy of the EIS once it has been prepared, and checks the statement against current legislation and the EIS guidelines. It also considers the advice and public review.[3] The Commission's review is generally guided by two issues: whether the EIS can assist in decision-making, and, if so, whether it is complete and accurate. The review concerns the adequacy of the EIS, not the environmental acceptability of the activity. Within two months of receiving the EIS, the Commission sends the results of this review to the competent authority, which makes the final decision.

The competent authority makes a decision based on the EIS, the advisors' comments, the Commission's review, and the results of the public hearing. It makes the results of this decision known, including how a balance was struck between environmental and other interests, and how alternatives were considered. The competent authority must subsequently monitor the project, based on information provided by the developer, and make the monitoring information publicly available. If actual impacts exceed those predicted, the competent authority must take measures to reduce or mitigate them.

EIA in the Netherlands does appear to influence decisions on projects. Van de Gronden (1994) believes that positive influences of the Dutch EIA system include: the withdrawal of unsound projects; the legitimization of sound projects; the selection of improved project locations; the reformulation of plans; a redefinition of the goals and responsibilities of projects' proponents. Forty EISs were prepared in 1988, 57 in 1989, and about 70 in 1990, rising to over 80 in 1994. By 1 July 1990, 159 EIA procedures had been begun for 191 activities (some EIAs cover more than one activity). The 191 activities include waste management (89), landfills (37), chemical waste processing (23), incineration installations (15), plans (14) and roads (13). Further information on the Dutch EIA system can be found in Koning (1990), van Haeren (1991), UNECE (1991) and Netherlands Commission for Environmental Impact Assessment (1996).

Case study: the integrated gasification combined-cycle (IGCC)
demonstration plant at Buggenum
In response to the policy for the diversification and competitive pricing of electricity supplies set out in the Dutch Electricity Plan 1989–98, the Dutch utilities' parent organization SEP[4] proposed to build a 285 MW demonstration IGCC plant at Buggenum in the south-east of the country, near the existing Maas power station (Jones 1992). The plant, which was expected to begin operating in late 1993, would

burn clean "syngas" produced by a coal gasifier located on site. It was expected to cost approximately NFL 1 billion, and will be the world's largest IGCC plant. The EIA process began when Demkolec BV, the company established by SEP to undertake the IGCC project, handed in a notice of intent to the Province of Limburg. The provincial authorities drew up EIA guidelines with advice from the EIA Commission and the public.

The EIA considered the proposed project's impacts on air quality, the quality and availability of water, noise, safety and landscape, as well as the effects of air contaminants. It evaluated alternatives to the main proposal:

No action alternative

A   3 units operating at Maas, reference year 1987
B   1 unit operating at Maas, 1992

Proposed project

C   1 unit operating at Maas and the IGCC, 1993
D   Maas decommissioned, the IGCC only, 2000

Process alternatives

E   improving desulphurization from 98 to 99.5 per cent
F   a cooling tower for the entire IGCC capacity
G   sealed storage of coal gasification slag and fly ash
H   future action: improved reduction of NO_x emissions.

The EIS was carried out within 12 months, at about 0.0004 per cent of the project cost. It was 292 pages long, and it discussed the purpose of and need for the project, the proposed project and its alternatives, the decision-making process, the existing environment, the environmental effects of the proposed project and of its alternatives, a comparison of the environmental effects of the project and of its alternatives, and uncertainties in the EIA process. It summarized the most important impacts of the proposed project and its alternatives in a table, part of which is shown in Table 11.2. It concluded that situations C and D, with E and then H would give the preferred alternative from an environmental point of view.

The EIS was submitted with the planning application to the Province of Limburg in August 1989. It was reviewed by the Province for adequacy and was made public in early October. A hearing was held in late October and public objections and advice were received for one month after the announcement. The EIA Commission examined the EIA for one additional month, and one objection to the project was resolved in the High Court within weeks. The entire EIA process after the submission of the EIS took three-and-a-half months, and the project was approved in April 1990.

Table 11.2 Part of table summarizing the most important environmental impacts of the demonstration IGCC project and its alternatives.

Scenario	Water	Landscape	Cost
A	Thermal discharge 690 MWt, little chemical influence	some impact	n/a
B	Thermal discharge 298 MWt, chemical influence < A	= A	n/a
C	Thermal discharge 590 MWt, additional NaOCl discharge	somewhat < A	n/a
D	Thermal discharge 285 MWt, better than C	less than A and C	n/a
E	= C/D	= C/D	NFL 1.1M/yr
F	Thermal discharge decreases depending on operation of cooling tower, no additional NaOCl discharge	greater impact	NFL 2.25M/yr
G	= C/D	= C/D	NFL 1.15M/yr
H	= C/D	= C/D	n/a

(Adapted from Demkolec 1989)

Canada

Canada has also set up a powerful system of environmental legislation, but under conditions different from those in the Netherlands. Its wealth of natural resources, which were originally plundered indiscriminately by the giant "trusts" in coal, steel, oil and railroads; its lack of strong planning and land-use legislation; and the conflicting needs of its powerful provincial governments – all prompted the development of a mechanism by which widespread environmental harm could be prevented. Recently, Canada and the USA have co-operated in joint ventures on the monitoring and protection of such assets as the Great Lakes (Ledgerwood et al. 1992).

Legislative framework
The responsibility for EIA in Canada is shared between the federal and the provincial governments and at both levels there have been recent changes. The federal Environmental Assessment and Review Process (EARP) established a broad framework and required EIA for federal-level programmes and activities. It was administered by the Federal Environmental Assessment Review Office (FEARO), which operated at arm's length from the federal environmental agency, Environment Canada, and on behalf of the Minister of the Environment. EIA procedures were laid out in the form of guidelines rather than in legislation (FEARO 1978), and it was only in 1979, with the Government Organization Act 1979, that the process was made mandatory. The EARP guidelines were judged to be legally binding in 1989. Additional support for the EARP was provided by the Fisheries Act, the Clean Air Act, etc. (FEARO 1978).

Following an initial check on whether an EIA is needed, EARP involved a two-stage process. Most projects had an initial environmental evaluation (IEE) only, but a few judged to have significant environmental effects went to the second stage of a full, formal, public EIA. Since 1973, thousands of activities have been subject to an initial assessment, hundreds have undergone IEE; and about three federal projects per annum have the full EIA. However, over the years there have been several criticisms of the federal EARP (Effer 1984, Bowden & Curtis 1988), including concern that public participation was limited, its Environmental Assessment Panels were controversial, and there was a lack of enabling legislation. In 1995 EARP was replaced by the Canadian Environmental Assessment Act, which entrenched in law procedures that were formerly administered under a guidelines order. FEARO was replaced by the Canadian Environmental Assessment Agency (CEAA), which, while reporting to the Minister of the Environment, operates independently of any federal department or agency, including Environment Canada.

The provinces have separate, and widely differing, EIA processes for projects under their own jurisdictions (Smith 1991). Most followed the federal guidelines approach, and similarly there are now examples of reform. In New Brunswick the basis for EIA is set out by the Clean Environment Act, and specified in the guidelines *Environmental impact assessment in New Brunswick*. In Alberta, EIA is required by the Clean Air Act, Clean Water Act and Land Surface Conservation and Reclamation Act, and procedures are laid out in the Alberta Environmental Assessment System Guidelines. Only Ontario has specific legislation on EIA, the Environmental Assessment Act 1975 (Effer 1984). Smith (1991) discerns three clear quality divisions in the provincial EIA systems. Those in the first division, including Saskatchewan, Newfoundland and Quebec, are characterized by a clear, inclusive definition of the environment, invoked by excellent institutional arrangements, on the basis of public participation in the assessment process. As adherence to these features softens, assessment provision declines in quality.

Federal procedures

Figure 11.2 summarizes Canada's federal EIA procedures. The EIA process under CEAA is similar to that under EARP. Although the new process is less discretionary, it does have valuable additions including the availability of mediation as a means of public review of projects with potentially significant impacts, the provisions for public scrutiny and consultation at the self assessment stage, and provisions relating to cumulative effects and to project follow-up. The federal process is in two stages. An initial self-assessment by the responsible agency proposing the action determines whether the action is likely to have a significant environmental impact. At this stage, agencies consider technical information, expert opinion, initial public reactions and any other surveys and studies carried out in the time available.[5] The purpose of this preliminary "start up" step in the process is to determine that the project under consideration:

- is a "project" as defined by the Act;
- is not excluded by the Act's Exclusion List regulation;
- involves a federal authority or action that triggers the need for an EIA under the Act.

The Exclusion List regulation identifies projects or classes of projects for which EIAS are not required because the adverse environmental effects are not regarded as

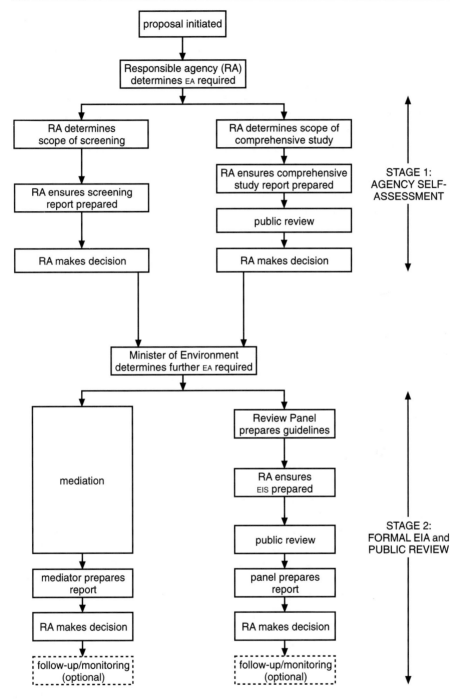

Figure 11.2 Canada's Federal (CEAA) EIA procedures.

significant. For example, simple renovation projects would be excluded. On the other hand, a Comprehensive Study List and an Inclusion List identify projects for which an EIA *is* required, by virtue of likely significant effects. For some projects on the list, numerical thresholds are applied to determine their effects' significance; in general, the projects are large and often generate considerable public concern – uranium mines, large military establishments and industrial plants for example.

Once it is determined that an EIA is needed, the next step is to decide which EIA track to follow. Most projects are handled as a screening EIA involving a brief review of available information and a short report. A small number of projects will have a comprehensive study EIA. This study replaces the EARP initial environmental assessment and involves a consideration of a wider range of factors than the screening EIA, a public review and a follow-up programme. The latter is a programme for verifying the accuracy of the EIA of a project and for determining the effectiveness of any mitigation measures taken to address adverse environmental effects. If the reports from either EIA track raise uncertainty as to whether significant negative effects are likely or whether they may or may not be justified, the Minister of the Environment can refer the project to a mediator or review panel. This is the second stage. Prior to the CEAA, a panel review was the only option for public review. The mediation option is an innovative alternative. It involves a voluntary process of negotiation in which an independent mediator appointed by the Minister, helps the interested parties to resolve their issues through a non-adversarial, collaborative approach to problem-solving. Following the completion of mediation or panel review, the responsible authority must decide whether to proceed with the project, and with what mitigation measures and follow-up programme.

While the commitment to follow-up programmes under the CEAA is encouraging, it is still only partial, as there are no absolute means of ensuring compliance. The issue is not new. Over time, many monitoring programmes have taken place, and some major research projects have been undertaken on the subject (Sadler 1988). Environment Canada and FEARO commissioned a series of EIS audits in the 1980s. These audits were themselves reviewed under the auspices of the Environmental Assessment Research Council (CEARC).[6] A conference held at Banff in 1985 discussed the findings of these and other studies (Sadler 1987).

Other notable dimensions of the amended EIA procedures include the provisions for public participation and for the inclusion of socio-economic and cultural effects. The new procedures improve notification to the public of the proposals and assessment reports, and public comments must be considered at various stages of the process. The new procedures also include a participant funding programme, which ensures that stakeholders have the opportunity to participate in panel reviews and mediation. The Act also requires the consideration not only of biophysical effects, but also the effects in socio-economic and cultural areas that flow directly from the environmental impacts. These include effects on:

- human health (physical, psychological, emotional, spiritual or mental health, and well-being);

- socio-economic conditions, including quality, "way of life" and home life;
- cultural and physical heritage, including things of archaeological, palaeon-
tological or architectural significance;
- the current use of land and resources for traditional purposes by aboriginal people.

The relationship of the federal and provincial EIA procedures has received considerable attention, and in 1992 a Framework for Environmental Assessment Har-
monisation was adopted as the basis for federal–provincial bilateral accords. The
Framework includes 16 principles, one of which is that both jurisdictions will
adhere to the provisions of the 1991 UN Convention on EIA in a Transboundary
Context. An agreement between Canada and Alberta was the first bilateral agreement under the new framework. But difficulties in harmonization include the fact
that some jurisdictions have not defined the role of EIA within decision-making, the
need to ensure that the EIA process does not constrain economic development,
uncertainties about the need for a multi-tier EIA system, and the lack of agreement
on public consultation procedures. Further information on Canada's EIA process can
be found, for example in Bowden & Curtis (1988), Smith (1991) and CEAA (1996).
Smith draws particular attention to a feature endemic to most EIA systems: change.
Any review can only be a snapshot at one point in time.

Case study: reforming EIA in British Columbia
This case study is of procedures in one province of Canada, and it provides a recent
example of proposals to reform the EIA system at the provincial level (Province of
British Columbia 1992). EIA procedures in British Columbia are orientated to projects
and project-related issues and impacts, and are currently addressed through three
formal review processes for energy projects, mine developments and major projects.
The strengths of these processes are seen to include a "one window" government
contact point for each sector, the use of inter-ministry committees and comprehensive approaches to review. A particularly clear feature of EIA practice in British
Columbia is the broad context attached to the term "environmental", which encompasses the biophysical, socio-economic, cultural, human health and safety factors
and other related aspects of human activity. This is partly explained by the significance of some of the projects for economic development in remote areas of the
province. By way of example, the Mount Milligan Ore Extraction Project, planned
to extract 22 million tonnes of gold and copper ore per annum at a location north of
Fort James, devotes two of its five substantial volumes to "community and regional
socio-economic impact assessment" and "socio-economic assessment of native communities" (Continental Gold Corporation 1991). These socio-economic studies include social as well as economic assessment, in considerable and disaggregated
detail. Some of the predictions bear some similarity to the monitoring information
on Sizewell B in the UK (discussed in Ch. 7): for example, some increases in
impaired driving, assaults and other criminal activity in the community are possible
because of the size of the construction labour force and the transient element that
may be attracted to the area in the hope of gaining employment on the project.

However, there has been growing concern about deficiencies in British Columbia's EIA system, including its definitions of categories of projects for review, the limited public participation and the lack of comparability between the different procedures. There is concern about potential bias in the procedures; proponents of projects submit their reports to the ministry responsible for promoting and regulating the industry. In addition, the province wishes to promote sustainability further, widen the scope of assessment to include cumulative impacts, integrate with federal legislation and integrate Native American people's participation. The relevant ministries have adopted an interesting way forward by producing a legislation discussion paper containing 45 recommendations for change. This was mailed to over 6,000 groups and individuals in the province; 600 written submissions were received, and many meetings, "open houses" and a workshop were held. Some recommendations and consultation responses are included in Table 11.3. This reveals some of the different and often conflicting perspectives – from the government, the private sector and the public – on the EIA process.

Table 11.3   Reforming EIA in British Columbia: some of the recommendations, and responses from consultation.

Recommendations	Example of reported consultation response
• A combination of category and threshold inclusion criteria should be used to determine which developments will be subject to EIA. There should be provision in the legislation for the minister(s) to require other projects to be subject to the EIA process where it is considered to be in the public interest.	• All projects, large and small, that have a serious significant effect on the environment must be given serious consideration for the EIA process (BC Wildlife Federation)
• Consistent with current practice, the legislation should state that it is the proponent's responsibility to identify and manage all direct environmental impacts associated with a project.	• The direct environmental impacts associated with a project are an integral part of project development and therefore the proponent's responsibility (Cominco Metals)
	• Making the project proponent responsible for the technical analysis on which the impact assessment is based is the most serious flaw in current impact assessment process. The impact assessment is not trusted by directly affected parties and members of the public. Thus, there is no commonly accepted factual foundation on which to base negotiations to achieve consensus (Irving Fox, University of BC)
• The legislation should specify that the proponent's responsibility for monitoring compliance be included in the project approval certificate and should enable	• The background paper indicates that the operator of a project should be responsible for monitoring impacts. This has the same weakness as assigning a

340

Table 11.3 (cont'd)

Recommendations	Example of reported consultation response
the minister(s) to withdraw the certificate for breach of its conditions.	fox to guard the chickens (Irving Fox, University of BC) • We agree that the proponent should have as much direct involvement and responsibility as possible in the assessment process. Therefore the onus for verifying and reporting compliance should remain with the proponent. (Noranda Inc.)
• Responsibility for chairing and directing the EIA process should be assigned to one of the following: • Min. of Environment, Lands and Parks • Min. of Environment, Lands and Parks and a lead agency; or • a neutral agency reporting to the Cabinet Committee on Sustainable Development.	• Wherever possible, we favour the use of a neutral agency to direct the EIA process – this would provide the best guarantee of independent and impartial review, consistent application of review requirements, and balanced consideration of a project's ecological, economic and social implications. (Alcan) • Environment, Lands and Parks – one only to avoid duplication of effort and conflicting priorities. (Canadian Earthcare Society)
• The legislation should outline public notification requirements, procedures for the release of documents, and public consultation. • The legislation should provide for the involvement of the public in issue identification and throughout the review process. • The government is formulating a policy on direct participant funding which will guide its application to EIA. Advice on how to deal with this important policy issue would be appreciated.	• The root of public frustration, animosity and civil unrest has been the exclusion of them from the resource development planning and management process. They will not be satisfied with legislation that merely states "the minister may decide to include the public in review processes". (J. Stelfox). • There should be no strings attached to participant assistance. (Sierra Club) • It is necessary to set limits on participant funding. (BC Hydro)

(*Source:* Adapted from *Province of British Columbia 1992*)

Australia

Like Canada, Australia also has a federal (Commonwealth) system with powerful individual states. Its environmental policies, including those on EIA, are generally not as powerful as those of Canada or the Netherlands.

Legislative framework
Australia's Environmental Protection (Impact of Proposals) Act 1974 requires EIA for actions that are carried out by the national government or require approval by

the government (e.g. railways and airports, defence facilities, activities requiring export licences) and are likely to have a significant environmental impact. The Act was implemented by Administrative Procedures of 1975, and substantially amended in 1987 by the Environment Protection (Impact of Proposals) Administrative Procedures. Other minor changes were made in 1995. EIA is also required by Australia's individual states; in recent years several of the states have pursued significant reform of their legislation:

- Australian Capital Territory's Environmental Assessment and Inquiries Act 1991;
- New South Wales's Environmental Planning and Assessment Act 1979; updated EPEA Regulations 1994;
- Northern Territory's Environmental Assessment Act 1982;
- Queensland's State Development and Public Works Organization Act 1971;
- South Australia's Development Act 1993; Amendments 1997;
- Tasmania's Environmental Management and Pollution Control Act 1994;
- Victoria's Environmental Effects Act 1978; Amendments 1994; Guidelines 1995;
- Western Australia's Environmental Protection Act 1986; Amendments 1993 and 1996.

Where a proposal affects both state and national decisions, arrangements have been made to facilitate and streamline EIA procedures.

Commonwealth procedures

Figure 11.3 summarizes Australia's Commonwealth EIA procedures. The EIA process begins when a developer prepares a "notice of intent". This includes a description of the proposed action, the environment that would be affected, the expected positive and negative impacts, any alternatives to the action, and the proposed environmental protection measures. This notice is submitted to the Environment Protection Agency (EPA) of the Department of the Environment, Sport and Territories (DEST), which determines the level of EIA needed:

- no further reports, provided specified conditions are met;
- a full EIS;
- a simpler and less comprehensive Public Environment Report (PER);
- examination by a Commission of Inquiry.

A PER is generally required when a proposal is expected to have only a few impacts or to be focused on a few specific issues, but when the issues still require consultation with the public. It briefly outlines the proposal, examines its environmental implications and describes the safeguards needed to protect the environment. Where an EIS or a PER is required, the EPA, in consultation with the developer and other organizations, prepares guidelines for the preparation of the document. A draft EIS or a PER is prepared, is announced in the *Commonwealth of Australia Gazette* and advertised in newspapers, and is made available for public review and comment (subject to commercial confidentiality) for at least 28 days. The Minister of the Environment may also call for "round table" discussions between the EPA, the developer and the public.

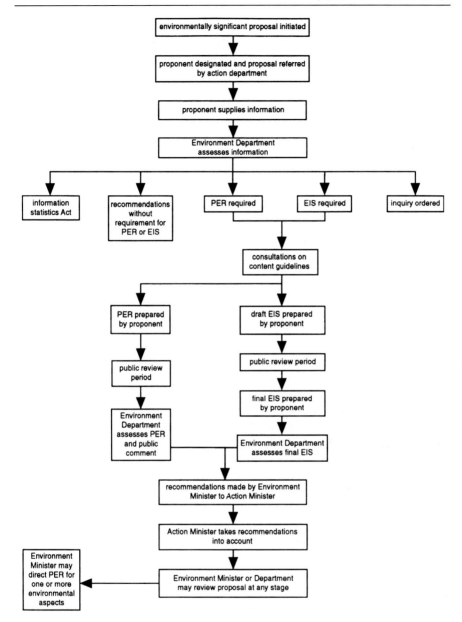

Figure 11.3 Australia's Commonwealth EIA procedures.

In the case of an EIS, comments by the public and relevant agencies are then sent to the developer, who revises the draft EIS; the developer prepares a final EIS and submits it to the EPA for assessment, and it is again made publicly available. In the case of a PER, no further revision is needed. The EPA then decides:

- whether the proposal meets the objectives of the Act;
- whether further environmental information is needed;
- any recommendations or conditions that should be associated with an approval of the proposed action.

The results of this examination (for an EIS) or of the first-stage review (for a PER) are presented to the Environment Minister as an environmental assessment report. The report may again be preceded by "round table" discussions. The minister may make recommendations to the competent authority, which must be taken into account when the competent authority makes its decision.

About 750 notices of intent for non-road projects are submitted annually in Australia; of these, about one-third require some form of environmental assessment. Of the latter, about 4 per cent have required a full EIA (Gilpin 1992, Formby 1987). Between 1975 and the end of 1991, 131 Commonwealth-level EISs had been prepared (DASETT 1992). These concern mineral exploration (about 35 per cent), transport development (25 per cent), military developments (10 per cent), communications, timber processing, and power generation and transmission. By 1994 only five inquiries had been held into environmental matters, including one for sand extraction at Fraser Island in 1975 and the Ranger Uranium Environmental Inquiry of 1977.

Australia's EIA procedures have been criticized for several reasons. Most of these are linked to government secrecy, which allows a great deal of ministerial and administrative discretion regarding, for example, whether an EIS is needed, the scope of the EIS, whether the draft EIS is to be made publicly available, whether a public inquiry should be conducted, whether additional information is required from the proponent, and whether monitoring is needed. They also restrict the legal standing of interest groups (Formby 1987).

There has also been growing concern about the variation in EIA procedures, and their implementation, between states in the Commonwealth of Australia. The Australian and New Zealand Environment and Conservation Council (ANZECC) established a working group to pursue the issues. A *National approach* on EIA, and *Background paper*, were produced in 1991 (ANZECC 1991). They identified key areas of agreement between states on the objectives of EIA in Australia, and outlined the principles of EIA for the following groups: assessing authorities, proponents, the public and the government. They also recommended a single national agreement for EIA between the Commonwealth, states and territories, with schedules to accommodate individual legislative arrangements. The national approach included proposals on many issues of concern in Australian EIA, including the need to integrate ecological and economic considerations, to consider cumulative and long-term impacts, for public participation much earlier in the process and for the post-development auditing of projects. An Intergovernmental Agreement on the Environment (1992) was subsequently released, and a section of this agreement relates to EIA. One example of the implementation of the agreement is the production of Commonwealth-wide guidelines and criteria for determining the need for and the level of EIA in Australia.

344

In parallel with the ANZECC studies, a series of Commonwealth government working groups produced a number of sector studies on ecologically sustainable development (ESD). Several of these studies addressed the role of EIA. For example, the Tourism Sector Working Group on ESD highlighted several recommendations on EIA procedures including: the clear definition of a triggering process for EIA that cannot be bypassed, the formal extension of the scope of EIA to include an assessment of the social and cultural impacts of proposed developments, and the integration of the principles of ESD. In 1992 *A national strategy for ecologically sustainable development* was published (Commonwealth of Australia 1992). Harvey (1992) discusses the relationship between ESD and EIA in these studies.

A major review of Commonwealth EIA processes was undertaken in 1994 producing a set of very useful reports on: cumulative impact and strategic assessment; social impact assessment; public participation; the public inquiry process; EIA practices in Australia; overseas comparative EIA practice (CEPA 1994). The 1994 review highlighted, among other issues, the need to reform EIA at the Commonwealth level – including a better consideration of cumulative impacts, social and health impacts, SEA, public participation and monitoring. There is a continuing concern about the variations in EIA between the states, and a questioning about whether or not EIA systems are converging as a result of the Intergovernmental Agreement on the Environment EIA principles and the Commonwealth accreditation of state EIA processes. For further discussion of the evolving Australian EIA systems we refer readers to Thomas (1996), Harvey & McCarthy (1997) and Harvey (1998).

Case study: the third runway at Sydney Airport

By the late 1980s, Sydney's Kingsford Smith Airport, Australia's main international gateway, was reaching its runway capacity (Gilpin 1992, FAC 1990). Two solutions to this problem had already been discussed for years: either supplementing the existing 4,000 m north–south runway and 2,500 m east–west runway with a 2,400 m north–south runway ("the third runway"), or building a new airport 70 km away at Badgery's Creek. The Federal Airports Corporation commissioned a group of engineering consultants to prepare an EIS for the third runway; because the proponent was a Commonwealth authority, the national EIA procedures were used. The EIS considered several alternatives to the main proposal of a third runway:

(a) no action;
(b) the development of the third runway and a staged development at Badgery's Creek;
(c) no third runway but directed development at Badgery's Creek;
(d) active traffic management of the two existing runways and a staged development at Badgery's Creek.

It also considered different spacings between the existing north–south runway and the proposed third runway (< 760 m, 760–1,525 m, > 1,525 m), and alternative ways of operating the three runways.

Public input into the EIA process was encouraged through the use of community access centres, telephone enquiry lines, newsletters and attitudinal research and by consulting interest groups. A draft EIS was released in September 1990. This seemed to take a rather limited view of the environment: in its summary of the comparison of alternatives, it compared options (b), (c) and (d) on the basis of capacity, cost, noise and other implications (operations, market trends, aviation sectors, timing and uncertainty). Copies of the draft EIS were put in public libraries and council offices, and were made available for purchase for A$25. Many submissions were received from members of the public, interest groups and local authorities during the ensuing three-month review period. Noise was a central issue. A final EIS was then prepared, reviewed by the Environment Department and considered by the national government. In 1991, the government decided to construct the third runway and also to start a staged development of a new international airport at Badgery's Creek.

EIA in Western Australia (WA)

The Western Australian EIA system provides an interesting example of a good state system that includes many innovative features. Central to the success of the Western Australian system is the role of the Environmental Protection Authority (EPA) (Wood & Bailey 1994). The EPA is an independent environmental adviser that recommends to the WA government whether projects are acceptable. It is independent of political direction. The EPA determines the form, content, timing and procedures of assessment and can call for all relevant information; the advice it provides to the Minister for the Environment must be published. The Environmental Protection Act overrides virtually all other legislation, and the environmental decision (with condtions) is central to the authorization of new proposals. Other permits must await the environmental approval, based on the EIA.

Proposals may be referred to the EPA by any decision-making authority, the proponent, the Minister for the Environment, the EPA or any member of the public. Unfortunately, the latter, public referral, has been greatly weakened by the 1996 amendments. The EPA determines the level of assessment, the most comprehensive being the Environmental Review and Management Programme (ERMP). Guidance is provided on scoping, the review document is produced by the proponent, and it is subject to public review. The EPA then assesses the environmental acceptability of the proposals on the basis of the review document, public submissions, proponents' response, expert advice and its own investigations. The resulting EPA report to the Minister for the Environment pronounces on the environmental acceptability or otherwise of the proposal and on any recommended conditions to be applied to ministerial approval.

The centrality of the EPA's review of the relevant environmental information to the Minister's decision, which itself has predominance, is the most remarkable aspect of the WA system, and one which highlights the significance of the EIA impact on decisions. The WA system also has a high level of public participation, especially in controversial EIAs. The central role of the EPA also ensures consistency. However, the limited integration of the EIA and planning procedures is a weaker feature of the WA procedures. The 1996 amendments are designed to secure better integration, improving the EIA of land-use schemes, but there is also a shift of control away from the EPA to the Ministry of Planning. This is symptomatic of attempts to weaken an effective system. In 1993, WA lost its pioneering Social Impact Unit, which had provided expert advice on social impacts, and there is a strong development lobby, in a state highly dependent on major mineral projects, to further "soften green laws".

Japan

In the 1970s, Japan responded to severe environmental degradation by adapting, developing and applying the newest technology for pollution control and energy efficiency. More recently, it has actively sought a more global role in environmental affairs and is putting itself forward as a leader in resolving global environmental problems[7] (Barrett 1991). However, Japan's large ministries (for example transport, construction and industry) tend to quash any environmental policies that are likely to harm the nation's economic development. Similarly, environmental policies that would require significant social changes, such as reduced car use, are unpopular (Barrett & Therivel 1991).

Legislative framework

The establishment of federal EIA legislation in Japan was discussed with increasing intensity from the early 1970s until 1983. A bill concerning EIA was first proposed by the Environment Agency in 1976, was discussed with other agencies, and was presented to the Diet (parliament) in 1981. However, after the bill was discussed at Diet level for two years, the Diet was dissolved and the bill nullified. Instead, the Cabinet adopted non-mandatory federal EIA guidelines (Implementation of Environmental Impact Assessment) in August 1984. Since then, the various ministries have established EIA guidelines for developments under their jurisdiction. Recently the Environment Agency has once again proposed an EIA bill, but this is being opposed by the other large ministries on the grounds that it may harm economic development and lead to lawsuits. About half of Japan's local authorities have established separate EIA regulations or guidelines, most of which are more stringent than the national guidelines.

National procedures

Figure 11.4 summarizes Japan's national EIA procedures. The Cabinet decision of 1984 requires the preparation of an EIA for certain listed projects. These include roads of four or more lanes more than 10 km long, dams with a water surface of more than 200 ha, airports with runways of 2,500 m or longer, and industrial estates, residential developments, urban development projects and land readjustment projects of 100 ha or more.

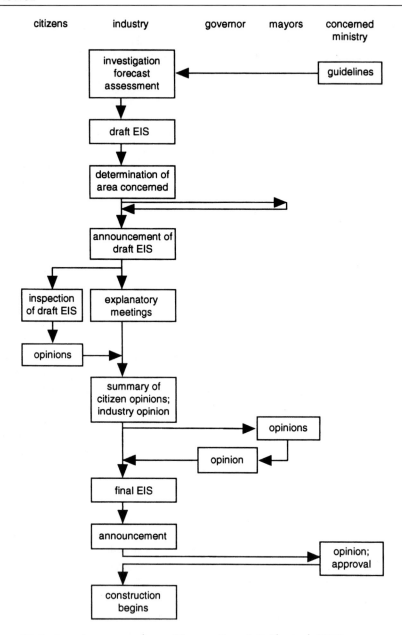

Figure 11.4 Japan's EIA procedures. (*Source:* Barrett & Therivel 1991)

For these listed projects, the developer prepares a draft EIS based on guidelines prepared by the responsible government ministry. Copies of the EIS are sent to the governor(s) and mayor(s) of the area the project will affect (as determined by the developer), and are made publicly available for one month. During that month, the developer also holds explanatory meetings for residents of the affected area "if possible". Residents can send written comments to the developer for one calendar month plus two weeks after the publication of the draft EIS; the developer summarizes these and passes them to the governor(s) and mayor(s).

Within three months of receiving the residents' opinions, the governor(s), in consultation with the mayor(s), comments on the EIS, focusing on pollution control and the conservation of the natural environment. The developer then prepares a final EIS, which includes revised information from the draft EIS, a summary of the residents' opinions and the governor's opinion, and measures to respond to these opinions. This final EIS is publicly available for one month. If permission is needed from a government ministry, the final EIS is presented with the application; before it grants a licence, the ministry must ensure that the EIS properly considers pollution control and nature conservation.

Local authority procedures

Local authority EIA procedures in Japan are broadly similar to national procedures, but they generally apply to more projects, have a broader scope and include more public participation. For instance, many local authorities require EIA for waste-treatment plants, recreational projects and water-supply projects, none of which is required at the national level. Some local authorities require the consideration of, for example, socio-cultural impacts, climate and the obstruction of sunlight (which is particularly important in Japan's densely populated urban areas). Again, these are not required by the national guidelines. Some local authorities require the developer to hold public hearings, and to collect and publicize monitoring information during the construction and operation of a project.

Approximately 70 federal-level EIAS are prepared annually in Japan. Of these, about half are for ports or harbours; most of the rest are for power stations. In addition, about 75 local authority EIAS are prepared annually. These vary more widely, including reclamation projects (approximately 25 per cent), residential developments (18 per cent), roads (13 per cent), power stations (9 per cent) and railways (7 per cent).

Japan's EIA procedures have been criticized for several reasons. They are non-mandatory and therefore non-enforceable. Some development projects that are bound to have significant environmental impacts – such as crude oil refineries, integrated chemical installations and nuclear power stations – are not subject to EIA. Alternatives to a proposed project do not have to be considered. The requirements for public consultation are not strong. Finally, the multiplicity of national government and local authority EIA procedures leads to duplication and confusion. EIA in Japan seems all too often to be used as a tool to justify development decisions and overcome local opposition.

Case study: the Trans-Tokyo Bay Highway

The Trans-Tokyo Bay Highway (TBH), when completed, will be a six-lane highway 15 km long (a 10 km bridge and a 5 km tunnel) that will connect the cities of

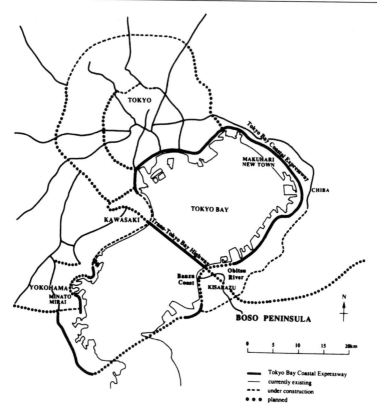

Figure 11.5 Tokyo Bay and new highway.

Kawasaki and Kisarazu by traversing the Tokyo Bay (Barrett & Therivel 1991). The TBH will connect with the Tokyo Bay Coastal Expressway to make a loop linking the greater Tokyo region, and it is expected to add further impetus to the development of that region's economy (see Fig. 11.5). Construction began in 1988, and it is estimated that it will take 10 years and cost Y1.15 trillion (approximately £6 billion). The Japan Highway Public Corporation (JHPC) was responsible for planning the highway and preparing its EIA. The TBH had been discussed since the early 1960s. In 1986, the TBH Company was established as a private company to design, construct and operate the highway.

The project affected 11 local authorities, and could have been subject to seven different EIA guidelines or regulations. Some of these were eliminated on the grounds that the local authorities were not sufficiently involved; in other cases the stronger EIA procedures were used. In the end, the EIA procedures of Kawasaki city, Chiba prefecture and the federal Ministry of Construction were used. The JHPC followed the (weaker) Ministry of Construction guidelines, and the local authorities followed their own procedures for public consultation and review.

Table 11.4 Environmental impacts considered in the EIA for the Trans-Tokyo Bay Highway.

	Air pollution	Water pollution	Noise	Vibration	Ground config.	Land plants	Land animals	Marine life	Scenery
Construction									
Man-made island	X	X						X	
Tunnel									
Bridge	X	X	X	X				X	
Roads	X		X	X		X	X		
Operation									
Man-made island		X						X	X
Tunnel									X*
Bridge		X			X			X	X
Roads						X	X		
Traffic	X	X	X	X		X	X	X	

* air ventilation towers
(*Source:* Barrett & Therivel 1991)

Table 11.4 shows the environmental impacts considered in the EIA for the TBH. This type of simple matrix is commonly used in Japanese EIAS. No attempt was made to specify the key impacts within this framework, or to discuss the potential for shipping accidents or for the ecosystem as a whole. Public participation in the EIA process was widespread. In total, 1,770 opinion statements were received concerning the EIS, of which 1,746 opposed the project. In addition, 154 people attended a public hearing at Kawasaki. Concern focused on future nitrogen dioxide (NO_2) levels, the loss of tidal flats and the continued decline of water quality in the bay.

The draft EIS was completed in mid-1986, and was then reviewed over the course of a year by the 11 local authorities. Kawasaki city alone took six months to analyze the EIS in specialist committees. As a result of this review process, the JHPC agreed to implement expensive measures to contain the emissions. The final EIS was 962 pages long, with 312 pages of supporting data. It was submitted to the Ministry of Construction in mid-1987. The ministry approved the development one month after receiving the EIS.

11.3 Emerging EIA systems

This section considers EIA systems that have recently been developed or are in the process of being developed. Sadler (1996) differentiates between two types of such emerging systems. First, about 70 "developing countries" have (mostly in the past ten years) enacted EIA legislation, usually within the framework of a more general

351

environmental law, but also as specific EIA legislation. This process is still continuing, particularly in Africa and South America. Second, "countries in transition" in Central and Eastern Europe are establishing EIA legislations, many of which are specifically in line with Directive 85/337 and the Espoo convention on transboundary impacts. This process is now almost complete.

Although the basic principles of these emerging EIA systems are similar to those of industrialized Western nations, they are generally applied in, and adapted to, quite different contexts:

- Many apply to countries in or near tropical areas, where environmental models, data requirements and standards from temperate regions may not apply.
- Socio-cultural conditions, traditions, hierarchies and social networks may be very different.
- The technologies used may be of a different scale, vintage and standard of maintenance, bringing greater risks of accidents and higher waste coefficients.
- Perceptions of the significance of various impacts may differ significantly.
- The institutional structures within which EIA is carried out may be weak and disjointed, and there may be problems of understaffing, insufficient training and know-how, low status and a poor coordination between agencies.
- EIA may take place late in the planning process and may thus have limited influence on project planning, or it may be used to justify a project.
- Development and aid agencies may finance many projects, and their EIA requirements may exert considerable influence (see Section 11.4).
- EIA reports may be confidential, and few people may be aware of their existence.
- Public participation may be weak, perhaps as a result of the government's (past) authoritarian character, and the public's role in EIA may be poorly defined.
- Decision-making may be even less open and transparent, and the involvement of funding agencies may make it quite complex.
- EIAS may be poorly integrated with development plans.
- Implementation and regulatory compliance may be poor, and environmental monitoring limited or non-existent (Bisset 1992, *EIA Newsletter* 1995, Hirji & Ortolano 1991, Kakonge & Imevbore 1993, Thanh & Tam 1992, Welles 1995).

Perhaps the greatest problem to overcome in these systems is a lack of political will, which means that EIA has little influence on project planning. Speaking of Asian countries, Welles (1995) notes that "Many countries have enacted EIA legislation, but the institutional commitment to EIA and the required level of technical and analytical skill to carry out such assessments is often lacking." Speaking of Africa, Okaru & Barannik (1996) suggest that "the presence of EA statutes – although a key ingredient in successful EAS – does not guarantee adequacy and enforceability or good EA practice". Similarly, Fisher (1992) notes of Eastern Europe that "There is no lack of goodwill and a flood of western assistance, and as a result, environmental policy documents and protection acts contain many very positive approaches. These intentions, however, are less evident in the economic sectors where they must be implemented."

Nevertheless, as all three of the subsequent case studies show, emerging EIA systems are developing rapidly, learning from existing systems, and adapting EIA techniques to their own needs. For instance, a communiqué from the 1995 African Ministerial Conference on Environment commits African environment ministers "to formalize the use of EIA within a legislative framework for development planning and decision-making at the project, programme and policy levels" and lists priority actions, including capacity-building, the promotion of co-operation between countries and the sensitization of policy and decision-making to the importance of EIA (Goodland et al. 1996).

This section considers three emerging EIA systems. Peru and China represent developing countries under Sadler's definition, and Poland is a "country in transition". These examples seek to illustrate the variety of emerging EIA systems as well as some of the points raised above.

Peru was chosen as a South American country about whose EIA system (of 1992) little has been written to date. The Peruvian system is notable for its use of a double system of preliminary and detailed EIA, for the problems of implementing a national environmental policy through a number of sector-specific ministries and for the confidential nature of the EIA findings.

China's EIA system was chosen because of the world-wide effect that any Chinese environmental policy is likely to have in the future. China's environmental policies are restricted by the need to harmonize them with plans for economic development. The Chinese EIA system's extensive use of mathematical modelling and lack of public participation are notable.

In contrast to the other two case studies, *Poland's* EIA system is part of its planning system. EIA can also be applied to existing installations as a form of environmental audit. Although Poland's EIA system is buttressed by strong government willpower and a wish to harmonize it with existing European EIA procedures, public participation is very limited and EIA quality is constrained by a lack of expertise and baseline data and by the strong impetus towards establishing a market economy.

The literature on emerging EIA systems is extensive, although that on any one country is limited (see Table 11.1). More general information can be found in, for instance, Angelsen et al. (1994), Beanlands (1994), Birley (1995), Biswas & Agarwala (1992), Goodland & Edmundson (1994), Goodland et al. (1996), Horberry & Muscat (1990), McLaren (1993), Wathern (1992), West et al. 1993, and in a range of publications by the OECD, ODA, UNEP and the World Bank.

EIA in Peru

Peru, the third largest country in South America, includes a thin dry strip of land along the coast, the fertile sierra of the Andean foothills and uplands, and the Amazon basin. Fishing, agriculture and mining are the main industries. A change in government in 1990 has led to the reconstruction of the country after years of economic difficulties, and an extensive privatization programme – including the privatization of many of the state-owned mines – has encouraged dramatic increases in foreign investment.

In September 1990, the Peruvian government enacted Decree 613–90, the Code of Environment and Natural Resources. This established EIA as a mandatory requirement for any major development project, but did not specify the EIA contents or legal procedures, or who the competent authority should be. The Ministry of Energy and Mining was the first ministry to put this decree into practice. In 1992, it approved the General Law of Mining by Supreme Decree no. 014-92-EM; this, in turn, contained regulations for environmental protection in mining activities, which were approved in 1993 by Supreme Decree 016-93-EM. The ministries for fishing, agriculture, and transport, communication, housing and building all established similar requirements in 1994 and 1995, but others (e.g. for tourism and industry) have not done so yet. The remainder of this section focuses on the mining sector.

According to Supreme Decree 016-93-EM, any developer who plans to exploit minerals, or to expand existing exploitation by 50 per cent or more, must carry out an EIA. The EIA has to be carried out by an institution authorized and registered with the Ministry of Energy and Mining. The decree lists the EIA contents in two parts. The first part is mandatory. If, after reviewing the EIA, the Ministry considers that the project will have significant environmental impacts, it can also require the EIA to address some of the aspects in the second part:

Part I   EIA contents

1   executive summary;
2   antecedents (e.g. applicable legislation);
3   introduction (project description and estimated cost);
4   project area description (general information about location, geological and biotic components, etc.);
5   project description (e.g. estimated volume of water used, waste-water and wastes produced, energy demand, employment, etc.);
6   predicted impacts (human health, flora and fauna, ecosystems, etc.);
7   control and mitigation (e.g. measures to control noise and dispose of wastes);
8   cost–benefit analysis.

Part II   Additional information

1   project alternatives (description, justification of chosen alternative);
2   affected environments (detailed studies of continental and marine waters, etc.).

Every EIA must include an environmental management plan which lists the project's "environmental obligations": the activities and programmes to be implemented before and during the project to guarantee the fulfilment of existing environmental standards and practices. The decree also includes a programme (the *Programa de Adecuación y Manejo Ambiental*) of actions and investments to incorporate new

354

technologies and alternative measures into mining activities so as to reduce emissions or discharges.

In September 1994, the Ministry of Energy and Mining also published more detailed guidelines (*Guía para elaborar EIA*) on how EIA should be carried out, which broadly cover the points above. Once an authorized body has carried out an EIA, the EIS is reviewed by the Ministry's environment directorate. In theory, this review should be carried out within 45 days, or else automatic consent is granted. In practice, since the ministry has 45 days "after receiving the EIS and/or additional information", and since the relevant directorate only has eight officers, who also have to deal with other matters, the officers often request additional information so that they have longer to review the EIS. Where a project is likely to be particularly harmful to the environment, the Ministry may ask the consultancy and developer to give a presentation and discuss contentious points. Until recently, only the EIA's non-technical summary was publicly available; the main body of the EIS was felt to be commercially sensitive and thus confidential. However, Ministerial Resolution 335-96-MEM/SG of July 1996 now requires that a public inquiry should be held before a decision is made.

Once a project is approved, the developer must carry out programmes of management, control and monitoring throughout the operations to ensure that the environmental management plan is adhered to. The developer has to contract a registered auditing consultancy to inspect its activities twice a year. The consultancy must prepare a report on activities at the site and submit it to the Ministry, which can apply sanctions for non-compliance.

Since 1993, when EIA became mandatory for mining activities, 77 EISs have been approved by the Minister of Energy and Mining. A survey of environmental consultants carried out by Iglesias (1996) showed that almost two-thirds of EISs are begun after the project planning is "more than 50% completed", with the sections on control, mitigation and cost–benefit appraisal being particularly difficult to carry out. Early indications are that the quality of EIS is quite high, but that the discussions of mitigation measures are weak or non-existent. The Ministry feels that the implementation of EIA is going well, with the exception of some gaps, which will be overcome (Iglesias 1996).

EIA in China

With its centrally planned economy, huge population, and relentless industrialization and urbanization, China is establishing environmental policies that attempt to balance the need for economic growth and environmental protection. The main impetus for the introduction of EIA in China was the adoption of the Environmental Protection Law (for trial implementation) in September 1979; this states that:

All enterprises and institutions shall pay adequate attention to the prevention of pollution and damage to the environment when selecting their sites, designing,

constructing and planning production. In planning new construction, reconstruction and extension projects, a report on the potential environmental effects
shall be submitted to the Environmental Protection Department and other
relevant departments for examination and approval before designing can be
started. (Article 6)

Over the years, guidelines for the implementation of this law have been prepared,
of which the central ones are the *Management rules on environmental protection of
basic projects* of 1981, which were revised and formalized as the *Management
guidelines on environmental protection of construction projects* of 1986, *Management procedures for environmental protection of construction projects* of 1990, and
*Management guidelines on strengthening loan projects for environmental impact
assessment* of 1993 (Welles 1995). The basis for carrying out EIA in China was
further strengthened when the Environmental Protection Law was revised and
issued without trial status in 1989. The National Environmental Protection Agency
administers EIA for projects of national economic or strategic significance, the
provincial environmental protection bureaus (EPBs) administer EIA for projects of
regional importance, and so on down through city and district EPBs for municipal
areas, and county, town and village EPBs outside municipal areas. Ortolano (1996)
estimates that there are about 2,500 EPBs in China, which between them receive
"tens of thousands" of EIAS every year.

Figure 11.6 summarizes China's EIA procedures. The guidelines require EIA for a
range of projects, including those in industry, transport, water conservation, forestry, commerce, health, culture and education, and tourism. The EIA process begins
when a developer asks a competent authority to determine whether or not a proposed action requires a full EIA. Most projects require only the preparation of an
environmental impact form, which describes the project and briefly states its environmental impacts. Large projects with significant impacts and smaller projects in
inappropriate locations require full EIAs. The competent authority personnel, sometimes assisted by outside experts, conduct a preliminary study, then make a ruling.
If an EIA is needed, those factors most likely to affect the environment are identified
and given an importance weighting. The EIA's management is then entrusted to
state-approved experts, who work to a brief prepared by the competent authority.

In 1995, about 250 organizations had "class A" licences which allowed them to
carry out EIA for any size of development, and about 400 had "class B" licences
applicable only to projects of regional and lower significance. Ortolano (1996)
suggests that this licensing system has both advantages and disadvantages. On the
one hand it aims to ensure high quality EIA work. On the other hand, there may be
concerns about conflicts of interest where licensed appraisers are closely linked to
the industry they are appraising or to the EPBs. Licensing may also restrict the take-
up of best-practice techniques from abroad and from non-licensed but innovative
organizations.

After the scoping stage, the licensed expert analyses the relevant impacts in
greater detail and compares them with relevant environmental quality standards.

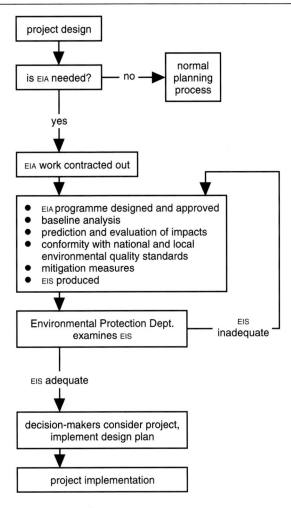

Figure 11.6 China's EIA procedures.

Baseline environmental assessments are carried out if the project is proposed for an area of low industrial activity where environmental standards are high. The impacts are then predicted, often using a systems approach and simulation techniques. They are evaluated for their impact on human health, ecosystems and, sometimes, social systems. Mitigation measures may be proposed. An EIS is then produced, which, according to the guidelines, needs to include the following information:

- the general legislative background;
- a description of the proposed development, including materials consumed and produced;
- the baseline environmental conditions and the surrounding area;

357

- the short-term and long-term environmental impacts of the project;
- proposals for monitoring;
- a cost–benefit analysis of the environmental impacts;
- an assessment of impact significance and acceptability;
- existing problems and proposals for addressing them.

The EIS or environmental impact form is submitted to the relevant EPB, which checks the proposal against environmental standards and makes a decision within two months, considering the comments of the competent authority and outside experts; for a controversial project, and for a project that crosses provincial boundaries, the document is submitted to the higher EPB for examination and approval. If the project is approved, conditions for environmental protection may be included, such as monitoring and verification procedures. The competent authority must submit a report that states how the project will be carried out and how any required environmental protection measures will be implemented. Once this has been approved by the provincial authorities, a certificate of approval is issued.

China's EIA system has no formal procedures for public consultation; the system is solely administrative, and makes no provisions for informing residents or for eliciting their views (e.g. in public meetings). All Chinese counties and cities have a kind of ombudsman's office that receives complaints on all types of matters, including environmental issues, but this is different from a formal system of public participation. Other criticisms of China's EIA system are that it places too much stress on mathematical modelling techniques rather than on practical assessment methods, that the biotic and socio-cultural components of EIA are often poorly assessed, and that many decision-makers seem to feel that they must put economic concerns before environmental issues (Wenger & Huadong 1990).

Another limitation of Chinese EIA practice is that institutional factors often cause EIAs to be carried out very late in the project planning process, or even after the projects have been built. In some cases, this is because EPBs learn about the projects only at a very late stage. In others the EPB is unwilling to antagonize other government departments or local leaders who strongly favour the proposed projects. In some cases, even where an EIA is prepared and environmental protection measures are agreed, "the mayor . . . steps in and asks [the EPB] . . . whether less money couldn't be spent on [pollution control] equipment" (Jahiel 1994), effectively cancelling out the project's environmental protection features. Sinkule & Ortolano (1995) cite a case where a factory did not permit the municipal EPB to inspect the factory, and the mayor's office subsequently instructed the EPB to stop conducting inspections. On the other hand, some EPBs have established co-operative relations with other organizations, so that other bureaux (such as the planning and economic commissions, the land management department and banks) do not act on an enterprise's application for a loan or permit unless the EPB has approved the EIA and the project (Sinkule & Ortolano 1995). In another case, an EPB established a form that enterprises proposing projects to other local agencies needed to fill out, which would notify the EPB that a project was being initiated (Jahiel 1994). Overall, how far EIA requirements are implemented is largely dependent on the degree of importance placed on EIA by the local government, especially the mayor's office.

However, EIA in China does seem to have broadened the traditional form of decision-making, which considered only economic criteria. Some projects (e.g. a coal-gas project at Lanzhou) have been stopped because of EIA, and others have been substantially modified. Further information on China's system of EIA can be found in Wenger & Huadong (1990), Sinkule & Ortolano (1995) and Ortolano (1996).

EIA in Poland

Poland, like the other "countries in transition" of Eastern and Central Europe, is subject to severe pollution, although in Poland this is confined to a limited number of sites (EBRD 1994). Since the overthrow of the Communist regime in late 1989, and since gaining associate membership of the European Union,[8] Poland has been undergoing enormous economic and social changes, which in turn have had environmental repercussions:

> There are no more economic plans and central planners, the currency is convertible and the best technology accessible, and the whole economy is being privatised. Moreover, administrative arrangements have been redesigned in order to create a strong central agency as an environmental watchdog . . . [but] old industry is still operating. The observed improvement of environmental records since 1989 is only a side effect of the recession . . . [and] EIA law in Poland still reflects two characteristic features of the Communist regime: an aversion to getting the general public involved in decision-making, and a reluctance to developing procedural rules for dispute settlement. This means that this legislation not only is not efficient enough from the "environmental" point of view, but also does not match the political and economic transformation towards an open and democratic society and a free market. (Jendroska & Sommer 1994)

EIA in Poland originated in the late 1970s, when the government's policy of borrowing from the West was backfiring, the Communist regime was providing only the most basic services, and environmental issues were being virtually ignored (Fisher 1992). The resulting deterioration in environmental quality caused widespread concern, leading to the enactment of the Environmental Protection Act (EPA) of 1980. This authorized environmental authorities to require the developers of proposed facilities and the managers of existing facilities to give an "opinion" on the environmental impacts of their facilities. Jendroska & Sommer (1994) discuss this early EIA system in greater depth.

The Town and Country Planning Act of 1984, which established a two-tier project authorization system, set the context for EIA in Poland. At the first tier, the local planning authority suggests possible sites for the proposed project in a "location

Table 11.5 Projects especially harmful to the environment and human health that require EIA in Poland.

A "project especially harmful to the environment and human health" is one which, during construction, operation or decommissioning, can involve:

- emissions of particulates and gases of more than 20,000 tonnes/year, or 5,000 tonnes in protected areas;
- discharges of waste-water into border waters; running waters in protected areas or areas of ecological hazard ("protected areas" hereafter); ≥ 100 m³/day into the Baltic Sea, lakes, reservoirs or the soil; or ≥ 5,000 m³/day into any other flowing water;
- the deterioration of water in protected areas or areas of high socio-economic value;
- the production or storage of hazardous waste;
- soil contamination, or a change in the use of agricultural or forest land of 100 hectares (50 hectares in protected areas);
- water abstraction from border waters; from groundwater of ≥ 5,000 m³/day (or > 2,000 m³ in protected areas); or from surface waters of ≥ 40,000 m³/day (or ≥ 20,000 m³ in protected areas);
- the production of electromagnetic fields of 0.1–300,000 MHz; exceeding field intensities or energy flux densities set in separate regulation;
- exceeding noise levels set in separate regulations.

indication", and at the second it considers the developer's detailed application for a specific site and hands down a "location decision". EIA is required at both tiers before any project can be authorized.

Both the EPA and the Planning Act have since undergone considerable amendments and refinements. The current EIA system is based on:

- Environmental Protection Act 1980, as amended in 1983, 1987, 1989, 1990, 1991 and 1993;
- Executive Order of the Environment Minister on development projects especially harmful to the environment and human health and on the requirements for environmental impact assessment, 23 April 1990;
- Executive Order of the Environment Minister on establishing the Commission for Environmental Impact Assessment, December 1989;
- Town and Country Planning Act 1984, as amended in 1989 and 1990.

Copies of these are given in EBRD (1994). Further amendments are still being considered. The following summary of Poland's EIA procedures brings together the requirements of these regulations. Three types of action require EIA in Poland. First, "projects especially harmful to the environment and human health" require EIA in all cases. Table 11.5 lists these projects: note that they are defined by their environmental impacts rather than by their nature or scale, as is the case with Directive 85/337. Second, "projects potentially dangerous to human health and the environment" may require EIA if they are likely to have significant environmental effects. No list of such projects is given (as is, for instance, in Annex II of Directive 85/337), although criteria for appraising significance are very similar to those of the Directive. The EIA process itself helps to determine whether a given project

Table 11.6 EIS contents required in Poland.

1   Examine air, land, soil, minerals, surface waters and groundwater, the marine environment, flora, fauna and their interrelations.
2   Consider the construction, operational and decommissioning (where appropriate) stages, and emergencies.
3   Use, where appropriate, quantitative and other data, including data collected during previous studies.
4   Present problems in a manner commensurate with their importance and facilitating their analysis.
5   Describe the project's technological and technical features.
6   Estimate the materials, energy and water consumed, and the amount and kind of waste and pollutants emitted.
7   Describe the methods to minimize environmental hazards, including the technical installations necessary and their technological and economic effectiveness.
8   Describe the size and management of the protective zone needed if the mitigation measures do not eliminate the project's adverse effects or keep them within prescribed limits.
9   Predict the possible environmental effects of the proposed (or existing) development.
10 Describe the land around the project and its current environmental quality, indicating parts that will be significantly affected.
11 Describe the possible levels of environmental impact.
12 Assess the likely significant impacts on the environment, human health and the natural beauty of the area, differentiating between direct, indirect and long-term impacts.
13 Include the data necessary to obtain permits for air pollution, water pollution, water consumption, noise, vibration and waste disposal.
14 State the conditions for using the environment.
15 Compare the applied technology with the best available technology in terms of environmental performance.
16 Assess how the technical design implements the mitigation measures incorporated in the "locational decision".
17 Refer to the environmental data gathered before and during the operation of the project.
18 Consider the data on water consumption, waste discharge, pollutant emissions and other adverse environmental effects gathered during the operation of the project.
19 Estimate the current and anticipated impacts on the environment, human health and the natural beauty of the area.

requires EIA or not. Third, existing projects require EIA as a type of environmental auditing, although this requirement has not yet been put into practice.

For proposed projects, the EIA process begins when a developer asks a local environmental authority (LEA) whether or not an EIA is needed. If it is, then the LEA draws up a list of suitable consultants to carry out the work. The developer chooses a consultant from the list, and a scoping meeting is carried out between the LEA, the developer, the consultant and the project designer. The EIS must include points 1–9 from Table 11.6. Once this EIS is complete, it is presented with an application to the LPA, to act as an input for the "location indication" tier of the planning decision. The LPA consults with the LEA (or with the national EIA committee if the project is exceptionally polluting or of national importance). The LEA reviews the EIS for its compliance with regulations and methodology, and it often seeks corrections. At this point consultation may be carried out, but it is not mandatory. If the EIS is

361

accepted, then the LPA can issue the "location indication", which lists alternative locations for the project within the local authority. In practice, this is the crucial stage for EIA, as the subsequent stage seems to be more of a formality.

Once the developers receive the "location indication", they choose a site and continue to design the project. The environmental consultants prepare the final EIS, which must include points 1–14 from Table 11.6. Both the planning application and EIS are given to the LPA, which again consults with the LEA about the EIA before making a "location decision" regarding the project. At this stage, construction consent must still be granted under the Building Law of 1974. This requires the preparation of yet another EIS to accompany the technical design of the project, this time including points 1–8 and 10–16 of Table 11.6. By the end of 1993, the EIA Commission had reviewed 22 cases, most of which had then been returned for further elaboration. It had withheld approval for five of these cases (EBRD 1994). In theory, existing facilities can also be required to prepare an EIS, which must include points 1, 2, 5–11, 13, 14, and 17–19 of Table 11.6. In practice, this has not yet happened.

Several points about Poland's EIA system are striking. First, unlike that of Peru or of China, it is linked to a pre-existing planning system. However, this system is having to deal with a large number of complex proposals, with little baseline data or experience with environmental appraisal. Secondly, the EIA Commission has an important role, but it is severely restricted by resource constraints. The Commission is a body of 75 experts which reviews applications for "location indications" and EISs for existing developments, publishes EIA-related information, and issues lists of EIA consultants. However, the Commission is able to review only about 10 per cent of the EISs that are worth reviewing (Jendroska & Sommer 1994), and it has been criticized for sometimes taking more than a year to review an EIS (Rzeszot & Wood 1992); in a climate of rapid economic expansion and fluctuating interest rates, these delays can seriously hinder project planning and throw EIA in a bad light. Similarly, the lack of adequately trained and approved EIA consultants means that those consultants that exist in Poland are very stretched. This in turn has led to poor-quality EISs for which further information has been required at the review stage, thus further slowing down the EIA process (Rzeszot & Wood 1992). The fact that EIA can be carried out only by approved consultants also tends to remove the developer from the EIA process, leading to a less integrated process, possibly with fewer project modifications.

The planning and EIA process is obviously cumbersome and redundant, with three EISs prepared for each project. Similarly, the lack of clear screening criteria means that full EIAs are prepared for projects which may not need them. Furthermore, the EIA system has the most effect at the "location indication" stage, but at this stage project design is generally not yet far enough advanced to allow much detailed environmental information to be provided in the EIS (Jendroska & Sommer 1994, Rzeszot & Wood 1992).

A serious limitation of Poland's EIA system has been its lack of procedures for public participation. Although the planning process generally encourages public participation, there is no requirement to consult the public on the EIA findings. Other problems include the fact that environmental experts in Poland are not used to carrying out interdisciplinary work, that the LEA is sometimes reluctant to set an EIA scope for fear the EIS may be rejected by the EIA Commission, with subsequent ramifications, and that EIA does not apply at the strategic level (Rzeszot & Wood 1992). However, many of these problems are likely to be ironed out through a future review of the existing laws. These may include:

- replacing the two-tier planning system with one tier;
- a longer list of "projects especially harmful to the environment and human health";
- a screening list of "projects potentially dangerous to human health and the environment";
- a requirement that public meetings should be held before an EIS is submitted to the LPA;
- the application of EIA to strategic levels as well as to the project level (EBRD 1994, Jendroska & Sommer 1994).

The three case studies illustrate some of the progress being made in the development of EIA systems worldwide. They also illustrate some of the continuing issues. These include, for example, a narrow exphasis in Peru and China on primarily biophysical impacts (particularly air and water), and, in the three countries, institutional constraints, limited public participation and the need for further capacity building. In this context, the roles of international funding institutions in EIA now merit some consideration.

11.4   International funding institutions and EIA

The range of international funding institutions and their EIA procedures

Several of the major international funding institutions (IFIS) have established EIA procedures to promote funding decisions that are environmentally sound. All these procedures seek to identify adverse environmental impacts early in a project's life-cycle, to assess the impacts and, within the limits of feasibility, to integrate measures to mitigate, minimize or compensate for environmental damage. However there are considerable variations in the organization and execution of the procedures and in the terminology used by the IFIS. Table 11.7 provides a brief summary of pre- and post-EIA requirements, project categories and the types of EIA guidelines available.

The European Bank for Reconstruction and Development (EBRD) provides general guidelines covering a variety of situations ranging from environmental audits to

363

Table 11.7 Summary of comparisons between EIA procedures of main IFIs.

International funding institutions	Pre-EIA requirements	Project categories	EIA guidelines; specific sectoral coverage	Post-EIA requirements
European Bank for Reconstruction and Development (EBRD)	Scoping study	Level A – list of specific major developments (EIA required) Level B – activities not included in level A (environmental analysis required) Level C – all activities that do not require levels A or B (no assessment required)	General EIA guidelines	Environmental monitoring
Asian Development Bank	Initial environmental examination	No explicit categorization	General EIA guidelines Specific guidelines – agriculture, health, social, coastal zone, industry, energy, power, transport, risks and hazards	Monitoring
African Development Bank	Initial environmental examination	I – projects with significant impacts II – projects with limited impacts III – projects with no adverse impacts All above categories considered in context of environmental sensitivity of the project location	General EIA guidelines	Environmental management; monitoring; auditing
World Bank	Screening	A – significant adverse impacts that may be sensitive, irreversible and diverse (EIA required) B – less significant impacts than A (no EIA, environmental analysis required) C – unlikely to have adverse impacts (no EIA or environmental analysis required)	General EIA guidelines Specific guidelines – health, social, coastal, energy, power, risks and hazards, human settlements, industry, waste and water	Analysis of alternatives; environmental management plan; training; environmental monitoring plan
Inter-American Development Bank	Environmental brief	I – project that improves environmental quality (no EIA required) II – no direct or indirect impact (no EIA required) III – moderate impact but with recognized solutions (semi-detailed EIA required) IV – significant negative impacts (EIA required)	General EIA guidelines	No explicit requirements
European Commission (e.g. of DGIB)	Initial screening	I – projects not expected to have significant adverse environmental impacts (no EIA required) II – projects with limited adverse impacts – semi-detailed EIA required III – projects with diverse and significant adverse impacts – detailed EIA study required	User guidance note; plus specific sectoral guidance (DG VIII Sourcebook)	Integrated with project life cycle management; monitoring and expert evaluation

project-specific EIA to regional strategic environmental assessments (SEAS). The EBRD stresses the importance of undertaking public consultation throughout the EIA process and has produced a manual specifically targeting this subject. The Asian Development Bank provides more specific guidelines for sectors such as agriculture, industry, energy and transport; the Bank also emphasizes the incorporation of a social dimension into assessments and has produced specific assessment guidelines for undertaking such analysis. An initial environmental examination (IEE) is undertaken for all development projects, following a pre-defined tabular format. The African Development Bank also has an initial environmental screening stage, and also encourages public participation and consultation with non-governmental organizations and interest groups at all stages of the project cycle. Following the completion of the EIA and associated mitigation measures, environmental management is required to enhance environmental quality (an important step beyond simply preventing environmental damage). Post-construction monitoring and auditing is also recommended to ensure that the mitigation measures and environmental damage is kept to a minimum.

World Bank EA procedures

"Environmentally sustainable development has become one of the most important challenges facing development institutions such as the World Bank in recent years. Accordingly, the Bank has introduced a variety of instruments into its lending and advisory activities. Environmental Assessment is one of the most important of these tools" (World Bank 1995). The World Bank and the European Investment Bank are the world's most significant development banks. In 1989, when the World Bank adopted Operational Directive (OD) 4.00, Annex A: Environmental Assessment (amended as OD 4.01 in 1991), EA became standard procedure for bank-financed investment projects. Over the years, the Bank has accumulated considerable EA experience; between 1989 and 1995 over a thousand projects were screened for their potential environmental impacts. As outlined in Table 11.7, screening can result in an assignment of a project to one of three environmental categories. Over a hundred projects were assigned to category A, for a full EA, over the 1989–95 period. Energy and power projects accounted for about 45 per cent of the category A projects, agriculture and transport together accounting for another 33 per cent.

The World Bank's EA process involves five stages: screening, scoping and terms of reference (TOR) development, preparing the EA report, EA review and project appraisal, and project implementation. (See Appendix 5 for further details of each stage.) Notable features of the process include a holistic environment definition, including physical, biological and socio-economic aspects, a high profile for public consultation and participation, and considerable focus on project implementation. However, the bank is not complacent about the quality and effective implementation of its procedures. While it has found an encouraging number of "good practice" cases in a variety of countries, the bank has identified five main challenges ahead to

365

Box 11.1   EIA – The road ahead

- *Moving EA "upstream"*
  Ideally, EA should be part of overall development planning at sectoral and regional levels. This allows for shaping of policy and investment strategies, including identification and early design of projects. By carrying out EA before major project decisions are made, alternatives can be considered more realistically, with greater possibility for influencing project design.

- *Public consultation*
  More effective consultation with local affected people enhances the development process. Decision making improves when local values and perceptions concerning development options and their environmental effects complement EA findings based on technical and economic analysis. Public consultation also increases local ownership of projects.

- *Integrating EA into the project work programme*
  EA findings and recommendations, including mitigation, monitoring and management plans, are useful only insofar as they are implemented. A key intermediate step is to convert these measures into agreed deliverables with specified costs, timing, responsibilities and funding. Establishing or strengthening mechanisms that allow this process to take place is essential to EA effectiveness.

- *Learning from implementation*
  Projects subject to EA that are under implementation need to be monitored to ensure environmental compliance, but also represent an opportunity to learn from experience how EA practice might be improved in the future. Establishing "feedback loops" to project preparation is an important potential environmental management tool.

- *Engaging the private sector*
  The private sector has become a major force in financing, designing or implementing development projects. A major, immediate challenge is to work with financiers and project sponsors in the private sector to ensure that projects are subject to EA of acceptable quality and the projects are compatible with country environmental strategies and plans. Significant progress has been made, but much more work needs to be done, especially in relation to the banking sector and capital markets, which are the main sources of investment finance in the development world, and in relating environmental planning at the macro level with private sector driven development.

(Source: World Bank, 1995)

---

make EA more effective (see Box 11.1). Davis (1996) provides an interesting World Bank paper on the crucial issue of improving public involvement in the process. Consultation and participation can be constrained by the lack of open government in the countries concerned, by the lack of social science expertise in EIA consultant teams and by the viewing of consultation as a hurdle by some bank project staff. He highlights certain Western European experience, such as local Agenda 21 exercises, Delphi techniques and citizen panels, which may have some relevance for future World Bank EA procedures.

European Commission EIA procedures for non-EU countries

The development of EC guidance on EIA for non-EC/EU countries was a response to the Lomé IV Convention, which provides the broad framework for development co-operation between the EC and ACP countries (Africa, the Caribbean and the Pacific). The convention explicitly recognizes the protection and enhancement of natural resources as a pivotal aspect of the economic development support that the EC can provide. The outcome was the EIA manual published in 1993 by DG VIII (External Relations and Development Co-operation with the ACP countries) (CEC 1993) and targeted at government authorities, delegations and the technical divisions of DG VIII. The manual is in two parts: *A users' guide* and a *Sectoral environmental assessment sourcebook*. The former provides project managers with direction on the main procedural steps of the process, plus practical guidance, such as model terms of reference (TOR) and checklists for a range of development types (e.g. mining, industry, tourism). The sourcebook includes guidelines for a wide range of development sectors. While the manual is potentially very useful, the Commission has increasing concern about its utility in practice, and about the harmonization of practice within the EC itself.

DGIA (External Relations: Europe and the New Independent States) and DGIB (External Relations: Southern Mediterranean, Middle East, Latin America, South East Asia) have been less proactive in terms of EIA procedures and their economic development actions. However, in 1997 DGIB produced a very useful DGIB *EIA Guidance Note* (CEC 1997). This builds on the DG VIII manual. It emphasizes the role of EIA through the project life-cycle (see Fig. 11.7), and has many other good features including an emphasis on public consultation, and the interrelationship between environmental change and social effects. It also involves a screening into three categories of project with the full EIA limited to category 3 projects. Progress on EIA for DGIA is more limited. In addition, the major funding agency of the EIB tends to draw indirectly on other agency/joint funder EIA procedures rather than strongly promote its own. Overall, EC EIA procedures for other countries are good in parts, but fragmented and perhaps weakly implemented in many cases. A review is currently in hand.

There are many other EIA guidelines for development aid projects and programmes. *A directory of impact assessment guidelines*, published by the International Institute for Environment and Development (Roe et al. 1995), provides a good summary. There are also guidelines from NGOs and from country-specific agencies; for example the UK Department for International Development (DFID, formerly ODA) produces *The manual of environmental appraisal* (ODA 1996). A major constraint on the implementation of all the good guidance emanating from funding agencies is the lack of indigenous EIA "capacity" in many countries. In conclusion, attention is drawn to a UNEP EIA training resources manual (UNEP 1997), developed to help fill the capacity gap. It includes both project EIA and strategic EIA and is available on the Internet at http:\\www.environment.gov.au/net/eianet.html.

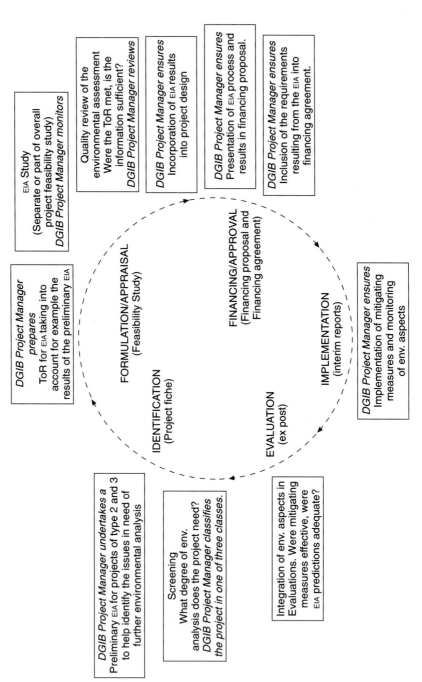

Figure 11.7 Environmental Assessment and the project cycle. (*Source:* CEC (1997) DGIB)

11.5   Summary

EIA systems can be found in many countries, as was shown in Figure 2.2. The nature of the systems varies – it is influenced by a particular country's resource base, the nature of its institutional environment and the development actions concerned; all have their strengths and weaknesses. However, an overview of at least some of these systems, as set out in this chapter, can provide valuable comparative experience. Chapters 12 and 13 draw on some of the ideas discussed here, and elsewhere, to identify possibilities for the future, focusing primarily on the UK system, but set in the wider European Union and global context.

References

Allotey, J. A. 1994. Environmental impact assessment in Ghana. *Environmental Assessment* **2**(1), 21–2.

Allotey, J. A. & Y. Amoyaw-Osei 1996a. Developing local capability in environmental impact assessment in Ghana. *Proceedings of the 16th Annual Conference of the International Association for Impact Assessment*, 551–6. Estoril: IAIA.

Allotey, J. A. & Y. Amoyaw-Osei 1996b. Developing environmental impact assessment procedures for a developing country – the experience of Ghana. *Proceedings of the 16th Annual Conference of the International Association for Impact Assessment*, 911–14. Estoril: IAIA.

Amir, S. 1992. EIA in Israel. *EIA Newsletter* **7**, 15–16.

Angelsen, A., F. Odd-Helge, U. R. Sumaila 1994. *Project appraisal and sustainability in less developed countries*, R. 1994:1. Fantoft-Bergen, Norway: Chr. Michelsen Institute.

ANZECC 1991. *Environmental impact assessment: a national approach; background paper to the national approach*. Canberra: Australian and New Zealand Environment and Conservation Council.

Au, E. W. K. 1996. Development of an EIA system in Hong Kong. *Environmental Link* **2**(1), 6–7.

Barrett, B. & R. Therivel 1991. *Environmental policy and impact assessment in Japan*. London: Routledge.

Barrett, B. F. D. 1991. Japan and the global environment: a case for leadership. *Japan Digest*, July, 29–35.

Baryeh, N., M. Silengo, M. Pugh-Thomas 1996. Improving environmental assessment in developing countries: the role of education and training. *Proceedings of the 16th Annual Conference of the International Association for Impact Assessment*, 191–5. Estoril: IAIA.

Beanlands, G. 1994. Summary of country policies and procedures – coherence of environmental assessment for international bilateral aid. Paris: OECD/DAC Working Party on Development Assistance and Environment.

Bianchi, G. & V. Rosova 1995. Public participation under environmental impact assessment in the Slovak Republic: case experience. *Acta Environmentalica Universitatis Comenianae* **4–5**, 183–6.

Birley, M. 1995. *The health impact assessment of development projects*. London: HMSO.

Bisset, R. 1992. Devising an effective environmental impact assessment system for a developing country: the case of the Turks and Caicos Islands. In *Environmental impact assessment in developing countries*, A. K. Biswas & S. B. C. Agarwala (eds), 214–34. Oxford: Butterworth-Heinemann.

Biswas, A. K. & S. B. C. Agarwala (eds) 1992. *Environmental impact assessment for developing countries*. Oxford: Butterworth-Heinemann.

Bowden, M. & F. Curtis 1988. Federal EIA in Canada: EARP as an evolving process. *Environmental Impact Assessment Review* **8**, 97–106.

Brachya, V. 1993. Environmental assessment in land use planning in Israel. *Landscape and Urban Planning* **23**, 167–81.

Branis, M. 1994. A system of certified environmental impact assessment experts in the Czech Republic. *Environmental Impact Assessment Review* **14**(2/3), 203–8.

Branis, M. & E. Kruzikova 1994. The Environmental Impact Assessment Act in the Czech Republic: origins, introduction and implementation. *Environmental Impact Assessment Review* **14**(2/3), 195–202.

Briffett, C. 1994. EIA in Singapore. *EIA Newsletter* **9**, 18.

Briffett, C. 1995. The effectiveness of environmental impact assessment in southeast Asia (MSc dissertation, Oxford Brookes University, Oxford).

Briffett, C. 1996. The effectiveness of environmental assessment in southeast Asia. *Proceedings of the 16th Annual Conference of the International Association for Impact Assessment*, 197–202. Estoril: IAIA.

Brito, E. N. & I. V. D. Moreira 1995. EIA in Brazil. *EIA Newsletter* **11**, 11–12.

Burgess, J. S., A. P. Spate, F. I. Norman 1992. Environmental impacts of station development in the Larsemann Hills, Princess Elizabeth Land, Antarctica. *Journal of Environmental Management* **36**, 287–299.

Cardenas, M. L. 1995. Programmatic EIA in the Philippines. *Environmental Link* **1**(3), 2–3.

CEAA (Canadian Environmental Assessment Agency) 1996. *Environmental assessment in Canada: achievements, challenges and directions*. Ottawa: CEAA.

CEARC (Canadian Environmental Assessment Research Council) 1986. *Philosophy and themes for research*. Ottawa: Minister of Supply and Services.

CEPA 1994. *Review of Commonwealth environmental impact assessment*. Canberra: CEPA.

Chaibva, S. 1996. EIA in Zimbabwe – past, present and future. *EIA Newsletter* **12**, 14–15.

Cherp, O. 1995. EIA in Russia: ten years of development. *EIA Newsletter* **11**, 5–6.

Cherp, O. 1996. Environmental impact assessment and sustainable development in Russia. *Russian Conservation News*, 6 February, 12–14.

Cherp, O. & M. Khotuleva 1996. Public participation in the EIA system in Russia. *EIA Newsletter* **12**, 8–9.

Chico, I. 1995a. EIA in Latin America. *Environmental Assessment* **3**(2), 69–71.

Chico, I. 1995b. Environmental management in Venezuela. *Environmental Assessment* **3**(3), 105.

Cizkova, H. & L. Zenaty 1993. EIA in the Czech Republic. *EIA Review* **8**, 16.

Coles, T. 1992. *Environmental impact assessment in Asia and Europe with special reference to Malaysia, Indonesia and the UK*. Lincoln: Institute of Environmental Assessment.

CEC (Commission of the European Communities) 1993. *Environment manual: user's guide; sectoral environmental assessment sourcebook*. Brussels: CEC DG VIII.

CEC 1997. *Environment impact assessment DGIB guidance note*. Brussels: CEC.

Commonwealth of Australia 1992. *A national strategy for ecologically sustainable development*. Canberra: Commonwealth of Australia.

Continental Gold Corporation 1991. *Mount Milligan project*. Placer Dome Inc.

Coombs, A. 1993. EA in Hong Kong. *Landscape Design* **224**, 35–6.

DASETT (Department of the Arts, Sport, the Environment, Tourism and Territories) 1992. *List of proposals on which environmental impact statements have been directed under the administrative procedures – Commonwealth Environmental Protection (Impact of Proposals) Act 1974*. Canberra: DASETT.

Davis, S. 1996. *Public involvement in environmental decision making*. Washington, DC: World Bank.

de Mello, M. 1995. The law on environmental impact assessment. *Environmental Policy and Law* **25**(1/2), 73.

Demkolec BV 1989. *EIS for the integrated gasification combined cycle (IGCC) demonstration plant at Buggenum*. Arnhem: Demkolec BV.

Devkota, S. R. 1996. Environmental impact assessment in Nepal. *Proceedings of the 16th Annual Conference of the International Association for Impact Assessment*, 209–210. Estoril: IAIA.

Dixon, J. & T. Fookes 1995. Environmental assessment in New Zealand: prospects and practical realities. *Australian Journal of Environmental Management* **2**(2), 104–11.

Douthwaite, R. J. 1996. EIA in Uganda, *EIA Newsletter* **12**, 15.

Dung-Gwom, J. Y. 1996. Environmental assessment in Nigeria. *EIA Newsletter* **12**, 15–16.

Effer, W. R. 1984. Ontario Hydro and Canadian environmental impact assessment procedures. In *Planning and ecology*, R. J. Roberts & T. M. Roberts (eds), 113–28. London: Chapman & Hall.

EPC (Environmental Protection Commission – People's Republic of China) 1986. *Management guidelines on environmental protection of construction projects of the People's Republic of China*. 26 March. Beijing: EPC.

EBRD (European Bank for Reconstruction and Development) 1994. *Environmental impact assessment legislation: Czech Republic, Estonia, Hungary, Latvia, Lithuania, Poland, Slovak Republic, Slovenia*. Netherlands: Wolters Kluwer.

Fearnside, P. M. 1994. The Canadian feasibility study of the Three Gorges dam proposed for China's Yangtze River: a grave embarrassment to the impact assessment profession. *Impact Assessment* **12**(1), 21–57.

FAC (Federal Airports Commission – Australia) 1990. *Summary draft EIS for the proposed third runway, Sydney (Kingsford Smith) Airport*.

FEARO (Federal Environmental Assessment Review Office) 1976. *Guidelines for preparing initial environmental evaluations*. Ottawa: FEARO.

FEARO 1978. *Guide for Federal screening*. Ottawa: FEARO.

FEARO 1979. *Revised guide to the Federal EARP*. Environmental Assessment Panel, Environment Canada. Ottawa: FEARO.

Fisher, D. 1992. *Paradise deferred: environmental policymaking in Central and Eastern Europe*. London: Royal Institute of International Affairs.

Formby, J. 1987. The Australian government's experience with environmental impact assessment. *Environmental Impact Assessment Review* **7**(3), 207–26.

Fowler, H. G. & A. M. Dias De Aguiar 1993. Environmental impact assessment in Brazil. *Environmental Impact Assessment Review* **13**(3), 169–76.

Gilpin, A. 1992. *Environmental impact assessment in Australia*. Unpublished report.

Gomez, M. A. B., B. L. Marzuez, F. Pilar Saldana 1996. Fundamental changes in EIA regulations. *Proceedings of the 16th Annual Conference of the International Association for Impact Assessment*, 211–17. Estoril: IAIA.

Goodland, R. & V. Edmundson 1994. *Environmental assessment and development.* Washington, DC: World Bank.

Goodland, R., J. R. Mercier, S. Muntemba 1996. *Environmental assessment (EA) in Africa: a World Bank commitment.* Washington, DC: World Bank.

Grönlund, S. 1996. Environmental impact assessment of the Kedainiai bypass and a new road from Kedainiai to Kaunus in Lithuania. *Proceedings of the 16th Annual Conference of the International Association for Impact Assessment*, 931–5. Estoril: IAIA.

Hacon, S. 1990. EIA in Brazil. *EIA Newsletter* **5**, 22–3.

Hahn, K. 1993. EIA experience in Korea. *EIA Newsletter* **8**, 17–18.

Han, E.-J., M.-J. Kim, J.-W. Lee, S. H. Kim 1996. Environmental impact assessment and environmental assessment. *Proceedings of the 16th Annual Conference of the International Association for Impact Assessment*, 219–24. Estoril: IAIA.

Harvey, N. 1992. The relationship between ecologically sustainable development and environmental impact assessment in Australia. *Environmental and Planning Law Journal* **9**(4).

Harvey, N. & K. Ferguson 1994. *Environmental impact assessment in South Australia: towards 2000.* Working Paper no. 4. Adelaide: University of South Australia, School of Planning and Building.

Harvey, N. & M. McCarthy (eds) 1997. *EIA for the 21st century: conference proceedings.* Adelaide: University of Adelaide Press.

Harvey, N. 1998. *EIA: procedures and prospects in Australia.* Melbourne: Oxford University Press.

Hirji, R. & L. Ortolano 1991. Strategies for managing uncertainties imposed by environmental impact assessment: analysis of a Kenyan river development agency. *Environmental Impact Assessment Review* **11**, 203–30.

Horberry, J. & G. Muscat 1990. EIA guidelines for developing countries. *EIA Newsletter* **5**, 20–1.

Hughes, H. R. 1996. Simultaneous preparation and review: a new approach to environmental assessments in New Zealand. *Impact Assessment* **14**(1), 97–103.

Husenicova 1996. Possibilities of the improvement of environmental impact assessment effectiveness in Slovak Republic through dynamic integration with urban planning and building law activities. *Proceedings of the 16th Annual Conference of the International Association for Impact Assessment*, 873–6. Estoril: IAIA.

Huskova, V. 1994. EIA in the Slovak Republic. *EIA Newsletter* **9**, 14.

Ibrahim, A. K. C. 1992. *An analysis of quality control in the Malaysian environmental impact assessment (EIA) process* (MSc dissertation, University of Manchester, Manchester).

Iglesias, S. 1996. *The role of EIA in mining activities: the Peruvian case* (MSc dissertation, Oxford Brookes University, Oxford).

IIED (International Institute for Environment and Development) & Institute of Resource Assessment 1995. *Environmental assessment in Tanzania: a needs assessment for training.* IIED Environmental Planning Issue no. 9. London: IIED.

Jahiel, A. R. 1994. *Policy implementation through organisational learning: the use of water pollution control in China's reforming socialist system* (PhD dissertation, University of Michigan, Ann Arbor).

Jendroska, J. & J. Sommer 1994. Environmental impact assessment in Polish law: the concept, development, and perspectives. *Environmental Impact Assessment Review* **14**(2/3), 169–94.

Jones, T. 1992. *Environmental impact assessment for coal.* IEAcr/46. London: IEA Coal Research.

Kakonge, J. O. 1994. Monitoring of environmental impact assessments in Africa. *Environmental Impact Assessment Review* **14**, 295–304.

Kakonge, J. O. 1995. Dilemmas in the design and implementation of agricultural projects in various African countries: the role of environmental impact assessment. *Environmental Impact Assessment Review* **15**(3), 275–85.

Kakonge, J. O. & A. M. Imevbore 1993. Constraints on implementing environmental impact assessment in Africa. *Environmental Impact Assessment Review* **13**(5), 299–308.

Kamukala, G. L. 1992. Application of the environmental impact assessment in the appraisal of major development projects in Tanzania. In *Environmental impact assessment for developing countries* A. K. Biswas & S. B. C. Agarwala (eds), ch. 15, 185–190. Oxford: Butterworth-Heinemann.

Kassenberg, A. 1994. EIA Commission in Poland. *EIA Newsletter* **9**, 13–14.

Kim, K. & D. H. L. Murabayashi 1992. Recent developments in the use of environmental impact statements in Korea. *Environmental Impact Assessment Review* **12**(13), 295–314.

Koning, H. 1990. EIA in the Netherlands. *EIA Newsletter* **5**, EIA Centre, University of Manchester.

Kozova, M. & I. Drdos 1995. Environmental impact assessment in the Slovak Republic: legislative context, methodology and steps in the process. *Acta Environmentalica Universitatis Comenianae* **4–5**, 153–69.

Kurasaka, H. 1996. Historical background and status quo of EIA systems in Japan. *Proceedings of the 16th Annual Conference of the International Association for Impact Assessment,* 229–34. Estoril: IAIA.

LaRovere, E. L. & M. M. L. Baraton 1996. Environmental auditing in Brazil: a new management tool. *Proceedings of the 16th Annual Conference of the International Association for Impact Assessment,* 887–91. Estoril: IAIA.

Ledgerwood, G., E. Street, R. Therivel 1992. *The environmental audit and business strategy.* London: Pitman / Financial Times.

Le Duc, A., Y. Tran Van, E. Nierynck 1997. *Proceedings of the first workshop on training in EIA: Vietnam.* Hanoi: Institute of Geography, National Centre for Natural Science and Technology.

Leu, W.-S., W. P. Williams A. W. Bark 1996. Development of an environmental impact assessment evaluation model and its application: Taiwan case study. *Environmental Impact Assessment Review* **16**, 115–33.

Li Xingji 1987. Discussing the methods and their selective principles adopted in the EIA. Paper presented at the International Environmental Impact Assessment Symposium, Beijing, October.

McLaren, D. E. 1993. Environmental considerations and public involvement in the assessment of the impacts of tourism development in Third World countries. *Impact Assessment* **11**(2), 175–202.

Ministry of Housing, Physical Planning and Environment 1989. *To choose or to lose: national environmental policy plan.* The Hague: Government of the Netherlands.

Ministry of Housing, Physical Planning and Environment 1990. *National environmental policy plan plus.* The Hague: Government of the Netherlands.

Modale, P. 1994. EIA in India. *EIA Newsletter* **9**, 17.

Mondok, Z. 1995. The new Environmental Protection Act and EIA in Hungary. *EIA Newsletter* **11**, 3–4.

Montz, B. E. & J. E. Dixon 1993. From law to practice: EIA in New Zealand. *Environmental Impact Assessment Review* **13**, 89–108.

Morgan, R. 1995. Progress with implementing the environmental assessment requirements of the Resource Management Act in New Zealand. *Journal of Environmental Planning and Management* **38**(3), 333–48.

Moshoeshoe, Z. & M. Malatsi 1996. The role of non-governmental organisations and community based organisations in strengthening the environmental assessment process in South Africa. *Proceedings of the 16th Annual Conference of the International Association for Impact Assessment*, 527–32. Estoril: IAIA.

Moreira, I. V. 1992. EIA in Latin America. In *Environmental impact assessment: theory and practice*, P. Wathern (ed.), 239–253. London: Unwin Hyman.

Mubvami, T. 1996. EIA in Zimbabwe: the quest for sustainable development in a developing country. *Environmental Education and Information* **15**(1), 33–42.

Nay Htun 1992. The EIA process in Asia and the Pacific region. In *Environmental impact assessment: theory and practice*, P. Wathern (ed.). London: Unwin Hyman.

Netherlands Commission for Environmental Impact Assessment 1996. *Country status reports on EIA*. The Hague: Netherlands Commission for EIA.

Ning, D., H. Wang, J. Witney 1988. Environmental impact assessment in China: present practice and future developments. *Environmental Impact Assessment Review* **8**, 85–95.

Nor, Y. M. 1991. Problems and perspectives in Malaysia. *Environmental Impact Assessment Review* **11**(2), 129–42.

ODA (Overseas Development Administration) 1996. *The manual of environmental appraisal*. London: ODA.

Ofori, S. C. 1991. Environmental impact assessment in Ghana: current administrative procedures – towards an appropriate methodology. *The Environmentalist* **11**(1), 45–54.

Okaru, V. & A. Barannik 1996. Harmonization of environmental assessment procedures between the World Bank and borrower nations. In *Environmental Assessment (EA) in Africa*, R. Goodland et al. (eds), 35–63. Washington, DC: World Bank.

Onorio, K. & R. K. Morgan 1995. In-country EIA training in the South Pacific: an interim review and evaluation of the South Pacific Regional Environment Programme's EIA program. *Impact Assessment* **13**(1), 87–99.

Ortolano, L. 1996. Influence of institutional arrangements on EIA effectiveness in China. *Proceedings of the 16th Annual Conference of the International Association for Impact Assessment*, 901–5. Estoril: IAIA.

Pavlickova, K., B. Gabriel, R. Viera 1995. Public participation in the environmental impact assessment process in the Slovak Republic. *Acta Environmentalica Universitatis Comenianae* **4–5**, 177–82.

Peters, D. 1994. Social impact assessment of the Ranomafana National Park Project of Madagascar. *Impact Assessment* **12**(4), 385–408.

Peterson, K. 1995. EIA in Estonia. *EIA Newsletter* **11**, 2–3.

Pisanty-Levy, J. 1993. Mexico's environmental assessment experience. *Environmental Impact Assessment Review* **13**(4), 267–72.

Province of British Columbia 1992. *Reforming environmental assessment in British Columbia: (i) A legislation discussion paper; (ii) A report on the consultation process*. Victoria, BC: Ministry of Environment, Lands and Parks, British Columbia.

Radnai, A. 1993. EIA in Hungary. *EIA Newsletter* **8**, 16–17.

Reed, S. B. 1994. EIA in Hong Kong. *EIA Newsletter* **9**, 16–17.

Roe, D., B. Dalal-Clayton, R. Hughes 1995. *A directory of impact assessment guidelines.* London: International Institute for Environment and Development.

Romanillos, J., J. Palerm, W. R. Sheate 1996. Environmental impact assessment in central and eastern Europe: lessons from the Czech Republic and Romania. *European Environmental Law Review*, January, 15–22.

Ross, W. A. 1994. Environmental impact assessment in the Philippines: progress, problems and directions for the future. *Environmental Impact Assessment Review* 14(3), 217–32.

Ruchti, S. 1993. EIA in Switzerland. *EIA Newsletter* 8, 19–20.

Rzsezot, U. 1995. EIA in Poland: the new regulations. *EIA Newsletter* 11, 4–5.

Rzsezot, U. & C. Wood 1992. Environmental impact assessment in Poland: an emergent process. *Project Appraisal* 7(2), 83–92.

Sadar, M. H. & Z. Si 1994. *Priority environmental issues in Asia: the need and importance of developing cooperative approaches.* Proceedings of the Canada–Asia Seminar, 21–22 June. Ottawa: Impact Assessment Centre, Carleton University.

Sadler, B. 1987. *Audit and evaluation in environmental assessment and management: Canadian and international experience.* Proceedings of the Conference on Follow-up/Audit of EIA Results, Banff Centre, Banff, Alberta, 13–16 October.

Sadler, B. 1988. The evaluation of assessment: post-EIS research and process development. In *Environmental impact assessment: theory and practice*, P. Wathern (ed.), 129–42. London: Unwin Hyman.

Sadler, B. 1996. *Environmental effectiveness in a changing world: evaluating practice to improve performance.* Final report of the international study on the effectiveness of environmental assessment, Canadian Environmental Assessment Agency.

Sanchez-Silva, R. & S. Cruz-Ulloa 1994. Environmental impact of an agricultural project in La Roca, Oaxaca, Mexico. *Impact Assessment* 12(1), 89–92.

Schoeffel, P. 1995. Cultural and institutional issues in the appraisal of projects in developing countries: South Pacific water resources. *Project Appraisal* 10(3), 155–61.

Sinkule, B. J. & L. Ortolano 1995. *Implementing environmental policy in China.* Westport, Connecticut: Praeger.

Smith, D. B. & M. van der Wansem 1995. *Strengthening EIA capacity in Asia: environmental impact assessment in the Philippines, Indonesia and Sri Lanka.* Washington, DC: World Resources Institute.

Smith, L. G. 1991. Canada's changing impact assessment provisions. *Environmental Impact Assessment Review* 11, 5–9.

SPREP (South Pacific Regional Environment Programme) 1993. *A guide to environmental impact assessment in the South Pacific.* Apia, Western Samoa: South Pacific Regional Environment Programme.

Sowman, M., R. Fuggle, G. Preston 1995. A review of the evolution of environmental evaluation procedures in South Africa. *Environmental Impact Assessment Review* 15(1), 45–67.

Takabe, M. 1994. EIA in Japan. *EIA Newsletter* 9, 17–18.

Ter-Nikoghosyan, V. 1996. EIA in Armenia. *EIA Newsletter* 12, 16–17.

Thanh, N. C. & D. M. Tam 1992. Environmental protection and development: how to achieve a balance? In *Environmental impact assessment in developing countries*, A. K. Biswas & S. B. C. Agarwala (eds), 3–15. Oxford: Butterworth-Heinemann.

Thomas, I. 1996. *EIA in Australia.* New South Wales: Federation Press.

Tongcumpou, C. & N. Harvey 1994. Implications of recent EIA changes in Thailand. *Environmental Impact Assessment Review* 14, 271–94.

Tortajada-Quiroz, H. C. 1996. Environmental assessment for water projects in Mexico. *Proceedings of the 16th Annual Conference of the International Association for Impact Assessment*, 923–26. Estoril: IAIA.

UNECE (United Nations Economic Commission for Europe) 1991. *Policies and systems of environmental impact assessment*. New York: United Nations.

UNEP (United Nations Environment Programme) 1994. *A sub-regional workshop on environmental impact assessment for Commonwealth countries of eastern and southern Africa – workshop report*. Environmental Economies Series Paper no. 10. Nairobi: UNEP.

UNEP 1997. *Environmental impact assessment training resource manual*. Stevenage: SMI Distribution.

UK Foreign and Commonwealth Office 1995. *Guide to EIA of activities in Antartica*. London: Foreign and Commonwealth Office.

Valappil, M., D. Devuyst, L. Hens 1994. Evaluation of the environmental impact assessment procedures in India. *Impact Assessment* **12**(1), 75–88.

Van de Gronden, E. D., J. Beentjes, F. von der Wonde 1994. *Use and effectiveness of EIA in decision-making*. Zoetermeer: Netherlands Ministry of Housing, Spatial Planning and the Environment.

van Haeren, J. 1991. EIA in the Netherlands. *EIA Newsletter* **6** (EIA Centre, University of Manchester).

Veleva, V. R. 1996. Public participation in EIA in Bulgaria. *EIA Newsletter* **12**, 9–10.

Vizayakumar, K. & P. K. J. Mohapatra 1991. Framework for environmental impact analysis – with special reference to India. *Environmental Management* **15**(3), 357–68.

Vrbensky, R. & F. Kolocany 1995. Environmental impact assessment in the Slovak Republic with particular focus on the position of the Ministry of Environment in this process. *Acta Environmentalica Universitatis Comenianae* **4–5**, 171–6.

Wallace, R. R. 1985. *Public input to government decision-making*. FEARO Occasional Paper no. 13. Calgary, Alberta: FEARO.

Wathern, P. 1992. *Environmental impact assessment: theory and practice*. London: Routledge.

Welles, H. 1995. EIA capacity-strengthening in Asia: the USAID/WRI model. *Environmental Professional* **17**, 103–16.

Wenger, R. B. & W. Huadong 1990. Environmental assessment in the People's Republic of China. *Journal of Environmental Management* **14**(4), 429–39.

Wenger, R., H. Wang, X. Ma 1990. Environmental impact assessment in the People's Republic of China. *Environmental Management* **14**(4), 429–39.

West, C., R. Bisset, R. Snowden 1993. *Developing countries EIAs*. London: IBC Technical Services.

Wiseman, K. 1996. Balancing theory and reality in the management of large EIAs: the Saldanha steel project, South Africa. *Proceedings of the 16th Annual Conference of the International Association for Impact Assessment*, 681–6. Estoril: IAIA.

Wood, C. 1993. Environmental impact assessment in Australia – can the old world learn from the new? *International Environmental Affairs* **5**(3), 256–73.

Wood, C. 1995. *Environmental impact assessment: a comparative review*. Harlow: Longman Scientific & Technical.

Wood, C. & J. Bailey 1994. Predominance and independence in environmental impact assessment: the Western Australia model. *Environmental Impact Assessment Review* **14**(1), 37–59.

World Bank 1991. *Operational Directive 4.01: Environmental assessment*. Washington, DC: World Bank.

World Bank 1995. *Environmental assessment: challenges and good practice.* Washington, DC: World Bank.

Xie, Y., D. Devuyst, L. Hens 1996. Trends in public participation in environmental impact assessment in some Asian developing countries. *Proceedings of the 16th Annual Conference of the International Association for Impact Assessment*, 949–53. Estoril: IAIA.

Notes

1 By mid-1990, 21 exemptions had been granted, of which seven were for projects covered by Annexes I or II of the Directive. As a result, the EC informed the Dutch government in a letter of April 1990 that changes were needed. In May 1991, the Dutch government published "The report of the government on the working of the EIA regulations" which proposed changes to address the issues raised by the EC, and a bill has been proposed that would amend the Environmental Protection Act to conform to the Directive.

2 The EIA Commission is an independent body which carries out research on the EIA system, and which advises on the scope and adequacy of each EIA. The core of the Commission is composed of a chairman, who is appointed by the Council of Ministers, two vice-chairmen and a full-time staff of about 25. In addition, about 200 members who are experts in EIA-related fields assist on a case-by-case basis.

3 The same group that carried out the scoping exercise usually also reviews the EIS.

4 Samenwerkende Electriciteits Produktiebedijuen.

5 The screening guidelines use a series of matrices, first to identify the activities that are likely to have an environmental impact, second to focus on specific problem areas. Federal agencies are expected to establish their own screening guidelines, including exclusion lists. These are lists of specific actions considered to be environmentally benign and are excluded from the requirements for EIA, for instance routine maintenance, the interior renovation of buildings, and surveys and inventories.

6 CEARC was established in 1984 to "advise on ways to improve the scientific, technical and procedural basis for environmental impact assessment" (CEARC 1986). It is composed of 12 members from the government and the academic and private sectors, who are appointed by the chairman of FEARO for three-year terms. CEARC promotes and reviews research related to improving the EIA system, in particular the integration of EIA with strategic planning, the incorporation of ecological and social sciences within EIA, the incorporation of social values in EIA evaluation and the strengthening of policy and institutional frameworks related to EIA. More recent research has focused on the EIA of government policies, the relation between EIA and economics, the role of EIA in sustainable development and the possibility of "intervenor funding", i.e. the funding of those participants who would be significantly affected by the proposal under review (Bowden & Curtis 1988).

7 Barrett (1991) notes:

> One Japanese journalist recently argued that no country other than Japan could realistically adopt such a rôle. The United States and the Soviet Union are "pollution superpowers" who have neglected the development of pollution control technology as they vied for military supremacy. Both countries must now devote a significant

proportion of their declining economic power in order to overcome the environmental threats they are currently facing. European countries, moreover, are not particularly interested in new environmental technologies and the ecological disaster now facing their eastern partners will keep the Europeans pre-occupied for some time. Meanwhile, developing countries, already burdened with debt, cannot afford to develop new technologies to counter the growing number of environmental problems they will have to face. The only country with the ability to effectively react to global environmental problems, therefore, is Japan.

8　This requires Poland to incrementally enact changes in its laws and statutes, to bring them in line with those of the EU.

Part 4

Prospects

CHAPTER 12

Improving the effectiveness of project assessment

12.1 Introduction

Overall, the experience of EIA to date can be summed up as being like the prover-bial curate's egg: good in parts. Current issues in the EIA process were briefly noted in Section 1.5: they include screening, scoping, EIA methodology, the roles of the participants in the process, EIS quality, monitoring and the extension of EIA to more strategic levels of decision-making. The various chapters on steps in the process have sought to identify best practice, and Chapter 8 provides an overview of the quantity and quality of UK practice from 1988 to 1997. Detailed case studies of good practice and comparative international experience provide further ideas for possible future developments. The limited experience in EIA among the main par-ticipants in the process – consultants, local authorities, central government, devel-opers and affected parties – explains some of the current issues.

However, less than ten years after the implementation of EC Directive 85/337, there is less scepticism in most quarters and a general acceptance of the value of EIA. There are still some fundamental shortcomings, and there is considerable scope for improving quality, but practice and the underpinning knowledge and under-standing are quickly developing; EIA is on a steep learning curve. The procedures, process and practice of EIA will undoubtedly evolve further, as evidenced by the comparative studies of other countries. The EU countries can learn from such experi-ence and from their own experience since 1988.

This chapter focuses on the prospects for project-based EIA. The following sec-tion briefly considers the array of perspectives on change from the various partici-pants in the EIA process. This is followed by a consideration of possible developments in some important areas of the EIA process and in the nature of EISs. The chapter concludes with a discussion of the parallel and complementary development of environmental management systems and audits. The nature and types of system and audit are explained, and their important relationships to EIA are discussed. Chapter 13 closes the book by widening the scope of EIA, from projects to programmes, plans and policies.

12.2 Perspectives on change

An underlying theme in any discussion of EIA is change. This has surfaced several times in the various chapters of this book. EIA systems and procedures are changing in many countries. Indeed, as O'Riordan (1990) noted (see Section 1.3), we should expect EIA to change in the face of shifting environmental values, politics and managerial capabilities. This is not to devalue the achievements of EIA to date. As the World Bank (1995) notes, "Over the past decade, EIA has moved from the fringes of development planning to become a widely recognised tool for sound project decision making."

The practice of EIA under the existing systems established in the EU Member States has also improved rapidly (see Ch. 8). This change can be expected to continue in the future, as the provisions of the amended Directive 97/11 are introduced and used. Changes in EIA procedures, like the initial introduction of EIA regulations, can generate considerable conflict between levels of government: between federal and state levels, between national and local levels and, in the particular case of Europe, between the EU and its Member States. They also generate conflict between the other participants in the process: the developers, the affected parties and the facilitators.

The *Commission of the European Communities* (CEC) is generally seen as positive and proactive with regard to EIA. The CEC welcomed the introduction of common legislation as reflected in Directive 85/337, the provision of information on projects and the general spread of good practice, but was concerned about the lack of compatibility of EIA systems across frontiers, the opaque processes employed, the limited access of the public, and lack of continuity in the process. It pressed hard for amendments to the Directive, and achieved some of its objectives in the amended Directive. The CEC is committed to reviewing and updating EIA procedures, which may involve further changes. Areas of attention include, for example, SEA (see Chapter 13), cumulative assessment, public participation, economic valuation, and EIA procedures for development aid projects. In contrast with the CEC, *Member States* tend to be more defensive and reactive. They are generally concerned to maintain "subsidiarity" with regard to activities involving the EU; this has been an issue with EIA, as reflected in the exchanges between the EC Commissioner for the Environment and the UK Government in 1991/92 (see Section 6.6). Governments are also sensitive to increasing controls on economic development in difficult economic times.

For example, within the *UK Government*, the Department of the Environment (DOE)(now DETR) has been concerned to tidy up ambiguities in the project-based procedures, and to improve guidance and informal procedures, but is wary of new regulations. However, it has commissioned and produced research reports on an EIA good practice guide, and on the evaluation and review of environmental information. Its response to many of the proposals in the amended Directive also reflect an acceptance of the value of EIA. *Local government* in the UK has begun to come to terms with EIA, and there is evidence that those authorities with considerable experience (e.g. Kent, Cheshire) learn fast, apply the regulations and guidance in user-friendly "customized" formats to help developers and affected parties in their areas, and are pushing up the standards expected from project proponents.

Pressure groups – exemplified in the UK by the Council for the Protection of Rural England (CPRE), the Royal Society for the Protection of Birds (RSPB) and Friends of the Earth (FOE) – and those parties affected by development proposals view project EIA as a very useful tool for increasing access to information on projects, and for advancing the protection of the physical environment in particular. They have been keen to develop EIA processes and procedures; see, for example, the reports by CPRE (1991, 1992). Many *developers* are less enthusiastic about changes in the regulations, but would welcome clarification on ambiguities – especially on whether EIAs are needed in the first place for their particular projects. For *facilitators* (consultants, lawyers, etc.), EIA has been a welcome boon; their interest in longer and wider procedures, involving more of their services, is clear.

Other participants in the process in the UK, such as the Institute of Environmental Assessment (IEA), the Association of Environmental Consultants, academics and some environmental consultancies, are carrying out ground-breaking studies into topics such as best-practice guidelines, the use of monetary valuation in EIA, and approaches to types of impact study. In addition the production of over 300 EISs a year in the UK is generating a considerable body of expertise, innovative approaches and comparative studies. EISs are also becoming increasingly reviewed, and it is hoped that bad practice will be exposed and reduced. Training in EIA skills is also developing.

12.3　Possible changes in the EIA process: the future agenda

An overview of possible changes

In the important *International study of the effectiveness of environmental assessment* for the IAIA, Sadler (1996) provides a summary of "best case" and "worst case" EA performance (see Box 12.1). He also provides a five-part agenda for action:

- "Going back to basics" involves building on well-established procedures, by providing, for example, more good practice guidance, explicit periods for the process, and the removal of duplication.
- "Upgrading EIA processes and activities" involves, in particular, better quality control, public involvement and addressing the issue of cumulative effects.

Box 12.1 Summary of international best and worst case EA performances

Best case performance

The EA process:

- facilitates informed decision making by providing clear, well-structured, dispassionate analysis of the effects and consequences of proposed actions;
- assists the selection of alternatives, including the selection of the best practicable or most environmentally-friendly option;
- influences both project selection and policy design by screening out environmentally unsound proposals, as well as modifying feasible action;
- encompasses all relevant issues and factors, including cumulative effects, social impacts, and health risks;
- directs (not dictates) formal approvals, including the establishment of terms and conditions of implementation and follow-up;
- results in the satisfactory prediction of the adverse effects of proposed actions and their mitigation using conventional and customized techniques; and
- serves as an adaptive, organizational learning process in which the lessons experienced are fed back into policy, institutional, and project design.

Worst case performance

The EA process:

- is inconsistently applied to development proposals with many sectors and classes of activity omitted;
- operates as a "stand alone" process, poorly related to the project cycle and approval process and consequently is of marginal influence;
- has a non-existent or weak follow-up process, lacking surveillance and enforcement of terms and conditions, effects monitoring, etc.;
- does not consider cumulative effects or social, health and risk factors;
- makes little or no reference to the public, or consultation is perfunctory, substandard and takes no account of the specific requirements of affected groups;
- results in EA reports that are voluminous, poorly organized and descriptive technical documents;
- provides information that is unhelpful or irrelevant to decision-making;
- is inefficient, time consuming and costly in relation to the benefits delivered; and
- understates and insufficiently mitigates environmental impacts and loses credibility.

(*Source*: Sadler 1996)

- "Extending SEA as an integral part of policy making" includes the development of methods, and extended applications.
- "Sharpening EA as a sustainability instrument" includes incorporating relevant sustainability indicators, the consideration of capacities, dealing with risks and uncertainty and linking EIA with other forms of assessment and other policy instruments, such as environmental accounting.

- "New opportunities and challenges" covers issues such as the trans-boundary management of common property resources (e.g. the Antarctic), global change and the decommissioning or replacement of major infrastructure items.

A pragmatic approach to change could subdivide the future agenda into pro-posals to *improve* EIA procedures, usually sooner and maybe more easily than proposals to *widen* the scope of EIA, which are likely to come later and probably be more difficult to implement.

Improvements to project EIA cover some of the changes heralded by the amended EC Directive, including developments in approaches to screening, the mandatory consideration of alternatives and a strong encouragement to undertake scoping at an early stage in the project development cycle. There is also more support for trans-parent procedures, and encouragement for consultation, for the explanation and publication of decisions and for the inclusion of cumulative impacts and risk assess-ment. There may be a case for further changes in the legal basis of project EIA, especially in the UK, where the wide array of regulations can cause the fragmenta-tion of the elements of a project linked EIA activity, as we revealed in Chapter 10. The methods of assessment could also benefit from further attention. Uncertainty about the unknown may mean the EIA process starts too late and results in a lack of integration with the management of a project's life-cycle. The EIA process and the resulting EISs may lack balance, focus on the more straightforward process of describing the project and its baseline environment and consider much less the identification, prediction and evaluation of impacts. The forecasting methods used in EIA are not explained in most cases (see Table 8.1). It is to be hoped that there will be advances in the application of concepts and techniques in operational prac-tice, in the areas of predicting the magnitude of impacts and determining their importance (including the array of multi-criteria and monetary evaluation tech-niques). A good "method statement", explaining how a study has been conducted – in terms of techniques, consultation, the relative roles of experts and others – should be a basic element of any EIS.

Widening the scope of EIA includes, in particular, the development of tiered assessment through the introduction of SEA (discussed in the next chapter). Another important extension of the scope of EIA includes "completing the circle" through the more widespread use of monitoring and auditing. Unfortunately this vital step in the EIA process will still not be mandatory under the amended EC Directive. More wide-ranging possibilities include the move to a "whole of environment" approach, with a more balanced consideration of both biophysical and socio-economic impacts. Such widening of scope should lead to more integrated en-vironmental assessment. There may also be a trend towards what might be termed "environmental impact design", with the use of EIA to identify environmental con-straints before the design process is begun.

The following sections discuss possibilities for some of these short- and long-term proposals, including allowing for cumulative impacts, building in better pro-cedures for public participation, widening the scope to include socio-economic impacts, developing integrated environmental assessment and moving towards environmental impact design.

Cumulative impacts

Many projects are individually minor, but collectively may impose a significant impact on the environment. Activities such as residential development, farming and household behaviour normally fall outside the scope of conventional EIA. The ecological response to the collective impact of such activities may be delayed until a threshold is crossed, when the impact may come to light in sudden and dramatic form (e.g. flooding). Odum (1982) refers to the "tyranny of small decisions" and the consequences arising from the continual growth of small developments. While there is no particular consensus on what constitutes cumulative impacts, the categorization by the Canadian Environmental Assessment Research Council (CEARC) (Peterson et al. 1987) is widely quoted, and includes:

- time-crowded perturbations – which occur because perturbations are so close in time that the effects of one are not dissipated before the next one occurs;
- space-crowded perturbations – when perturbations are so close in space that their effects overlap;
- synergisms – where different types of perturbation occurring in the same area may interact to produce qualitatively and quantitatively different responses by the receiving ecological communities;
- indirect effects – are those produced at some time or distance from the initial perturbation, or by a complex pathway; and
- nibbling – which can include the incremental erosion of a resource until there is a significant change / it is all used up.

"Cumulative impact assessment is predicting and assessing all other likely existing, past and reasonably foreseeable future effects on the environment arising from perturbations which are time-crowded; space-crowded; synergisms; indirect; or, constitute nibbling" (Commonwealth Environmental Protection Agency (CEPA) 1994). The need to include cumulative impact assessment in EIA has been long recognized. In the Californian Environmental Quality Act of 1970, significant impacts are considered to exist if "the possible effects of a project are individually limited but cumulatively considerable". More recent legislative reference is found in the 1991 Resource Management Act of New Zealand, which makes explicit reference to cumulative effects, and now also in the amended Directive 97/11/EC, which refers to the need to consider the characteristics of projects having regard to "the cumulation with other projects".

However, it is in the practical implementation of the consideration of cumulative impacts that the problems and deficiencies become clear, and cases of good practice and useful methodologies are limited. In Australia, assessments have largely been

386

carried out by regulatory authorities, rather than by project proponents, and have focused on regional air quality and the quality and salinity of water in catchment areas (CEPA 1994). Figure 12.1 provides an example of a simple perturbation impact model developed by Lane and Associates (1988). It is basically an "impact tree" which links (a) the principal causes driving a development with (b) the main perturbations induced with (c) the primary biophysical and socio-economic impacts, and (d) the secondary impacts. The figure shows some of the potential cumulative impacts associated with a number of area-related tourism developments.

Public participation

The lack of effective public participation in EIA is a major weakness in the UK, in most of Europe and in many other countries. It tends to occur late, if at all, and is often tokenistic and limited to minimum requirements and to the lowest rungs on Arnstein's ladder (see Section 6.2). There is an unequal balance of participants between the "impactors" and the "impactees". We hear much of "expert speak" but often very little of "people speak". Yet the public have much to contribute: they may offer a superior knowledge of local conditions; they bring their own values as stakeholders; and they contribute a non-scientific discourse to a process which is often *too* scientific. Lack of effective public participation is not only inequitable and inefficient; it may also be very counter-productive, as frustrated and unequal participants resort to other means, including direct action.

More effective public participation needs both the will and the methods. In Europe, there are signs from the European Union, and more widely, that the will is strengthening, as evidenced by the following declaration from the Third Conference of European Environment Ministers (Sofia, Bulgaria, October 1995):

> We believe it is essential that, in accordance with Principle 10 of the Rio Declaration, States should give the public the opportunity to participate at all levels in decision-making processes relating to the environment, and we recognize that much remains to be done in this respect.
>
> We call upon countries in the region to ensure that they have a legal framework and effective and appropriate mechanisms to secure public access to environmental information, to facilitate and encourage public participation, inter alia through environmental impact assessment procedures, and to provide effective public access to judicial and administrative remedies for environmental harm.
>
> We invite countries to ensure that in relevant legislation effective public participation as a foundation for successful environmental policies is being introduced.

New methods are needed to empower people in EIA to participate genuinely and constructively. These could include deliberative techniques, such as focus groups, Delphi panels and consultative committees, and appropriate resourcing, perhaps through intervenor funding.

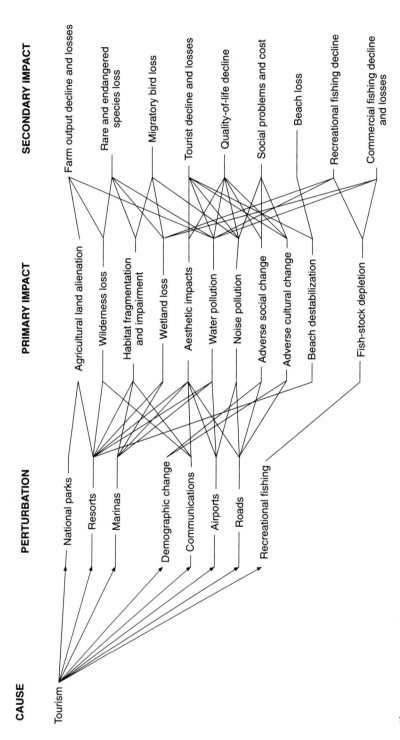

Figure 12.1 Cumulative impacts – perturbation impact model. (*Source:* Lane, P. and associates 1988)

CAUSE PERTURBATION PRIMARY IMPACT SECONDARY IMPACT

Tourism

National parks
Resorts
Marinas
Demographic change
Communications
Airports
Roads
Recreational fishing

Agricultural land alienation
Wilderness loss
Habitat fragmentation and impairment
Wetland loss
Aesthetic impacts
Water pollution
Noise pollution
Adverse social change
Adverse cultural change
Beach destabilization
Fish-stock depletion

Farm output decline and losses
Rare and endangered species loss
Migratory bird loss
Tourist decline and losses
Quality-of-life decline
Social problems and cost
Beach loss
Recreational fishing decline
Commercial fishing decline and losses

Socio-economic impacts

Widening the scope of EIA to include socio-economic impacts much better is seen as a particularly important item for the agenda. While there are varying interpretations of the scope of socio-economic or social impacts, a number of recent reports have highlighted the importance of this area (see, for example, CEPA 1994, IAIA 1994). Social impact assessment (SIA) has been defined by Bowles (1981) as "the systematic advanced appraisal of the impacts on the day to day quality of life of people and communities when the environment is affected by development or policy change". Most development decisions involve trade-offs between biophysical and socio-economic impacts. Also, development projects affect various groups differently; there are invariably winners and losers. Yet the consideration of socio-economic impacts is very variable in practice, and often very weak. There is useful practice and associated legislative impetus for SIA in some countries, for example the USA, Canada and some states of Australia. Some of the procedures of international funding institutions also give a high profile to such impacts. But in Europe the profile is lower, and the consideration of socio-economic impacts has continued to be the poor relation (Glasson 1995). Even when socio-economic impacts are included, the socio-cultural aspect (with impacts such as severance, alienation, social polarization, crime and on health) may still be very marginal.

The fuller and better consideration of socio-economic impacts raises issues and challenges, for example about the types of impact, their measurement, the role of public participation, and their position in EIA. One categorization of socio-economic impacts is into: (a) quantitatively measurable impacts, such as population changes, and the effects on employment opportunities or on local financial implications of a proposed project, and (b) non-quantitatively measurable impacts, such as effects on social relationships, psychological attitudes, community cohesion, cultural life or social structures (CEPA 1994). Such impacts are wide ranging, many are not easily measured, and direct communication with people about their perceptions of socio-economic impacts is often the only method of documenting such impacts. There is an important symbiotic relationship between developing public participation approaches and the fuller inclusion of socio-economic impacts. SIA can establish the baseline of groups which can provide the framework for public participation to further identify issues associated with a development proposal. Such issues may be more local, subjective, informal and judgemental than those normally covered in EIA, but they cannot be ignored. Perceptions of the impacts of a project and the distribution of those impacts often largely determine the positions taken by various groups on a given project and any associated controversy.

Integrated environmental assessment

Hopefully such widening of scope will lead to integrated environmental assessment, with decisions based on the extent to which various biophysical, social and economic

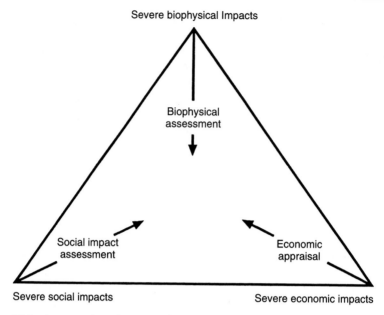

Severe biophysical Impacts

Biophysical
assessment

Social impact
assessment

Economic
appraisal

Severe social impacts

Severe economic impacts

Figure 12.2 Integrated environmental assessment.

impacts can be traded (Fig. 12.2). For example, decision-makers might be unwilling to trade critical biophysical assets (e.g. a main river system and the quality of water supply) for jobs or lifestyle, but willing to trade less critical biophysical assets. Integrated environmental assessment or IEA (Bailey et al. 1996, Davis 1996) differs from traditional EIA in that it is consciously multi-disciplinary, does not take citizens' participation or the ultimate users of EIA for granted and recognizes the critical role of complexity and uncertainty in most decisions about the environment. Hence it tolerates a much broader array of methods and perspectives (quantitative and qualitative; economic and sociological; computer modelling and oral testimony) for evaluating and judging alternative courses of action. However, integration is not without its problems, including limitations on the transferability of assessment methods (see *Project Appraisal* 1996).

Another equally important perspective is of the integration of relevant planning, environmental protection and pollution procedures. At the one extreme the UK has over 40 regulations for EIA, grafting the procedures into an array of relevant planning and other legislation; there is also parallel environmental protection and pollution legislation. At the other, there is the New Zealand "one-stop shop" *Resource Management Act*. A better integration of relevant procedures represents another challenge for most EIA systems.

390

Extending EIA to project design: towards environmental impact design

An important and positive trend in EIA has been its application at increasingly early stages of project planning. For instance, whilst the DOT's 1983 *Manual of environmental appraisal* applied only to detailed route options, its 1993 *Design manual for roads and bridges* requires a three-stage approach covering, in turn, broadly defined route corridors, route options and the chosen route (see Chapter 10). British Gas also now uses three levels of environmental analysis for its pipelines, from broad feasibility studies to detailed design (Parkinson 1996). This application of EIA to the early stages of project planning helps to improve project design and to avoid the delayed and costly identification of environmental constraints that comes from carrying out EIA once the project design is completed.

McDonald & Brown (1995) suggest that the project designer must be made part of the EIA team:

> Currently, most formal administrative and reporting requirements for EIA are based on its original role as a stand alone report carried out distinct from, but in parallel with the project design ... We can redress [EIA limitations] by transferring much of the philosophy, the insights and techniques which we currently use in environmental assessments, directly into planning and design activities.

A further evolution of this concept is to use EIA to identify basic environmental constraints before the design process is begun, but then allow designers freedom to design innovative and attractive structures as long as they meet those constraints. Holstein (1996) calls this postmodern approach "environmental impact design" (EID),[1] and distinguishes it from EIA's traditionally conservative, conservation-based focus. The following paragraphs explain Holstein's view of EID.

> EIA as presently practised deconstructs a site: it takes an environment apart to highlight the different interacting components within it (e.g. soil, water, flora). EIA suggests that the site has another (environmental) function other than that for which it is being developed. Yet this relationship to deconstruction is only superficial because EIA is conservation based; it makes little challenge to the fixed hierarchies of modernism that underpin it, such as development-induced growth and technological subservience. Environmental design within EIA is too often merely a by-product of assessment or is even handed back to the developer to have another shot at the design themselves. It makes little use of artistic-based metaphors to provide any re-enchantment or return to human landscape values, it makes no attempt to rip apart environmental function and form, and creates no demand for the kind of relative individualism needed to reflect cultural sustainability to an uninterested-unless-aroused population (all characteristics of postmodernism). Through this passivity of EIA, time, space, communication, leadership – all the key elements of good flowing design are lost.

This said, initially it might be argued that true postmodernism is simply beyond the remit of an EIA which exists for objective assessment rather than artistic purposes. The above description should be called environmental impact design. EID emphasises the artistic contribution to EIA; it requires a different set of approaches (and probably personnel) than pure EIA, as well as creativity and elements of cultural vision. To an extent some of the principles of EID are already being undertaken in EIA, in the mitigation sections of EISs, and especially within environmental divisions of the larger developers (e.g. the utilities) who often seem to see the formal EIA process as merely a lateral extension to their own design policies. Even so, rarely is it recognised as an artistic activity.

The key difference between EIA and EID lies in the concept of "unmodifiable design". Traditionally, EIAS are carried out on projects in which most of the structural elements have already been finalised. In more EID-oriented approaches, there is less unmodifiable design and thus more scope for introducing environmentally sound design as mitigation measures. An even more radical path would be a postmodern EIA which aims to begin with so few unmodifiable design ideas that the EIA essentially becomes the leading player in design.

The process would begin by reducing (deconstructing) this idea of unmodifiable design back to a single design principle, one which maintains the distinction between form and function (e.g. not "we must build an incinerator" but "we must build something else"). Thus the whole development would be focused around a separate concept from which it was intended. For nuclear power stations an initial design principle might be "we must build a structure which cannot be accessed for fifty years", for a hydroelectric dam it might be "we must flood an area the size of ten football pitches", and for an incinerator it might be "we must build a structure which reaches a height of 70 m".

This deconstructed design principle offers a reversal in the standard hierarchy of engineer–designer to designer–engineer, from scientist–artist to artist–scientist. For when faced with such an open-ended invitation, the prospect of building an incinerator would actually represent a golden invitation to all the sculptors, public artists, architects and designers who are normally so frustrated by the confines of planning regulations on height, size and shape. From this design principle any active imagination could automatically come up with a dozen uses and shapes for the structure, which would later of course have to be engineered around the project (e.g. the incinerator) in some way. This is where the EIA would be the controlling process; because for every design idea that came forth – from radio aerials to revolving restaurants to rollercoasters – the first role an EIA would have is to assess how potentially environmentally damaging the use would be in the context of the original function of the structure (e.g. an incinerator EIA would probably conclude that, due to the emissions, it might be too unhealthy to have human activity near the top of the structure). And in this way postmodern EIA, instead of acting as a conservation-based force on predetermined scientific engineering, acts as a conservation-based force on a flowing artistic design (adapted from Holstein 1996).

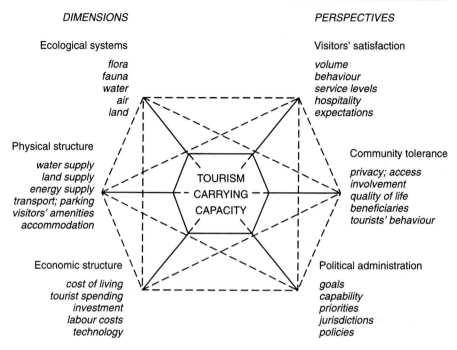

Figure 12.3   Carrying capacity – a tourism example. (*Source:* Glasson et al. 1995)

Complementary changes: enhancing skills and knowledge

The previous discussions indicate that EIA practitioners need to develop further their substantive knowledge of the wider environment. There is an important role for "state of the environment reports" and the development of "carrying capacity and sustainability indicators" – if not interpreted too narrowly. For example, carrying capacity is multi-dimensional and multi-perspective (see Fig. 12.3 for an example for tourism impact assessment). Carrying capacity is also an elastic concept, and the capacity can be increased through good management.

Practitioners also need to develop both "technical" and "participatory" approaches, such as the focus group, the Delphi approach and the mediation approaches noted earlier. EIA has been too long dominated by the "clinical expert" with the detached quantitative analysis. Notwithstanding, there is still a place for the sensible use of the rapidly developing technology – including expert systems, participatory techniques and text-oriented analysis (e.g. non-numerical unstructured data – NUD*IST – pulling out issues from focus group transcripts). There is also a need for more capacity building of EIA expertise, plus relevant research, including, for example, more comparative studies and longitudinal studies (following impacts over a longer life-cycle – moving towards adaptive EIA).

393

12.4 Extending EIA to project operations: environmental management systems and environmental audits

Less inspiring but considerably more implemented is the application of environmental management systems (EMS) and environmental auditing. EMS, like EIA, is a tool which helps organizations to take more responsibility for their actions, by determining their aims, putting them into practice and monitoring whether they are being achieved. However, in contrast with the orientation of EIA to future development actions, EMS involves the review, assessment and incremental improvement of an *existing* organization's environmental effects. EMS can thus be seen as a continuation of EIA principles into the operational stage of a project.

EMS has evolved from environmental audits, which were first carried out in the 1970s by private firms in the USA for financial and legal reasons, as an extension of financial audits. Auditing later spread to private firms in Europe as well and, in the late 1980s, to local authorities in response to public pressure to be "green". In the early 1990s environmental auditing was strengthened and expanded to encompass a total quality approach to organizations' operations through EMS. EMS is now seen as good practice and has mostly subsumed environmental auditing.

This section reviews existing standards on EMS, briefly discusses the application of EMS and environmental auditing by both private companies and local authorities and concludes by considering the links between EMS and EIA.

Standards and regulations on EMS

Three EMS standards apply in the UK: British Standard 7750 of early 1992 (updated and revised in early 1994), the EC Eco-Management and Audit Scheme (EMAS) of 1993, and the International Standards Organisation's ISO series 14000. The three are compatible with one another, but differ slightly in their requirements.

BS7750 is a direct evolution from the well-known British Standard 5750 on quality systems. Whereas BS5750 addresses an organization's products and services, BS7750 focuses more on by-products such as wastes and emissions. Both standards establish criteria for improving the organization's management system. Fulfilment of BS7750 entails the following steps:

- a commitment by the organization to undertake the audit;
- a preparatory review and assessment of relevant regulatory requirements, environmental effects, environmental management practices and procedures, and feedback from investigations of previous incidents and non-compliance;
- the formulation of an environmental policy;
- a full inventory and assessment of the organization's activities and environmental effects;

394

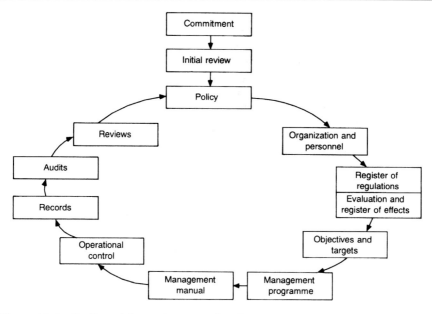

Figure 12.4 Outline of the environmental audit process.

- an assessment of whether the activities conform to relevant regulations and requirements;
- the formulation of targets and objectives;
- the development of an environmental management programme and supporting manual;
- the application of the management plan in the organization's operations and record-keeping;
- an audit to test whether the organization achieves its targets and objectives, which feeds back into environmental policy formulation to form a cycle.

Figure 12.4 summarizes this process. Although BS7750 requires that these steps must be fulfilled, it allows organizations to determine what environmental standards they plan to achieve and how they are to be achieved. The standard requires outside verification only to the extent that relevant environmental management records must be established and maintained, and can thus be checked. The records do not need to be made public. Achievement of BS7750 is voluntary, but it is likely to be increasingly seen as a sign that an organization is carrying out good environmental management. In particular, BS7750 organizations are likely to require their subsidiary or supplying firms to achieve the standards as well.

The *EC's Eco-management and Audit Scheme (EMAS)* was adopted by EC Regulation 1836/93 in July 1993, and became operational in April 1995. It is also voluntary and can apply on a site-by-site basis. It incorporates the elements of BS7750, but goes beyond them in terms of public disclosure and review of the auditing information. Thus, in addition to the stages outlined in Figure 12.4, EMAS requires:

- the periodic publication of a non-technical public statement which outlines how the organization has met its environmental objectives, including quantified data on emissions, waste generation, the raw material used, the energy and water consumed and other aspects related to the site's operation;
- a review and validation of the policy statement and EMS by an accredited EMAS verifier.

Once these steps are fulfilled, the organization can apply for EMAS registration from the relevant competent body (in the UK this is the Secretary of State for the Environment, Transport and the Regions).

Although EMAS was originally oriented towards larger private companies, it can also apply to local authorities and smaller companies. The DOE/Welsh Office's Circular 2/95 discusses EMAS for local authorities, and the Small Company Environmental and Energy Management Assistance Scheme was established in November 1995 to help companies with fewer than 250 employees to carry out EMAS.

The *International Standard Organisation's ISO 14000* series was first discussed in 1991, and a comprehensive set of EMS standards was published in September 1996. These include ISO14001 on EMS specifications, ISO14002 on EMS for small and medium-sized companies, ISO14004 on general EMS guidance, and ISO14010–14014, which give guidance on environmental auditing and review. The ISO requirements are similar to, and compatible with, BS7750 and EMAS. In addition, ISO14000 makes a useful distinction between what an organization controls and what it influences.

Implementation of EMS and environmental auditing

Although early environmental audits carried out by private sector firms varied widely depending on the audit's purpose – they included acquisition/divestiture audits' risk audits, compliance audits and audits of individual sites – the practice of EMS is becoming more consistent as a result of the new standards. Private companies see EMS as a way to reduce their costs through good management pactices such as waste reduction and energy efficiency. They also see EMS as good publicity and, less directly, as a way of boosting employees' morale. However, private companies still have problems implementing EMS owing to commercial confidentiality, legal liability, cost, and lack of commitment. Smaller companies are especially affected by the cost implications of establishing EMS systems, and have been slower than the larger companies in applying it to their operations.

EMS in local authorities initially had a more consistent format than in the private sector, primarily because they all aimed to do the same thing: provide information on baseline conditions in the relevant area and suggest ways in which they could change their operations to become "greener". Generally, these audits included, in varying levels of depth:

- a state of the environment report, which reviews baseline environmental conditions in an area, preferably in conjunction with a regularly updated environmental data base;
- a policy impact assessment, which evaluates the local authorities' policies and practices and suggests actions to improve matters where necessary;
- an EMS like that of private firms, to implement, monitor and review the audit findings.

Although a survey by the County Planning Officers' Society in late 1990 showed that 87 per cent of county councils had carried out or were planning to carry out some form of environmental audit, by 1992 only about a dozen audits had actually been completed (Grayson 1992). The use of EMS by local authorities has been limited by cutbacks in central government funding, government reorganization, growing public concerns about economic rather than environmental issues and the realization that environmental data become rapidly dated unless constantly (and expensively) updated. All these factors mean that local authorities are re-evaluating whether EMS is the best way of fulfilling their environmental responsibilities. In particular, those authorities that began EMS in the past have found it difficult to implement the EMS's recommendations because of lack of funding and motivation. Again, the establishment of EMS standards, interpreted as guidelines for local authorities by Jacobs and Levett (1993), may help to reverse this rather discouraging trend. Auditing advice to local authorities includes that by the Association of County Councils et al. (1990), FOE (1990) and the Local Government Management Board (1991).

Links between EMS and EIA

The growth in EMS is important to EIA for several reasons. First, it is clear that in the process of environmental auditing both private- and public-sector organizations will increasingly generate environmental information that will also be useful when carrying out EIAS. Local authorities' state of the environment reports provide data on environmental conditions in the area that can be used in EIA baseline studies. Generally, a state of the environment report will contain information on such topics as local air and water quality, noise, land-use, landscape, wildlife habitats and transport. Unfortunately, unless this information is regularly revised it quickly becomes outdated. It is also often collected only on a large-scale (e.g. county-wide) basis, and so may not be suitable for any specific site. However, state of the environment reports do generally identify sources for environmental data that can be contacted for the most up-to-date information. Similarly, the reports may be useful when determining suitable locations for new developments, by identifying sites that are particularly environmentally sensitive and should clearly be avoided, or those that are environmentally robust and more suitable for development. A local authority's policy impact assessment is useful in clarifying its views on environmental matters and highlighting future policy directions that may influence the planning decision and the future operation of the project. The policy impact assessments of local authorities or government departments are also likely to generate a need for EIAS of policies, plans and programmes; this is discussed in Chapter 13.

Private firms' environmental audit findings have mostly been kept confidential to date. An audit is likely to be useful for EIA only if a firm with one intends to open a similar facility. However, the new auditing standards are likely to have a major effect in making auditing information more accessible both to the public and to other organizations. This information could include the levels of wastes and emissions produced by different types of industrial process, the types of pollution abatement equipment and operating procedures used to minimize these by-products, and the effectiveness of the equipment and operating procedures. This type of information will be useful for determining the impact of similar future developments and mitigation measures. Some of these audits are also likely to provide models of "best practice", which other firms can aspire to in their existing and future facilities.

Most interestingly, however, project EIAS are increasingly used as a starting point for their projects' EMSS. For instance, emission limits stated in an EIA can be used as objectives in the company's EMS, once it is operational. The EMS can also test whether the mitigation measures discussed in the EIA have been installed and whether they work effectively in practice.

Overall, environmental auditing is likely to increase the level of environmental monitoring, environmental awareness and the availability of environmental data. All of this can only be of help in EIA.

12.5   Summary

As in a number of other countries discussed in Chapter 11, the practice of EIA for projects in the UK, set in the wider context of the EU, has progressed rapidly up the learning curve. Understandably however, practice has highlighted problems as well as successes. The resolution of problems and future prospects are determined by the interaction between the various parties involved. In the European Union the introduction of the amended EIA Directive in 1999 will help to improve some steps in the EIA process, including screening, scoping, the consideration of alternatives, and consultation. However, some key issues remain unresolved, including the lack of support for mandatory monitoring. This chapter has identified an agenda for other possible changes, including cumulative impacts, public participation, socio-economic impacts, integrated environmental assessment and environmental impact design. Some of these will be easier to achieve than others, and there will no doubt be other emerging issues and developments. EIA is a dynamic area, and systems and procedures will continue to evolve in response to the environmental agenda and to our managerial and methodological capabilities.

There is an urgent need to "close the loop", to learn from experience. While the practice of mandatory monitoring is still patchy, there is some notable progress in the development of environmental management and auditing systems. Assessment can be aided by the recent development of environmental auditing for existing organizations, be they private-sector firms or local authorities. The information from such auditing could provide a significant change in the quality and quantity of baseline data for EIA.

As EIA activity spreads, so more groups will become involved. Capacity building and training is vital both in the EIA process, which may have some commonality across countries, and in procedures that may be more closely tailored to particular national contexts. EIA practitioners also need to develop their substantive knowledge of the wider environment and to improve both their technical and participatory approaches in the EIA process.

References

Association of County Councils, District Councils and Metropolitan Authorities 1990. *Environmental practice in local government.* London: Association of County Councils.

Bailey, P. et al. 1996. *Methods of integrated environmental assessment: research directions for the European Union.* Stockholm: Stockholm Environmental Institute.

Bowles, R. T. 1981. *Social impact assessment in small communities.* Toronto: Butterworth.

British Standards Institution 1992. *British Standard 7750: specification for environmental management systems.* Linford Wood, Milton Keynes: BSI.

CEPA 1994. *Review of Commonwealth environmental impact assessment.* Canberra: CEPA.

CRPE 1991. *The environmental assessment directive – five years on.* London: Council for Protection of Rural England.

CRPE 1992. *Mock directive.* London: Council for Protection of Rural England.

Davis, S. 1996. *Public involvement in environmental decision making: some reflections on West European experience.* Washington, DC: World Bank.

FOE 1990. *Environmental audits of local authorities: terms of reference.* London: Friends of the Earth.

Glasson, J. 1995. Socio-economic impacts 1: Overview and economic impacts. In *Methods of EIA*, P. Morris and R. Therivel (eds). London: UCL Press.

Glasson, J., K. Godfrey, Goodey B. 1995. *Towards visitor impact management.* Aldershot: Avebury.

Grayson, L. 1992. Environmental auditing: a guide to best practice in the UK and Europe. London/Letchworth: British Library / Technical Communications.

Holstein, T. 1996. Reflective essay: postmodern EIA or how to learn to love incinerators (written as part of MSc course in Environmental Assessment and Management, Oxford Brookes University, Oxford).

IAIA (International Association for Impact Assessment) 1994. Guidelines and principles for social impact assessment. *Impact Assessment* **12**, Summer.

Jacobs, M. & R. Levett 1993. *A guide to eco-management and audit scheme for UK local government*. London: HMSO.

Lane, P. and Associates 1988. *A reference guide to cumulative effects assessment in Canada*, vol 1. Halifax: P. Lane and Associates / CEARC.

Local Government Management Board 1991. *Environmental audit for local authorities: a guide*. Luton: Local Government Management Board.

McDonald, G. T. & L. Brown 1995. Going beyond environmental impact assessment: environmental input to planning and design. *Environmental Impact Assessment Review* **15**, 483–95.

Odum, W. 1982. Environmental degradation and the tyranny of small decisions. *Bio Science* **32**, 728–9.

O'Riordan, T. 1990. EIA from the environmentalist's perspective. *VIA* **4**, March, 13.

Parkinson, P. 1996. Best practice for the environmental assessment of pipelines (paper presented at Conference on Advances in Environmental Impact Assessment), 9 July. London: IBC UK Conferences.

Peterson, E. et al. 1987. *Cumulative effects assessment in Canada: an agenda for action and research*. Quebec: Canadian Environmental Assessment Research Council (CEARC).

Project Appraisal 1996. Special edition: Environmental Assessment and Socio-economic Appraisal in Development. **11**(4).

Sadler, B. 1996 *International study in the effectiveness of environmental assessment*. Ottawa: Canadian Environmental Assessment Agency.

Simmons & Simmons 1993. *Environmental Newsletter* **16**.

Turner, T. 1995. *City as landscape: a post-postmodernist view of design and planning*. London: E&FN Spon.

Wood, C. & C. Jones 1992. The impact of environmental assessment on local planning authorities. *Journal of Environmental Planning and Management* **35**(2), 115–27.

World Bank 1995. *Environmental assessment: challenges and good practice*. Washington, DC: World Bank.

Note

1   This term was originally coined for a slightly different context by Turner (1995).

CHAPTER 13

Widening the scope: strategic environmental assessment

13.1   Introduction

One of the most recent trends in EIA is its application at earlier, more strategic stages of development – at the level of policies, plans and programmes. In the USA, since the enactment of the NEPA, this so-called strategic environmental assessment (SEA) has been carried out as an extension of project EIA in a relatively low-key manner. However, in the EC it has recently come to be viewed as a valuable technique for achieving sustainable development, and a directive on SEA is being discussed. The UK government is also taking steps to appraise the environmental impacts of its policies. SEA is likely to be an area of strong growth in the years ahead, and this in turn will influence – and improve – the process of project EIA.

This chapter discusses the need for SEA and some of its limitations. It reviews the "best practice" status of SEA in other countries and discusses the most recent legislative advances in SEA in the USA, New Zealand, the Netherlands, the EC and the UK. It then focuses on two case studies, one of a local authority development plan, the other of a flood defence strategy. The chapter concludes with proposals for links between SEA and sustainable development. By necessity this chapter must radically simplify many aspects of SEA. The reader is referred to Therivel et al. (1992), English Nature (1996), Sadler and Verheem (1996) and Therivel and Partidario (1996) for a discussion in greater depth.

13.2   Strategic environmental assessment (SEA)

Definitions

Strategic environmental assessment can be defined as "the formalized, systematic and comprehensive process of evaluating the environmental impacts of a policy,

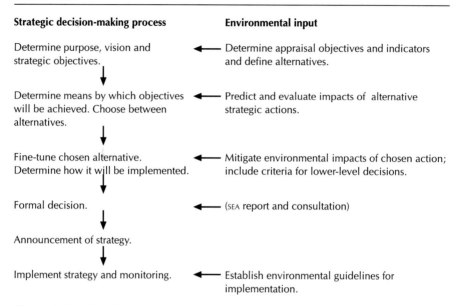

Strategic decision-making process	Environmental input
Determine purpose, vision and strategic objectives.	← Determine appraisal objectives and indicators and define alternatives.
Determine means by which objectives will be achieved. Choose between alternatives.	← Predict and evaluate impacts of alternative strategic actions.
Fine-tune chosen alternative. Determine how it will be implemented.	← Mitigate environmental impacts of chosen action; include criteria for lower-level decisions.
Formal decision.	← (SEA report and consultation)
Announcement of strategy.	
Implement strategy and monitoring.	← Establish environmental guidelines for implementation.

Figure 13.1 Links between SEA and the decision-making process.

plan or programme and its alternatives, including the preparation of a written report on the findings of that evaluation, and using the findings in publicly accountable decision-making" (Therivel et al. 1992). It is, in other words, the EIA of policies, plans and programmes (PPPS), but we must keep in mind that the process of evaluating environmental impacts at a strategic level is not necessarily the same as evaluating them at a project level. Figure 13.1 shows the links between PPP-making and SEA. Although policies, plans and programmes are generally all described as strategic in this and other texts, they are not the same things, and may themselves require different forms of environmental appraisal. A policy is generally defined as an inspiration and guidance for action, a plan as a set of co-ordinated and timed objectives for the implementation of the policy, and a programme as a set of projects in a particular area (Wood 1991). Here, policies, plans and programmes will be referred to as PPPS unless otherwise noted. PPPS may be sectoral (e.g. transport, mineral extraction), spatial (e.g. national, local), or indirect (e.g. education, research and development, privatization).

In theory PPPS are tiered; a policy provides a framework for the establishment of plans, plans provide frameworks for programmes, programmes lead to projects. For instance, the UK government's road policies, set out in its White Papers on roads, give rise to suggested road schemes, which are then incorporated in a national roads programme. This in turn forms a basis for the proposal of specific routes, for which project EIAS are prepared. In practice, as will be discussed later, these tiers are amorphous and fluid, without clear boundaries. The EIAS for these different PPP tiers can themselves be tiered, as we show in Figure 13.2, so that issues considered at higher tiers need not be reconsidered at the lower tiers.

402

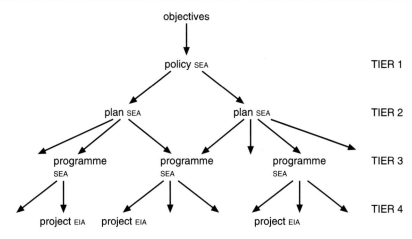

Figure 13.2 Tiers of SEA and EIA.

The need for SEA

Various arguments have been put forward for a more strategic level of EIA, most of which *relate to problems with the existing system of project EIA*. Project EIAS react to development proposals rather than anticipate them, so they cannot steer development towards environmentally robust areas or away from environmentally sensitive sites.

Project EIAS do not adequately consider the cumulative impacts[1] caused by several projects, or even by one project's subcomponents or ancillary developments. For instance, under present UK regulations different EIAS may be prepared for a power station, the gas pipeline providing the power station's fuel, the facilities for receiving and processing the gas, and the transmission lines carrying electricity away from the power station. Separate EIAS can also be prepared for different sections of one road. Small individual mineral extraction operations may not need an EIA, but the total impact of several of these projects may well be significant. At present in the UK there is no legal requirement to prepare comprehensive cumulative impact statements projects of these types.

Project EIAS cannot fully address alternative developments or mitigation measures, because in many cases these alternatives will be limited by choices made at an earlier, more strategic level. In many cases a project will already have been planned quite specifically, and irreversible decisions taken, by the time an EIA is prepared.

Project EIAS cannot consider the impacts of potentially damaging actions that are not regulated through the approval of specific projects. Examples of such actions include farm management practices, privatization and new technology. Project EIAS often have to be carried out very quickly because of financial constraints and the timing of planning applications. This limits the amount of baseline data that can be collected and the quality of analysis that can be undertaken. For instance, the planning periods of many projects have required their ecological impact assessments to be carried out in the winter months, when it is difficult to identify plants and when many animals either are dormant or have migrated. The amount and type of public consultation undertaken in project EIA may be similarly limited.

By being carried out earlier in the decision-making process and encompassing all the projects of a certain type or in a certain area, SEA can ensure that alternatives are adequately assessed, cumulative impacts are considered, the public is fully consulted, and decisions concerning individual projects are made in a proactive rather than reactive manner. As will be discuss later, SEA is also seen as a central step in the *achievement of sustainable development*.

Problems with SEA

On the other hand, the implementation of SEA is fraught with both technical and procedural problems. On the *technical* side, many potential future developments spread over a large area can lead to great analytical complexity. Information about existing and projected environmental conditions and about the nature, scale and location of future development proposals is usually very limited, so the impacts of these developments cannot be predicted precisely. The large number and variety of alternatives to be considered further complicates the process, as do requirements for public participation. Until recently, there was a general lack of information about SEA, and a dearth of case studies in which SEA had been successfully applied, particularly to policies, so there have been few models to suggest how to carry out SEA.

More intractable than these technical and information problems are those inherent in the *policy-making process*. Many PPPs are nebulous, and they evolve in an incremental and unclear fashion, so there is no clear point of time when their environmental impacts can best be assessed: "the dynamic nature of the policy process means issues are likely to be redefined throughout the process, and it may be that a series of actions, even if not formally sanctioned by a decision, constitute policy" (Therivel et al. 1992).

PPPs do not have clear boundaries at which they stop and other policies begin. For instance, it is impossible to distinguish fully between policies for transport, energy and land-use, as they all affect one other. Furthermore, the actual effects of policies are strongly influenced by how the policies are interpreted when they are implemented. The government's emphasis on deregulating many government activities also means that increasing numbers of interest groups, with increasingly diverse aims, are involved in formulating policy. As a result of these factors, policies may be more fragmented and less well understood and can often have unintended and unpredictable outcomes.

Finally, and most importantly, policy-making is a political process. Decision-makers will weigh up the implications of a PPP's environmental impacts in the wider context of their own interests and those of their "constituents".

13.3 Evolving systems of SEA

Despite these problems, SEA *is* carried out in certain countries, and looks likely to be increasingly used worldwide. Existing systems of SEA can be divided into those established through legislation, through administrative orders (or Cabinet directives), and through advisory guidelines. For instance, SEA legislation has been established in the USA, the Netherlands, New Zealand and Western Australia; administrative orders or Cabinet decisions have been promulgated in Canada, Denmark and Hong Kong; and guidelines have been published by the UK, the EC and the World Bank. A number of other countries have established partial SEA systems or are actively researching the feasibility of such a system, including Australia, Austria, Belgium, Germany and Finland. SEA systems are developing rapidly, and others are likely to be set up in the near future. Here the SEA legislations of the USA, New Zealand and the Netherlands are summarized, both because their legislative status is stronger than that in other countries, and because they demonstrate a range of possible approaches to SEA.

The USA

In the USA, hundreds of "programmatic environmental impact statements" (PEISs) have been prepared by government agencies under the National Environmental Policy Act of 1970, primarily as an extension of project EIA to the programme and plan level. Individual government agencies have established regulations and guidelines to implement these requirements, including the Department of Housing and Urban Development's ground-breaking *Areawide environmental assessment guidebook* (USHUD 1981). Agencies must prepare PEISs for the following actions, if these are likely to significantly affect the quality of the human environment: agency proposals for legislation; the adoption of rules, regulations, treaties, conventions or formal policy documents; the adoption of formal plans that guide the use of federal resources; the adoption of groups of connected actions that implement a policy (40 CFR 1508.18[b]). In 1994, 128 PEISs were prepared by federal agencies, including 25 for plans to reuse military bases, 19 for river basin plans, 17 for public land management plans and 28 for national park or national forest management plans (Bass & Herson 1996).

About one-third of the USA's 50 states have their own EIA regulations, but only a few of these also cover PPPs. Of these, the SEA system established by the California Environmental Quality Act of 1986 (State of California 1986) is the most well developed. Like project EIAS, such a "programme environmental impact report (EIR)" and "master EIR"[2] must include a description of the action, a section on the baseline environment, an evaluation of the action's impacts, a reference to alternatives, an indication of why some impacts were not evaluated, the organizations consulted, the responses of these organizations to the EIS, and the agency's response to the responses. Between September 1994 and September 1995, 96 EIRs were prepared in California, including 15 for city or county comprehensive plans, 14 for state-wide hunting and fishing plans for particular species, nine for regional transport plans, and six for community plans (Bass & Herson 1996).

New Zealand

In contrast to the USA, where SEA has clearly evolved from EIA provisions, in New Zealand it is seen as a tool for achieving sustainability as part of an integrated planning and assessment process. To help achieve its objective of promoting the sustainable management of natural and physical resources, the Resource Management Act of 1991 requires all PPPs at national, regional and district level to be evaluated to determine the likely costs and benefits of alternative means of achieving the PPPs and so as to be "satisfied that any such [PPP] (i) Is necessary in achieving the purpose of [the] Act; and (ii) Is the most appropriate means of exercising this function" (Article 32(c)). Because environmental issues and information are used as an integral part of the policy process, formal SEA has only rarely been used in New Zealand to date (Sadler & Verheem 1996).

The Netherlands

In the Netherlands, a two-pronged SEA system applies, one tier reminiscent of the USA's EIA-based system, the other more like New Zealand's integrated system. First, under the EIA Act of 1987, plans for waste management, and the supply of drinking-water, energy and electricity, and some land-use plans, require SEA. These SEAs must include full public participation, independent expert review at both the scoping and review stages, the consideration of alternatives, and monitoring. By 1995, SEAs had been prepared for two national and 15 regional waste-management plans, one national and seven regional sludge-management plans and seven other PPPs (Sadler & Verheem 1996). Second, since 1995 an environmental test (the "e-test") has been required for all Cabinet decisions with significant environmental impacts. As part of this test, an "environmental section or paragraph" must be prepared by the lead agency, which aims to fully integrate environmental and sustainability concerns into national policy-making.

In summary, many SEA systems are still in a state of evolution and refinement, and only a few are well established. It is against this background of the relatively limited application of SEA and development of SEA methodologies that the draft EC Directive on SEA and existing UK guidance on SEA must be seen.

13.4   SEA at the European level

Although the original version of EC Directive 85/337 was intended to apply also to PPPs, the final version applies only to projects. However, the EC's Fourth Action Programme on the Environment of 1987 (CEC 1987) stated that EIA "will also be extended, as rapidly as possible, to cover policies and policy statements, plans and their implementation, procedures, programmes . . . as well as individual projects", and the Fifth Action Programme of 1992 (CEC 1992) reiterated this aim within its broad framework for achieving sustainable development.

In response to these requirements, the EC's Directorate General XI (DGXI on Environment) has been working on several fronts to establish SEA requirements both for the EC's own activities and for those of Member States. These include:

- an internal procedure whereby (a) all DGs must examine their PPPs' environmental repercussions at the time when the PPPs are first conceived, and (b) any proposed PPP which is likely to have a significant environmental impact is marked with a "green star", and the impact discussed and justified in the accompanying documentation;
- a requirement for Member States to appraise the environmental impacts of plans and projects (only) which could have a significant impact on Special Areas of Conservation or Special Protection Areas under EC Directive 92/43 on habitats;
- a requirement that Structural Fund applications are accompanied by an "environmental profile";
- a draft SEA Directive.

Both DGXI and DGVII (on transport) have also commissioned considerable further research on SEA, for instance on existing SEA methodologies, case studies of SEA, the costs and benefits of SEA, and SEA in the transport sector (English Nature 1996). SEA of Structural Fund applications and the proposed SEA Directive are discussed here in greater detail; not much information exists about the "green star" system, and the requirements of the Habitats Directive have rarely been used to date. Further information about SEA in the EC can be found in English Nature (1996) and Lee & Hughes (1995).

SEA of Structural Fund applications

The main way in which the EC provides financial support for regional development is through the allocation of Structural Funds. An application for funding takes the form of a regional development plan drawn up by a government office in partnership with businesses, local authorities and others. The plans establish strategic objectives and the framework for the region's future economic development, and indicate how projects will be selected in the future. In the UK, Structural Funds apply to Objective 1 areas (where development lags behind the rest of the European Union), Objective 2 areas (industrial areas in decline) and Objective 5b areas (disadvantaged rural areas).

In 1993, the EC revised its "Framework" Regulation 2081/93, which governs the operation of the Structural Funds, so that applications for the 1994–9 tranche of Structural Funds had to be accompanied by an "environmental profile" comprised of:

> an appraisal of the environmental situation of the region concerned and an evaluation of the environmental impact of the strategy and operations . . . in terms of sustainable development in agreement with the provisions of Community law in force; the arrangements made to associate the competent environmental authorities designated by Member States in the preparation and implementation of the operations foreseen in the plan, and to ensure compliance with Community environmental rules. (Articles 8(4), 9(8), 11(a)(5))

Several DGs jointly prepared a non-mandatory aide memoire in 1993, to provide guidance on what should be included in such an environmental profile. The *aide-mémoire* suggests that the profile should include:

1   a description of key environmental issues, including:

 (a)   designated areas and other zones of special environmental interest;
 (b)   acute pollution problems, for instance where EC standards for air or water are being breached, or where a public health hazard or irreversible environmental damage is occurring;
 (c)   areas where "serious stress on the ecosystem" occurs, for instance areas of poor soil quality or deforestation;

2   a description of the environmental impact of the regional development plan, including any changes predicted to the issues highlighted in 1(b) and 1(c), proposed mechanisms for incorporating environmental protection in lower-tier plans and programmes, and details of environmental monitoring systems, including the use of indicators;
3   a discussion of the legal and administrative framework, including the role of the environmental authority in the development plan, the mechanisms for designating and protecting special zones, and the co-ordination of the development plan and environmental policies through land-use planning and control.

The UK's latest round of 25 applications, which were prepared in 1994–5, included draft "environmental profiles". These were then refined in discussions with DGXI. The resulting Community Support Framework or Single Programming Document contains a revised environmental profile, as well as a list of measures for preventing harm to, or enhancing, the environment (Bradley 1996).

Proposed EC directive on SEA

After several further versions of the proposal, a draft SEA Directive was agreed in December 1996 (CEC 1997). This would apply to

> land use plans and programmes which are subject to preparation and adoption by a competent authority or which are prepared by a competent authority by an act of legislation, and which are part of the land use decision-making process for the purpose of setting the framework for subsequent development consent decisions, and which contain provisions on the nature, site, location or operating conditions of projects. This definition includes plans and programmes in sectors such as transport (including transport corridors, port facilities and airports), energy, waste management, water resource management, industry (including extraction of mineral resources), telecommunication and tourism.

In the UK context, this means that development plans – structure plans, local plans and unitary development plans – and regional planning guidance would certainly be subject to the Directive's provisions. What is less clear, and still the subject of discussions, is to what extent sectoral plans and programmes would also require SEA. Policies are exempt from SEA under the draft Directive.

The draft Directive would require the lead agency responsible for the plan or programme (PP) to assess its impacts on human beings, fauna, flora, soil, water, air, climate, landscape, material assets and the cultural heritage. The SEA would need to include a discussion of:

- the contents and objectives of the PP;
- the environmental characteristics of any area likely to be significantly affected by the PP;
- the existing environmental problems, especially those related to areas of particular environmental importance;
- the relevant environmental protection objectives at international, EC or Member State level, and how these were considered during the PP's preparation;
- the PP's likely significant environmental effects;
- any alternative ways of achieving the PP's objectives considered during its preparation and reasons why these were not adopted;
- any mitigation measures for the PP;
- any difficulties encountered in compiling this information.

The environmental authorities and public would comment on the SEA findings, and these comments, the SEA itself and the comments of any Member States affected by trans-boundary effects would be taken into account before the PP were adopted.

The draft SEA Directive is clearly based heavily on the EIA Directive's consent-based approach, and is considerably weaker than previous proposals in terms of the range of PPPs it applies to. This has been in response to a very chilly reaction to earlier proposals by other DGs and the Member States. In the UK, for instance, central government is concerned that for most PPPs there is no clear moment when a decision is made, and that any consent-based SEA system could only apply to a limited number of PPPs that have a clear decision-making stage. Reportedly the response in other countries has been no more encouraging. It is unlikely that an SEA Directive will be adopted soon, although interest in the subject remains high.

13.5   SEA in the UK

Despite its lack of enthusiasm for an externally imposed SEA requirement, the UK government has promoted the environmental appraisal[3] of its own policies (fairly unsuccessfully), of local authorities' development plans (very successfully) and of QUANGO and other organizations' PPPs (increasingly successfully).

SEA of central government PPPS

The UK government's White Paper on the Environment of September 1990 (DOE 1990) promised that it would carry out "a review of the way in which the costs and benefits of environmental issues are assessed within the Government". A year later the DOE published a guidebook entitled *Policy appraisal and the environment* and distributed copies to central government mid-level managers (DOE 1991). The guide-book's procedures are not mandatory, but they aim to help civil servants consider the environmental repercussions of their decisions and to promote a "cultural change" in how civil servants formulate policies. *Policy appraisal and the environment* suggests that the department or agency from which a PPP originates should carry out the policy appraisal. Policy appraisal should apply to a wide range of PPPs, since generally any PPP "which concerns changes in the use of land or resources, or which involves the production or use of materials or energy, will have some environmental impact". According to the guidebook, policy appraisal should involve the following steps:

- summarizing the policy issue;
- listing the objectives;
- identifying the constraints;
- specifying the options;
- identifying the costs and benefits;
- weighing up the costs and benefits;
- testing the sensitivity of the options;
- suggesting the preferred option;
- setting up any monitoring necessary;
- evaluating the policy at a later stage.

At the central government level, these guidelines have been of limited effectiveness in promoting SEA. A DOE (1994) publication entitled *Environmental appraisal in government departments*, which summarizes central government studies carried out in response to *Policy appraisal and the environment*, does discuss a range of studies, but these were mainly cost–benefit analyses, not SEAs, and the booklet's publication was widely seen as a *pro forma* exercise. A survey conducted by the CPRE (1996) noted that in 1995 no government department had conducted an environmental appraisal of its policies.

SEA of local authorities' development plans

In contrast, a real "SEA-change" in environmental appraisal in the UK was begun by the (soon to be revised) publication of Policy Planning Guidance Note no. 12, *Development plans and regional planning guidance* (DOE 1992). PPG12's ostensible purpose is to explain the provisions of the Planning and Compensation Act 1991 as they relate to development plans. However, it also notes that the preparation of development plans "can contribute to the objective of ensuring that development and growth are sustainable" and that development plans should take environmental considerations into account through systematic appraisal. PPG12 refers local authorities to *Policy appraisal and the environment* for guidance on this appraisal process.[4]

In response to this guidance, some local authorities began to carry out environmental appraisals, albeit using much simpler techniques than those advocated by the government. Lancashire and Kent County Councils were two of the early pioneers. Lancashire's first appraisal was based on a matrix, which listed the 164 policy statements of its structure plan on the vertical axis and 11 environmental components on the horizontal axis. The impact of each statement on each component was recorded in the relevant cell, using numerical scores from +2 (sizeable benefit) to −2 (sizeable cost). The penultimate column of the matrix summed up each policy's score to form a "sustainability score", which gave an indication of the policy's impact on environmental resources. The final column indicated whether a revision of the policy was felt necessary because of high negative scores or poor wording of the policy. The structure plan's various policy areas could then be ranked by how well they performed on the sustainability score (Lancashire County Council 1992). The approach has the advantage of simplicity, but also has strong limitations, as Pinfield (1992) notes. The use of a single matrix is rudimentary and subjective. The methodology assumes that all environmental components and policies should be given the same weighting, and that scores from one cell can be added to scores from another.

The Kent appraisal, which followed quickly after Lancashire's early study, introduced some innovative thinking on the environmental appraisal of development plans (Kent County Council 1993). The Kent approach was set in the context of advice from the Local Government Management Board (1992): "A concept of sustainable development transforms a local authority's approach from a series of ad hoc steps to a strategy and from the need for controls alone to a need for policies. The local authority has to plan, to co-ordinate and to manage for sustainable development."

411

The first step in the Kent approach, as for Lancashire, was to evaluate the structure plan's policies in a matrix, in terms of their impact on the environment and their contribution to sustainability. The environmental components were grouped into three scales:

- *local* (including any impact on the quality of people's lives, townscape, noise etc.);
- *county-wide* (including any impact on air and water quality, ecology etc.);
- *global* (including any impact on renewable and non-renewable resources etc.).

The county's 130 structure plan policies were then scored against the various criteria, using a five-point scale of ticks and crosses (see Figure 13.3). The appraisal clearly indicated the environmental impacts of various policies; for example, economic policies were generally positive on local criteria but less so on county and global criteria; some transport policies were seen to generate considerable deterioration in environmental quality at all levels. A summary figure showed the general emphasis of policies. This approach also had its limitations; for example, environmental issues should also be considered in relation to social and economic needs, public input is needed, and policy interaction should be built in. As with the Lancashire study, it was only the first pioneering stage in the environmental appraisal of development plans, but it was an influential stage.

These early appraisal exercises provided the basis for the DOE (1993) publication *Environmental appraisal of development plans: a good practice guide*. The guide proposes a three-step SEA process:

- Characterize the environment: identify and assess the environmental stock that could be affected by the development plan. The guide suggests 15 environmental components, divided into those that relate to global sustainability, to natural resources, and to local environmental quality.
- Scope the plan: ensure that it covers an appropriate range of environmental concerns by comparing its policies to the requirements of existing government advice and other relevant guidelines;
- Appraise the plan's content: determine whether its objectives and policies are internally consistent, and assess their likely environmental effects, possibly using matrices.

Since then, many UK local authorities have carried out, or begun to carry out, SEAS of their development plans. By early 1997, most local authorities in England and Wales, and a few Scottish authorities, had begun to carry out environmental appraisals, and more than 180 had completed at least one complete round of appraisal (Therivel 1995, 1997). The DOE's good practice guide has been very well accepted, and most local authorities now followed its three steps. In many cases, local authorities have simply gone through the steps recommended by the DOE on the final plan, without using the findings to improve their strategies or policies further. However, other authorities have used their SEAS to revise their plans thoroughly. A more comprehensive, sustainability-related model of SEA emerges from these "good

Example of Kent Structure Plan Strategic Environmental Assessment Matrix

Environmental Criteria / Policy	Local Environment					Sound County Environmental Policies					Global Sustainable Policies				
	Human Beings	Townscape	Cultural Heritage	Noise	Open Space/Access	Ecology	Air Quality	Water	Land/Ground pollution	Landscape	Renewable natural resources	Non renewable natural resources	Non renewable energy resources	Energy	Sustainability/Global change
Policy T1				✓	✓	✓			✓					✓	✓
T2	✓			XX	XX	XX	XX		XX	X	X			XX	X
T3	✓	✓			✓		✓		✓					✓	✓
etc															

Key:

Local Environment
- Human Beings
- Townscape
- Cultural Heritage
- Noise
- Open Space/Access to Countryside

Sound County/Environmental Policies
- Ecology
- Air Quality
- Water
- Land/Ground pollution
- Landscape

Global Sustainable Policies
- Renewable natural resources
- Non renewable natural resources
- Non renewable energy resources
- Energy
- Sustainability/Global change

Policy Evaluation
- ✓✓ Concerted environmental action
- ✓ Some environmental action
- Neutral – no effect
- X Some deterioration in environmental quality
- XX Major environmental deterioration

Figure 13.3 Example of Kent Structure Plan strategic environmental assessment matrix. (Adapted from Kent County Council 1993)

practice" environmental appraisals, which, in its most comprehensive form, could involve:

- setting sustainability objectives (possibly linked to Agenda 21);
- setting plan objectives;
- setting environmental targets and/or carrying capacities;

413

- comparing alternative locational strategies;
- describing the baseline environment;
- identifying environmental criteria;
- "scoping";
- preparing a compatibility matrix;
- preparing matrices of policies v. environmental criteria;
- preparing matrices of policy areas v. environmental criteria;
- preparing matrices of proposals v. environmental criteria;
- preparing a written description of policy impacts;
- preparing a written description of proposal impacts;
- monitoring. (Therivel 1996)

As SEA becomes more widespread, it is likely that these techniques will be used more commonly, and that SEA will be used more effectively to improve the plan.

SEA of QUANGO and other PPPs

Sectoral SEAS have also increasingly been carried out by certain UK QUANGOs and other bodies. These studies – many of which might be considered to be only partial SEAS (owing, for instance, to their lack of public consultation or their limited consideration of mitigation measures) – include those for the National Forest in the Midlands, development in the Yorkshire Dales National Park, and transport schemes for Hull, the greater Edinburgh area, the Pennines and the Channel Tunnel Rail Link (English Nature 1996). The Environment Agency in particular has been active in preparing SEAS for its flood alleviation and water management schemes in recent years, based on techniques used by its predecessor, the National Rivers Authority (Fry 1996). However, only a few of such SEAS are carried out annually.

In sum, although UK guidance on SEA is relatively well developed, there is considerable disparity between the national government's emphasis on cost–benefit appraisal and the much more qualitative and subjective methods used by local authorities and other bodies. The following two sections consider two case studies: a local authority's "sustainability appraisal" of its development plan and an Environment Agency SEA for flood defence works.

13.6 Case study: the sustainability appraisal of Hertfordshire County Council's Structure Plan

Hertfordshire County Council's sustainability appraisal of its structure plan is widely seen as a particularly good UK example of SEA. The SEA has been an integral part of the plan-making process rather than an add-on; it is innovatively based on a "vision" for Hertfordshire and considers sustainability issues, not just environmental concerns.

Hertfordshire is a mixed urban and rural area north of London, and covers about 164,000 hectares. In 1993, Hertfordshire County Council's planning department started to prepare a new county structure plan[5] from scratch, using sustainability as the starting point. This process – and particularly the emphasis on sustainability – was based on the results of several previous studies, including a 1992 review of the state of the county's environment and a 1993 environmental strategy for the county.

As a first stage in the SEA process, Hertfordshire's planners wrote a discussion document entitled *Future directions for Hertfordshire*, which critically evaluated approaches to sustainability and some of the principles underlying Local Agenda 21. This document was used as a basis for a series of "sounding" exercises with local groups, which aimed to get the community's views on the new approach proposed for the future planning of the county, and on how sustainable development could be achieved in it. Based on this consultation, five strategy objectives were developed for the plan:

- to enable activities and development to be carried out in the most sustainable way;
- to improve people's quality of life;
- to encourage people to make sustainable choices;
- to allow the same degree of choice for the future;
- to contain consumption of, and damage to, natural resources.

Based on these objectives, a "vision" for Hertfordshire was prepared, which described aspirations for the county. Sixteen sustainability aims (A–O in Figure 13.4) – which included socio-economic, cultural and spiritual as well as environmental aims – were also developed. Thirty indicators were then developed by which these aims could be measured. These correspond quite closely to those in the DOE (1993) guidance on how to characterize the environment (see Section 13.5). For instance, the aim of reducing the total demand for resources (aim A in the figure) can be expressed and measured in terms of the indicators of reducing trip length, reducing the number of motorized trips, reducing the consumption of fossil fuels and minerals and improving the standards of maintenance and design in the continuous renewal of buildings.

The remit of structure plans is limited to strategic land-use matters, but Hertfordshire's sustainability aims addressed a much wider area of influence and dealt with sustainability in a much more comprehensive manner. Thus, for the purposes of the structure plan, the sustainability aims were also expressed in the form of structure plan objectives which translate sustainability into terms related to land-use. For instance, sustainability aim A of reducing the demand for resources could be partly translated into policies for concentrating development in towns, increasing accessibility and reducing the gap between the use and production of energy in the county. Each of the structure plan's objectives that evolved from this exercise helps to achieve one or more of the sustainability aims, and all of the sustainability aims are addressed by one or more of the plan's objectives. The sustainability aims thus

Policy 13: CONSTRUCTION OF NEW ROADS

Where new roads are constructed to relieve adverse environmental conditions resulting from the effects of traffic on existing roads, these will principally be aimed at relieving these conditions and will not necessarily be designed to accommodate all projected growth on the new road. Where new road building is undertaken every effort will be made to ensure that the physical impact of the road and its effect on the landscape and environment is minimized. Safety considerations for those using the road by whatever means will be given high priority in the design.

Achievement of Sustainability Aims	Commentary	Compatibility with Other Policies
Key o no significant effect ✓ beneficial effect ✓? likely beneficial effect ✗ adverse effect ? uncertainty of prediction or knowledge ? A) Reduce overall demand for resources ? B) Make the most efficient use of non-renewable resources o C) Increase the use of renewable resources o D) Increase the reuse and recycling of resources o E) Maintain biological diversity ? F) Apply aims A–D in relation to energy efficiency ? G) Mitigate the possible effects of greenhouse gases o H) Increase the rate of carbon fixing o I) Reduce the effects of pollution o J) Maintain the capacity of land to renew itself ✗ K) Maintain critical national and local assets ✗ L) Maintain stocks of less critical assets ✓ M) Improve the overall quality of life o N) Ensure that needs for shelter and economic support are met o O) Increase community awareness and involvement o P) Improve equality of opportunity	See commentary on Policies 38 and 39. It is important to clarify that the policy relates to new road construction which is intended primarily to relieve adverse environmental conditions. This is the main objective of County road schemes but not Department of Transport schemes for trunk roads, which are intended primarily to increase capacity. This is why the adverse effects which are noted concern County but probably not DOT schemes. **Appraisal summary** Incompatibility with some other policies and adverse effects on some sustainability aims may arise. New road proposals may achieve certain environmental gains (e.g. in regard to improving traffic conditions in towns), but at an unavoidable cost in terms of damage elsewhere, mainly to rural and urban fringe areas.	Key ✓ compatibility ✗ incompatibility ? uncertainty ✓ 1. Whole settlement strategies ✓ 2. Town centres ✓ 3. Green belt ✓ 4. Development: main settlements ✓ 5&6. Development strategy ✓ 7. Distribution of new dwellings ✓ 8. Affordable housing ✓ 9. Design and form of new devel. ✗ 10. Reduction in growth of car usage ✓ 11. Primary routes: traffic and improvements ? 12. Improvements to other roads ? 14. Pedestrian, bus and cycle networks ? 15. Environmental areas (for transport planning) ? 16. Public transport: infrastructure development ✓ 17. Assessment of development: transport impacts ✓ 18. Car parking provision ✓ 19. Rail and water freight depots ? 20. Protection of critical env'al assets ✓ 21. Chilterns AONB ✗ 22. Cumulative impacts on environmental stock ✓ 23. Landscape regions ✓ 24. Mineral deposits: non-sterilisation of reserves ✓ 25. Restoration of damaged land ✓ 26. Open spaces in towns: protection and provision ✓ 27. Tree cover ✓ 28. Water: catchment management plans ✓ 29. Renewable energy: provision of facilities ✓ 30. Aggregates: secondary and primary sources ✓ 31. Waste ✓ 32. Employment: devel. proposals ✓ 33. Employment: key sites ✓ 34. Retail development ✓ 35. Sports and recreation development ✓ 36. Rights of way network ✓ 37. Regional parks ? 38. Trunk roads programme: major investment ✗ 39. Other roads: County transport schemes

Figure 13.4 An example of policy appraisal matrix. (*Source:* Hertfordshire County Council 1994b)

Table 13.1 Structure of the Hertfordshire consultation draft structure plan SEA.

---

1 Context and purpose of appraisal (4 pages)
2 Characterizing the environment (4)
3 Scoping the plan (3)
4 Appraisal of plan content (3)
5 Summary of main conclusions (6)
6 Next step in environmental appraisal of the plan (2)

References
A Policy analysis forms covering draft policies 1 to 39 (80)

(*Source:* Hertfordshire County Council 1994b)

provided the framework for devising and refining the plan policies and programmes. However, in that the vision and sustainability aims provide a picture of how Hertfordshire's future could be, they also apply to other areas of county council activity and the activities of other agencies.

Based on the plan objectives, 39 specific draft policies were developed; these are summarized in the right-hand column of Figure 13.4. The vision, sustainability aims, structure plan objectives and draft plan policies were published in an informal consultation document in May 1994 (Hertfordshire County Council 1994a). They were then tested through another soundings process carried out by CAG Consultants with Land Use Consultants, who also commented on Hertfordshire's work to date. A leaflet on the subject was made widely available, and a six-month road show was taken around shopping centres. Slightly more than 1 per cent of Hertfordshire's population was contacted, as well as environmental groups. This consultation resulted in some changes in emphasis, for instance giving more importance to jobs.

After these stages, which were primarily carried out by the planners who had prepared the consultation documents and written the draft policies, an attempt was made to get an independent view at a "formal" SEA stage, which was carried out in-house by a planner who was not originally associated with the plan. The structure of the resulting SEA report is shown in Table 13.1.

The assessor first appraised three approaches to the plan's development strategy: new settlements, the peripheral expansion of towns, and urban regeneration. Partly as a result of the SEA, the option of urban regeneration was promoted. For each of the plan's 39 policies, a policy appraisal was then carried out which considered (a) whether the policy would have any adverse effects on the achievement of the plan's sustainability aims and (b) whether any adverse or problematic impacts resulted from any incompatibility between it and other policies. Figure 13.4 shows this appraisal for one of the policies: the left-hand column covers (a), the right-hand column (b), and the central column provides explanatory text.

Overall, Hertfordshire's draft policies were found to be broadly beneficial to the plan's aims, and their environmental impacts compatible with each other. However, the appraisal did identify some policies with adverse impacts on sustainability, and/

417

or showing incompatibility with other policies. For instance, there was some concern that the plan merely fulfilled housing demand and did not address the *need* for housing. Similarly, road development causes adverse environmental impacts, but the plan had to include policies for such development.

The SEA results were published in December 1994. The next SEA stage was to involve a SEA of the deposit draft plan, since this would incorporate the changes to the plan reflecting the consultation draft SEA and public consultation. These results, which were due out in late 1996 or early 1997, would then feed into the next stage of plan-making.

Hertfordshire County Council has taken an innovative approach by applying sustainability principles to the land-use planning process. The SEA process gave the planning team new ideas, raised questions, suggested answers and involved the public. It identified many issues that are not usually included in a standard structure plan: rather than deal with "traditional" specific policies (e.g. on employment or housing), the new policies focus on sustainable development, whole settlement strategies and issues about the quality of life and assets. These principles, if agreed and incorporated in the final adopted plan, will be implemented through local plans and the development control process: only then can the plan's effectiveness be gauged (Rumble & Therivel 1996).

13.7   Case study: SEA for the River Nene flood defence programme

Several of the eight Environment Agency (previously National Rivers Authority) regions have been undertaking SEAs for their flood defence programmes since the early 1990s. One of the most recent of these SEAs is that for the River Nene from Peterborough to the Wash, which was undertaken by the Anglian region in 1995. This SEA is notable for its objectives-led approach, which is similar to Hertford-shire's "vision-led" approach. It differs from the Hertfordshire example in that most of the work was carried out by the consultants Posford Duvivier Environment, rather than internally. It is also essentially a completed study rather than part of an ongoing cycle of consultation and plan development, and the SEA report is presented as a separate document from the main report (the "strategic study"), which covers engineering and economic aspects of the flood defence strategy, rather than being integrated within it.

The River Nene strategic study covers an 8 km stretch of the River Nene from Peterborough to the Dog-in-a-Doublet sluice, where the river is fluvial, and a 39 km stretch from the sluice to the river's outfall at the Wash, where the river is tidal. The surrounding land is flat and predominantly agricultural. The study relates closely to other studies for the area, including a 1990 study about erosion along the tidal River Nene, management plans for the Wash estuary and shoreline, a strategic study for

Table 13.2 Structure of River Nene flood defence programme SEA.

1	Executive summary (3 pages)
2	Introduction (5)
3	Environmental baseline (14)
4	Consultation (4)
5A	Strategic environmental objectives (3)
5B	Strategic flood defence options (3)
5C	Strategic development of options: riverbank protection (20)
5D	Strategic development of options: construction impacts (1)
6	Monitoring recommendations (1)
7	Management framework for the strategy (2)

the Wash outfalls, the Lower River Nene catchment management plan and a study of options for water resources along the Lower River Nene (Fry 1996). The integrated objectives of the River Nene strategic study are:

- to examine the flood defences along the River Nene between the Wash and Peterborough and to prepare a report on the strategy to be adopted for the future maintenance of these defences;
- to include a preliminary programme of works for the next five-year period;
- to economically justify any works proposed by cost–benefit analysis;
- to prevent any loss of life and damage to property from flooding by protecting the integrity of the Lower Nene river system, where appropriate, by protecting the river banks from erosion;
- to conserve and enhance the environment;
- to maintain the existing channel cross section;
- to preserve the integrity of the Nene Washes in their present equilibrium condition (NRA 1995).

The study's strategic environmental objectives, which form the basis of the SEA, are to ensure that the strategic option is environmentally as well as technically and economically acceptable, and establish an environmental baseline to ensure that environmental constraints and opportunities are taken into account in the detailed appraisals arising from the study.

The SEA report structure is shown in Table 13.2. The report describes the area's environmental baseline in terms of nature conservation sites, ecological characteristics, landscape, archaeology and heritage, land-use, water quality, fisheries, navigation, recreation and amenity, and the local community. As part of the process of gathering baseline data and analysing management options, the consultants contacted 21 outside organizations as well as a number of NRA officers to reveal their concerns and objectives about the future use of the lower River Nene: copies of the resulting 18 letters are included in an annex of the SEA report, and the results of the consultation were incorporated into the SEA. For instance, several organizations highlighted the need to take a "whole river" approach to the Lower Nene's future management; although this was not in the strategic study's remit, consultation and reference to the other studies for the Nene tried to ensure that such an approach was achieved.

The strategic study considered five management approaches for flood defence, based on a 50-year planning horizon: do nothing (no works undertaken, no repairs), a minimum investment (e.g. repairing a breach only once it had occurred), maintain the existing defences with minimum capital investment, sustain the existing standard of defence (e.g. protection work to prevent further erosion, restoring the integrity of the flood banks) and improve the existing defences by raising the flood banks. The study concluded, on safety and economic grounds, that the preferred option was to sustain the existing standards of defence.

Based on this preferred option, the SEA considered the environmental advantages, disadvantages and applications of various engineering options for protecting the tidal and fluvial stretches of the river and possible improvements to these options:

Tidal River Nene:

(a) a limestone "toe" in the riverbed, with a layer of limestone placed on the slope by a machine ("revetment");
(b) a limestone "toe" with concrete blocks on the slope;
(c) sheet piling in the riverbank, with the eroded bank slope backfilled.

Fluvial River Nene:

(a) as option (a) above;
(b) as option (c) above, but on a smaller scale;
(c) willow branches woven around a willow stake to create a wall, with the eroded bank slope backfilled ("spiling");
(d) planting along the riverbank;
(e) sheet piling installed so that it is almost always submerged, with a roll of coir fibres placed behind the piling and the remaining area planted with reeds;
(f) as (e), but with a low limestone "toe" instead of piling;
(g) timber rather than steel piling.

The SEA summarized its analysis of these engineering options in a table (see Table 13.3), which "can be used to identify the most environmentally appropriate option(s) for a specific location" (NRA 1995). Further measures to enhance the environment were also listed, including re-seeding and planting on embankments, enhancing fisheries and fencing off grazing zones, as were measures to minimize the environmental impacts of any construction works.

The SEA establishes a framework within which subsequent flood defence projects can be planned. These projects are promoted through five-year strategies. Over the next five years, for each problem area along the river, the most environmentally acceptable engineering option will be chosen, based on the table, as long as it fits technical and cost criteria. A monitoring programme will also be put in place to identify existing environmental interest, assess whether the environmental impact of new projects is as anticipated, and provide information to improve subsequent projects.

Table 13.3 Part of a table showing the suitability of fluvial riverbank protection options to a range of environmental criteria.

Environmental criteria	Fluvial riverbank protection option				
	stone toe & revetment	stone toe & concrete blocks	sheet piling	spiling	planting along the bank
Fringing vegetation					
Suitability of option to:					
● promote marginal growth below water line	**	**	*	*	***
● promote marginal growth above water line	***	***	*	***	****
● retain continuity of river habitats	**	**	*	**	***
● provide microhabitat for fish and invertebrates	***	***	*	***	***
Berm					
Suitability of option to:					
● provide opportunity to extend berm	***	*	***	**	*
Landscape					
Suitability of option to:					
● create varied profile	**	**	*	****	****
Grazing					
Suitability of option to:					
● withstand grazing and trampling damage	***	***	***	**	*
● withstand grazing and trampling damage given provision of water access and for fencing	N/A	N/A	N/A	***	***
Navigation					
Suitability of option to:					
● withstand damage by boatwash	****	****	****	**	**
● provide moorings	*	*	****	*	*
Angling					
Suitability of option to:					
● withstand damage	***	***	***	***	*
● withstand damage given boards, fishing stands etc.	N/A	N/A	N/A	N/A	***

Key: **** good; *** suitable; ** marginal; * poor
(*Source:* Based on NRA 1995)

The Environment Agency sees SEA as a technique to enable it to fulfil its duties and use public resources more effectively. SEA is seen to promote consistency by taking environmental considerations into account systematically, and to improve the efficiency of project EIA by tackling appropriate issues at a strategic level rather than having to repeat studies for individual projects (Fry 1996).

13.8 SEA and sustainable development

In addition to resolving many of the difficulties associated with project EIA, SEA can be a central step in the achievement of sustainable development. To date, SEA practice has mostly expanded EIA techniques and principles to more strategic actions: the greater incorporation of environmental concerns in PPP decision-making, the mitigation of PPPs' adverse environmental impacts, greater transparency and public involvement, all used for PPPs with a clear authorization stage, in an attempt to lead to more sustainable practices. The USA system and the proposed EC directive are examples of such a "consent-related" or "incremental" SEA approach (see Fig. 13.5).

However, SEA can also be a central step in the achievement of sustainable development. Therivel et al. (1992) suggest that a "trickle-down" approach to sustainable development could involve:

- a commitment to the objective of sustainable development;
- a determination of the parameters within which sustainable development is to be achieved (e.g. area, resource, time);
- a determination of carrying capacity within these parameters;
- SEA of all relevant PPPs using alternative development scenarios which do not exceed the carrying capacity;
- choice of one scenario that optimizes socio-economic factors;
- EIA of individual projects within the constraints set by the SEA;
- a monitoring programme that would give feedback to modify any/all of the above steps.

Sadler & Verheem (1996) are less deterministic and probably more practical. They suggest that a sustainability-based approach to SEA involves refocusing SEA processes towards sustainability assurance rather than towards impact minimization, and they propose a "proactive, forward looking approach that focuses on maintaining environmental bottom lines". This could be achieved through precautionary principles, the no-regrets policy and the application of simple sustainability tests.

The Netherlands, New Zealand and (the case study in) Hertfordshire are some of the few cases where sustainability-oriented SEA has been carried out to date, and they provide contrasting approaches to how this has been done. The Dutch system aims to achieve sustainable development in a clearly top–down, objectives-led manner. The report *Concern for tomorrow* (National Institute for Health and Environ-

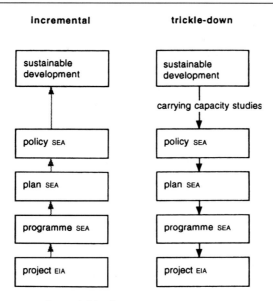

Incremental   **trickle-down**

Figure 13.5   Incremental vs. trickle-down SEA system.

ment 1988) established carrying capacities for the Netherlands, which determined the changes needed to achieve sustainable development. For instance the report noted that the emissions of various air pollutants would need to be reduced by 70–90 per cent to achieve sustainability. The National Environmental Policy Plan of 1989 (MHPPE 1989) in turn set out specific actions for achieving these changes, including the requirement that policy instruments should be assessed for their contribution to effecting sustainable development.

In contrast, New Zealand has not established quantitative objectives to be reached, but instead aims to integrate sustainability into all levels of PPPs, essentially obviating the need for a formal SEA stage. The Hertfordshire example combines these two approaches: it has established a non-quantitative "vision", a type of objective based on sustainable development, is integrating this into all levels of planning and is trying to influence other organizations to help achieve the vision. These examples suggest that SEA, to help achieve sustainable development, must become increasingly objectives-led and integrated into PPP-making. It should be used to co-ordinate all the activities in an area, and should be based on the area's carrying capacity rather than simply try to minimize impacts (Partidario 1996).

However, a number of factors limit this process. The political systems of many countries focus on short-term objectives that react to specific events or problems, rather than on long-term preventative goals such as sustainable development. Effective parameters for sustainable development are virtually impossible to set. At a regional or national level, data collection and political agreements may be feasible, but will constantly be influenced by outside factors.[6] At a global level, data may be consistent (for example, trans-boundary pollution will be accounted for) but the

423

amount of data needed is overwhelming, and political agreement may be impossible. Even if parameters are agreed upon, determining the carrying capacity within
those parameters is very complex: carrying capacity is affected by such factors as
personal consumption, technical innovation, and trade in resources within and
outside the region. In many cases, the determination of carrying capacity requires
much more environmental information than is presently available. The generation
of alternative development scenarios and the prediction of the impacts of these
scenarios have all the limitations discussed in Section 13.2. And, as discussed in
Chapter 7, monitoring is only in its infancy.

Despite these limitations, SEA is still likely to be one of the most direct and
effective ways of ensuring that human activities are carried out at a level that is
environmentally sustainable. Many countries are already carrying out research on
how to overcome these problems, and the initiatives of New Zealand and the Netherlands are particularly valuable models in this sense. In particular, they show that
the concept of sustainable development can be effectively implemented even if the
relevant methodologies are still relatively underdeveloped, as long as they are carried out in the spirit of "best practice". *Concern for tomorrow* pointed to a broad
direction and order of magnitude of change, so that its implementation could be
rapidly begun, even while further, more precise studies were being carried out.
Other countries could expand their research on EIA and carrying capacity to develop
their own objectives-led, integrated SEA systems, and begin to act on the preliminary
findings of these studies, even as they refine the concepts through further studies.

13.9 Summary

Strategic environmental assessment is the EIA of policies, plans and programmes.
SEA is often considered to be a tiered or nested process, in which the EIAs of
individual projects are carried out within the framework established by the SEA of a
programme (a group of similar projects), which in turn takes place within the
framework of an SEA of a plan (co-ordinated and timed objectives), and before that
of a policy (guidance for action).

SEA is still quite new. SEA systems have been established in the USA (particularly
in California), the Netherlands, New Zealand, and to a lesser extent some other
countries, but these are generally limited to plans and programmes (not policies)
and are not yet well developed. The EC is working towards developing a directive
on SEA. The UK Department of the Environment is encouraging the SEA of both
national and local government PPPs through a range of guidance, and good practice
examples of SEA are emerging.

However, SEA can do more than merely expand EIA to the more strategic levels of
decision-making. Objectives-led, integrated approaches to SEA are becoming seen
as best practice. As shown in particular by the Dutch National Environmental Policy

Plan of 1989, SEA can be a crucial step towards achieving sustainable development. SEA is the link between the EIA of projects, as it is presently being carried out, and the achievement of a level of human activities that maintains the quality of the environment.

References

Bass, R. & A. Herson 1996. Strategic environmental assessment in the United States: policy and practice under the National Environmental Policy Act and the California Environmental Quality Act. *Proceedings of the 16th annual conference of the International Association for Impact Assessment*, 21–6. Estoril: IAIA.

Bradley, K. 1996. SEA and the Structural Funds. In *The practice of strategic environmental assessment*, R. Therivel & M. R. Partidario (eds). London: Earthscan.

CEC (Commission of the European Communities) 1987. *Fourth action programme on the environment*, COM(86)485 final. Brussels: CEC.

CEC 1991. Draft proposal for directive on the environmental assessment of policies, plans and programmes, XI/194/90-EN-REV.4. Brussels: CEC.

CEC 1992. *Towards sustainability: a European Community programme of policy and action in relation to the environment and sustainable development*, COM(92)23 Final. Brussels: CEC.

CEC 1997. Proposal for a Council directive on the assessment of the effects of certain plans and programmes on the environment. OJ no. 6 129/14, 25 April. Brussels: CEC.

CPRE (Council for the Protection of Rural England) 1996. *Greening government – from rhetoric to reality*. London: CPRE.

DOE (Department of the Environment) 1990. *This common inheritance: Britain's environmental strategy*, CM 1200. London: HMSO.

DOE 1991. *Policy appraisal and the environment*. London: HMSO.

DOE 1992. *Development plans and regional planning guidance. Policy Planning Guidance Note no. 12*. London: HMSO.

DOE 1993. *Environmental appraisal of development plans: a good practice guide*. London: HMSO.

DOE 1994. *Environmental appraisal in government departments*. London: HMSO.

English Nature 1996. *Strategic environmental assessment and nature conservation*. Peterborough: English Nature.

Fry, C. 1996. *Strategic environmental assessment in the water sector*, MSc dissertation, Oxford Brookes University.

Hertfordshire County Council 1994a. *Structure plan review: future directions, consultation document*. Hertford: Hertfordshire CC.

Hertfordshire County Council 1994b. *Structure plan review: environmental appraisal stage 1*. Consultation document. Hertford: Hertfordshire CC.

Kent County Council 1993. *Kent structure plan 3rd review consultation draft: strategic environmental assessment of consultation draft policies*. Maidstone: Kent County Council.

Lancashire County Council 1992. *Lancashire structure plan technical report no. 13 – environmental appraisal of the plan*. Preston: Lancashire CC.

Lee, N. & J. Hughes 1995. *Strategic environmental assessment: legislation and procedures in the Community*. Manchester: University of Manchester.

Local Government Management Board 1992. *The UK local government agenda for the Earth Summit*. Luton: LGMB.

MHPPE (Ministry of Housing, Physical Planning and Environment) 1989. *National environmental policy plan – to choose or to lose*. The Hague: Netherlands Ministry of Housing, Physical Planning and Environment.

National Institute for Health and Environment 1988. *Concern for tomorrow*. Bilthoven, the Netherlands: Government of the Netherlands.

NRA (National Rivers Authority, Anglian Region) (1995) *River Nene strategy study: Peterborough to the Wash*, vol. 2, *Strategic environmental assessment*. Peterborough: NRA.

Partidario, M. R. 1996. Strategic environmental assessment – key issues emerging from recent practice. *Environmental Impact Assessment Review* **16**.

Pinfield, G. 1992. Strategic environmental assessment and land use planning. *Project Appraisal* 7(3), 157–63.

Rumble, J. & R. Therivel 1996. SEA of Hertfordshire County Council's structure plan. In *The practice of strategic environmental assessment*, R. Therivel & M. R. Partidario (eds). London: Earthscan.

Sadler, B. & R. Verheem 1996. *Strategic environmental assessment: status, challenges and future directions*. The Hague: Netherlands Ministry of Housing, Spatial Planning and the Environment.

State of California 1986. *The California Environmental Quality Act*. Sacramento: Office of Planning and Research.

Therivel, R. 1995. *Environmental appraisal of development plans 1992–1995*. Working Paper no. 160. Oxford: Oxford Brookes University, School of Planning.

Therivel, R. 1997. Strategic environmental assessment of development plans in Great Britain. *Environmental Impact Assessment Review* **17**.

Therivel, R. & M. R. Partidario (eds) 1996. *The practice of strategic environmental assessment*. London: Earthscan.

Therivel, R., E. Wilson, S. Thompson, D. Heaney, D. Pritchard 1992. *Strategic environmental assessment*. London: Earthscan.

USHUD (Department of Housing and Urban Development) 1981. *Areawide environmental assessment guidebook*. Washington, DC: Office of Policy Development and Research.

Wood, C. 1991. EIA of policies, plans and programmes. *EIA Newsletter* **5**, 2–3.

Notes

1   See Chapter 12, Section 12.3 for a discussion of cumulative impacts.

2   Programme EIRS and master EIRS differ slightly. PEIRS apply to government actions that are related geographically, as part of a chain or in connection with rules, regulations, plans or other general criteria governing the conduct of a continuing programme, and to activities carried out under the same authority and having generally similar environmental effects (14 California Code of Regulations 15168[a]). Master EIRS are a more recent phenomenon, which apply to city or county general plans, specific plans, large projects consisting of several smaller projects, the rules or regulations that will be implemented by projects, development agreements, urban redevelopment projects, multi-stage highway

or mass transit projects, transport plans and plans for the re-use of federal military bases. MEIRS are essentially a hybrid of PEIRS and project EIAS (Bass & Herson 1996).

3   Environmental appraisal is basically a more flexible, less rigorous version of full SEA.

4   In Wales, Planning Policy Guidance has traditionally run in parallel with English guidance. However, no equivalent to PPG12 has been ratified in Wales, and the recent restructuring of the Welsh local government system and PPGS has cast doubt on when and how Welsh guidance on SEA will emerge. In Scotland, the Scottish Office's National Planning Policy Guidance 1, *The planning system*, notes that structure plans should be "consistent with broader environmental objectives and sustainable development", and that all plans "should be regularly reappraised to ensure that policies are consistent with broader environmental objectives". The Scottish Office have prepared a Planning Advice Note on the environmental appraisal of development plans.

5   A structure plan is one form of local authority development plan in the UK, the other types being district local plans and unitary development plans. These plans in turn are generally composed of broad strategies for the location of development, which are then interpreted as more specific policies. In structure plans, these policies focus on strategic issues, while local plan policies are more location-specific. The SEA of development plans is somewhat easier than other forms of SEA, in that a development plan in the UK goes through a distinct life-cycle, lasting broadly 5–10 years, and has clear stages at which it is published and an SEA can be carried out.

6   For instance, the Dutch NEPP includes two very different costing estimates for its implementation: one if other countries aim towards sustainable development, the other if they do not.

APPENDIX 1

The text of Council Directive 97/11/EC

of 3 March 1997 amending Directive 85/337/EEC on the assessment of the effects of certain public and private projects on the environment[1]

Article 1

1. This Directive shall apply to the assessment of the environmental effects of those public and private projects which are likely to have significant effects on the environment.

2. For the purposes of this Directive:

"project" means:
– the execution of construction works or of other installations or schemes,
– other interventions in the natural surroundings and landscape including those involving the extraction of mineral resources;

"developer" means:
the applicant for authorization for a private project or the public authority which initiates a project;

"development consent" means:
the decision of the competent authority or authorities which entitles the developer to proceed with the project.

3. The competent authority or authorities shall be that or those which the Member States designate as responsible for performing the duties arising from this Directive.

4. Projects serving national defence purposes are not covered by this Directive.

5. This Directive shall not apply to projects the details of which are adopted by a specific act of national legislation, since the objectives of this Directive, including that of supplying information, are achieved through the legislative process.

Article 2

1. Member States shall adopt all measures necessary to ensure that, before consent is given, projects likely to have significant effects on the environment by virtue, inter alia, of their nature, size or location are made subject to a requirement for development consent and an assessment with regard to their effects. These projects are defined in Article 4.

2. The environmental impact assessment may be integrated into the existing procedures for consent to projects in the Member States, or, failing this, into other

procedures or into procedures to be established to comply with the aims of this Directive.

3. Without prejudice to Article 7, Member States may, in exceptional cases, exempt a specific project in whole or in part from the provisions laid down in this Directive.

In this event, the Member States shall:

(a) consider whether another form of assessment would be appropriate and whether the information thus collected should be made available to the public; (b) make available to the public concerned the information relating to the exemption and the reasons for granting it; (c) inform the Commission, prior to granting consent, of the reasons justifying the exemption granted, and provide it with the information made available, *where applicable*, to their own nationals.

The Commission shall immediately forward the documents received to the other Member States.

The Commission shall report annually to the Council on the application of this paragraph.

2a. Member States may provide for a single procedure in order to fulfil the requirements of this Directive and the requirements of Council Directive 96/61/ EC of 24 September 1996 on integrated pollution prevention and control[1].

Article 3

The environmental impact assessment shall identify, describe and assess in an appro-

priate manner, in the light of each individual case and in accordance with Articles 4 to 11, the direct and indirect effects of a project on the following factors:
– human beings, fauna and flora;
– soil, water, air, climate and the landscape;
– material assets and the cultural heritage;
– the interaction between the factors mentioned in the first, second and third indents.

Article 4

1. Subject to Article 2 (3), projects listed in Annex I shall be made subject to an assessment in accordance with Articles 5 to 10.

2. Subject to Article 2 (3), for projects listed in Annex II, the Member States shall determine through:

(a) a case-by-case examination, or
(b) thresholds or criteria set by the Member States

whether the project shall be made subject to an assessment in accordance with Articles 5 to 10.

Member States may decide to apply both procedures referred to in (a) and (b).

3. When a case-by-case examination is carried out or thresholds or criteria are set for the purpose of paragraph 2, the relevant selection criteria set out in Annex III shall be taken into account.

4. Member States shall ensure that the determination made by the competent authorities under paragraph 2 is made available to the public.

Article 5

1. In the case of projects which, pursuant to Article 4, must be subjected to an environmental impact assessment in accordance with Articles 5 to 10, Member States shall adopt the necessary measures to ensure that the developer supplies in an appropriate form the information specified in Annex IV inasmuch as:

(a) the Member States consider that the information is relevant to a given stage of the consent procedure and to the specific characteristics of a particular project or type of project and of the environmental features likely to be affected;

(b) the Member States consider that a developer may reasonably be required to compile this information having regard inter alia to current knowledge and methods of assessment.

2. Member States shall take the necessary measures to ensure that, if the developer so requests before submitting an application for development consent, the competent authority shall give an opinion on the information to be supplied by the developer in accordance with paragraph 1. The competent authority shall consult the developer and authorities referred to in Article 6(1) before it gives its opinion. The fact that the authority has given an opinion under this paragraph shall not preclude it from subsequently requiring the developer to submit further information.

Member States may require the competent authorities to give such an opinion, irrespective of whether the developer so requests.

3. The information to be provided by the developer in accordance with paragraph 1 shall include at least:

– a description of the project comprising information on the site, design and size of the project,
– a description of the measures envisaged in order to avoid, reduce and, if possible, remedy significant adverse effects,
– the data required to identify and assess the main effects which the project is likely to have on the environment,
– an outline of the main alternatives studied by the developer and an indication of the main reasons for his choice, taking into account the environmental effects,
– a non-technical summary of the information mentioned in the previous indents.

4. Member States shall, if necessary, ensure that any authorities holding relevant information, with particular reference to Article 3, shall make this information available to the developer.

Article 6

1. Member States shall take the measures necessary to ensure that the authorities likely to be concerned by the project by reason of their specific environmental responsibilities are given an opportunity to express their opinion on the information supplied by the developer and on the request for development consent. To this end, Member States shall designate the authorities to be consulted, either in general terms or on a case-by-case basis. The information gathered pursuant to Article 5 shall be forwarded to those authorities. Detailed arrangements for consultation shall be laid down by the Member States.

2. *Member States shall ensure that any request for development consent and any information gathered pursuant to Article 5 are made available to the public within a reasonable time in order to give the public concerned the opportunity to express an opinion before the development consent is granted.*

3. The detailed arrangements for such information and consultation shall be determined by the Member States, which may in particular, depending on the particular characteristics of the projects or sites concerned:

– determine the public concerned,
– specify the places where the information can be consulted,
– specify the way in which the public may be informed, for example by bill-posting within a certain radius, publication in local newspapers, organization of exhibitions with plans, drawings, tables, graphs, models,
– determine the manner in which the public is to be consulted, for example, by written submissions, by public enquiry,
– fix appropriate time limits for the various stages of the procedure in order to ensure that a decision is taken within a reasonable period.

Article 7

1. *Where a Member State is aware that a project is likely to have significant effects on the environment in another Member State or where a Member State likely to be significantly affected so requests, the Member State in whose territory the project is intended to be carried out shall send to the affected Member State as soon as possible and no later than when informing its own public, inter alia:*

(a) a description of the project, together with any available information on its possible transboundary impact;

(b) information on the nature of the decision which may be taken,

and shall give the other Member State a reasonable time in which to indicate whether it wishes to participate in the Environmental Impact Assessment procedure, and may include the information referred to in paragraph 2.

2. *If a Member State which receives information pursuant to paragraph 1 indicates that it intends to participate in the Environmental Impact Assessment procedure, the Member State in whose territory the project is intended to be carried out shall, if it has not already done so, send to the affected Member State the information gathered pursuant to Article 5 and relevant information regarding the said procedure, including the request for development consent.*

3. *The Member States concerned, each insofar as it is concerned, shall also:*

(a) arrange for the information referred to in paragraphs 1 and 2 to be made available, within a reasonable time, to the authorities referred to in Article 6(1) and the public concerned in the territory of the Member State likely to be significantly affected; and

(b) ensure that those authorities and the public concerned are given an opportunity, before development consent for the project is granted, to forward their opinion within a reasonable time on the information supplied to the competent

authority in the Member State in whose territory the project is intended to be carried out.

4. The Member States concerned shall enter into consultations regarding, inter alia, the potential transboundary effects of the project and the measures envisaged to reduce or eliminate such effects and shall agree on a reasonable time frame for the duration of the consultation period.

5. The detailed arrangements for implementing the provisions of this Article may be determined by the Member States concerned.

Article 8

The results of consultations and the information gathered pursuant to Articles 5, 6 and 7 must be taken into consideration in the development consent procedure.

Article 9

1. When a decision to grant or refuse development consent has been taken, the competent authority or authorities shall inform the public thereof in accordance with the appropriate procedures and shall make available to the public the following information:

– the content of the decision and any conditions attached thereto,
– the main reasons and considerations on which the decision is based,
– a description, where necessary, of the main measures to avoid, reduce and, if possible, offset the major adverse effects.

2. The competent authority or authorities shall inform any Member State which has been consulted pursuant to Article 7, forwarding to it the information referred to in paragraph 1.

Article 10

The provisions of this Directive shall not affect the obligation on the competent authorities to respect the limitations imposed by national regulations and administrative provisions and accepted legal practices with regard to commercial and industrial confidentiality, including intellectual property, and the safeguarding of the public interest.

Where Article 7 applies, the transmission of information to another Member State and the receipt of information by another Member State shall be subject to the limitations in force in the Member State in which the project is proposed.

Article 11

1. The Member States and the Commission shall exchange information on the experience gained in applying this Directive.

2. In particular, Member States shall inform the Commission of any criteria and/or thresholds adopted for the selection of the projects in question, in accordance with Article 4(2).

3. Five years after the entry into force of this Directive, the Commission shall send the European Parliament and the Council a report on the application and effectiveness of Directive 85/337/EEC as amended by this Directive. The report

shall be based on the exchange of information provided for by Article 11(1) and (2).

4. On the basis of this report, the Commission shall, where appropriate, submit to the Council additional proposals with a view to ensuring further coordination in the application of this Directive.

Article 12

1. Member States shall bring into force the laws, regulations and administrative provisions necessary to comply with this Directive by 14 March 1999 at the latest. They shall forthwith inform the Commission thereof.

When Member States adopt these provisions, they shall contain a reference to this Directive or shall be accompanied by such reference at the time of their official publication. The procedure for such reference shall be adopted by Member States.

2. If a request for development consent is submitted to a competent authority before the end of the time limit laid down in paragraph 1, the provisions of Directive 85/337/EEC prior to these amendments shall continue to apply.

Article 13

This Directive shall enter into force on the twentieth day following that of its publication in the Official Journal of the European Communities.

Article 14

This Directive is addressed to the Member States.

Done at Brussels, 3 March 1997.
For the Council
The President
M. DE BOER

---

[1] Consolidated version.
(1) oj No L257, 10. 10. 1996, p. 26

ANNEX I: Projects subject to Article 4 (1)

1. Crude-oil refineries (excluding undertakings manufacturing only lubricants from crude oil) and installations for the gasification and liquefaction of 500 tonnes or more of coal or bituminous shale per day.

2. – Thermal power stations and other combustion installations with a heat output of 300 megawatts or more, and
– nuclear power stations and other nuclear reactors including the dismantling or decommissioning of such power stations or reactors (^{1}) (except research installations for the production and conversion of fissionable and fertile materials, whose maximum power does not exceed 1 kilowatt continuous thermal load).

3. (a) Installations for the reprocessing of irradiated nuclear fuel.
   (b) Installations designed:
– for the production or enrichment of nuclear fuel,
– for the processing of irradiated nuclear fuel or high-level radioactive waste,
– for the final disposal of irradiated nuclear fuel,
– solely for the final disposal of radioactive waste,
– solely for the storage (planned for more than 10 years) of irradiated nuclear fuels or radioactive waste in a different site than the production site.

4. – Integrated works for the initial smelting of cast-iron and steel;
– Installations for the production of non-ferrous crude metals from ore, concentrates or secondary raw materials by metallurgical, chemical or electrolytic processes.

5. Installations for the extraction of asbestos and for the processing and transformation of asbestos and products containing asbestos: for asbestos-cement products, with an annual production of more than 20 000 tonnes of finished products, for friction material, with an annual production of more than 50 tonnes of finished products, and for other uses of asbestos, utilization of more than 200 tonnes per year.

6. Integrated chemical installations, i.e. those installations for the manufacture on an industrial scale of substances using chemical conversion processes, in which several units are juxtaposed and are functionally linked to one another and which are:
(i) for the production of basic organic chemicals;
(ii) for the production of basic inorganic chemicals;
(iii) for the production of phosphorous-, nitrogen- or potassium-based fertilizers (simple or compound fertilizers);

(iv) for the production of basic plant health products and of biocides;

(v) for the production of basic pharmaceutical products using a chemical or biological process;

(vi) for the production of explosives.

7. (a) Construction of lines for long-distance railway traffic and of airports (2) with a basic runway length of 2 100 m or more;

(b) Construction of motorways and express roads (3);

(c) Construction of a new road of four or more lanes, or realignment and/or widening of an existing road of two lanes or less so as to provide four or more lanes, where such new road, or realigned and/or widened section of road would be 10 km or more in a continuous length.

8. (a) Inland waterways and ports for inland-waterway traffic which permit the passage of vessels of over 1 350 tonnes;

(b) Trading ports, piers for loading and unloading connected to land and outside ports (excluding ferry piers) which can take vessels of over 1 350 tonnes.

9. Waste disposal installations for the incineration, chemical treatment as defined in Annex IIA to Directive 75/442/EEC (4) under heading D9, or landfill of hazardous waste (i.e. waste to which Directive 91/689/EEC (5) applies).

10. Waste disposal installations for the incineration or chemical treatment as defined in Annex IIA to Directive 75/442/EEC under heading D9 of non-hazardous waste with a capacity exceeding 100 tonnes per day.

11. Groundwater abstraction or artificial groundwater recharge schemes where the annual volume of water abstracted or recharged is equivalent to or exceeds 10 million cubic metres.

12. (a) Works for the transfer of water resources between river basins where this transfer aims at preventing possible shortages of water and where the amount of water transferred exceeds 100 million cubic metres/year;

(b) In all other cases, works for the transfer of water resources between river basins where the multi-annual average flow of the basin of abstraction exceeds 2 000 million cubic metres/year and where the amount of water transferred exceeds 5% of this flow.

In both cases transfers of piped drinking water are excluded.

(1) Nuclear power stations and other nuclear reactors cease to be such an installation when all nuclear fuel and other radioactively contaminated elements have been removed permanently from the installation site.

(2) For the purposes of this Directive, "airport" means airports which comply with the definition in the 1944 Chicago Convention setting up the International Civil Aviation Organization (Annex 14).

(3) For the purposes of the Directive, "express road" means a road which complies with the definition in the European Agreement on Main International Traffic Arteries of 15 November 1975.

13. Waste water treatment plants with a capacity exceeding 150 000 population equivalent as defined in Article 2 point (6) of Directive 91/271/EEC (6).

14. Extraction of petroleum and natural gas for commercial purposes where the amount extracted exceeds 500 tonnes/day in the case of petroleum and 500 000 m^3/day in the case of gas.

15. Dams and other installations designed for the holding back or permanent storage of water, where a new or additional amount of water held back or stored exceeds 10 million cubic metres.

16. Pipelines for the transport of gas, oil or chemicals with a diameter of more than 800 mm and a length of more than 40 km.

17. Installations for the intensive rearing of poultry or pigs with more than:
(a) 85 000 places for broilers, 60 000 places for hens;
(b) 3 000 places for production pigs (over 30 kg); or
(c) 900 places for sows.

18. Industrial plants for the
(a) production of pulp from timber or similar fibrous materials;
(b) production of paper and board with a production capacity exceeding 200 tonnes per day.

19. Quarries and open-cast mining where the surface of the site exceeds 25 hectares, or peat extraction, where the surface of the site exceeds 150 hectares.

20. Construction of overhead electrical power lines with a voltage of 220 kV or more and a length of more than 15 km.

21. Installations for storage of petroleum, petrochemical, or chemical products with a capacity of 200 000 tonnes or more.

(4) OJ No L 194, 25. 7. 1975, p. 39. Directive as last amended by Commission Decision 94/3/EC (OJ No L 5, 7. 1. 1994, p. 15).
(5) OJ No L 377, 31. 12. 1991, p. 20. Directive as last amended by Directive 94/31/EC (OJ No L 168, 2. 7. 1994, p. 28).
(6) OJ No L 135, 30. 5. 1991, p. 40. Directive as last amended by the 1994 Act of Accession.

ANNEX II: Projects subject to Article 4 (2)

1. Agriculture, silviculture and aquaculture

(a) Projects for the restructuring of rural land holdings;

(b) Projects for the use of uncultivated land or semi-natural areas for intensive agricultural purposes;

(c) Water management projects for agriculture, including irrigation and land drainage projects;

(d) Initial afforestation and deforestation for the purposes of conversion to another type of land use;

(e) Intensive livestock installations (projects not included in Annex I);

(f) Intensive fish farming;

(g) Reclamation of land from the sea.

2. Extractive industry

(a) Quarries, open-cast mining and peat extraction (projects not included in Annex I);

(b) Underground mining;

(c) Extraction of minerals by marine or fluvial dredging;

(d) Deep drillings, in particular:
– geothermal drilling,
– drilling for the storage of nuclear waste material,
– drilling for water supplies,
with the exception of drillings for investigating the stability of the soil;

(e) Surface industrial installations for the extraction of coal, petroleum, natural gas and ores, as well as bituminous shale.

3. Energy industry

(a) Industrial installations for the production of electricity, steam and hot water (projects not included in Annex I);

(b) Industrial installations for carrying gas, steam and hot water; transmission of electrical energy by overhead cables (projects not included in Annex I);

(c) Surface storage of natural gas;

(d) Underground storage of combustible gases;

(e) Surface storage of fossil fuels;

(f) Industrial briquetting of coal and lignite;

(g) Installations for the processing and storage of radioactive waste (unless included in Annex I);

(h) Installations for hydroelectric energy production;

(i) Installations for the harnessing of wind power for energy production (wind farms).

4. Production and processing of metals

(a) Installations for the production of pig iron or steel (primary or secondary fusion) including continuous casting;

(b) Installations for the processing of ferrous metals:
(i) hot-rolling mills;
(ii) smitheries with hammers;
(iii) application of protective fused metal coats;

(c) Ferrous metal foundries;

(d) Installations for the smelting, including the alloyage, of non-ferrous metals, excluding precious metals, including recovered products (refining, foundry casting, etc.);

(e) Installations for surface treatment of metals and plastic materials using an electrolytic or chemical process;

(f) Manufacture and assembly of motor vehicles and manufacture of motor-vehicle engines;

(g) Shipyards;

(h) Installations for the construction and repair of aircraft;

(i) Manufacture of railway equipment;

(j) Swaging by explosives;

(k) Installations for the roasting and sintering of metallic ores.

5. Mineral industry

(a) Coke ovens (dry coal distillation);

(b) Installations for the manufacture of cement;

(c) Installations for the production of asbestos and the manufacture of asbestos-products (projects not included in Annex I);

(d) Installations for the manufacture of glass including glass fibre;

(e) Installations for smelting mineral substances including the production of mineral fibres;

(f) Manufacture of ceramic products by burning, in particular roofing tiles, bricks, refractory bricks, tiles, stoneware or porcelain.

6. Chemical industry (Projects not included in Annex I)

(a) Treatment of intermediate products and production of chemicals;

(b) Production of pesticides and pharmaceutical products, paint and varnishes, elastomers and peroxides;

(c) Storage facilities for petroleum, petrochemical and chemical products.

7. Food industry

(a) Manufacture of vegetable and animal oils and fats;

(b) Packing and canning of animal and vegetable products;

(c) Manufacture of dairy products;

(d) Brewing and malting;

(e) Confectionery and syrup manufacture;

(f) Installations for the slaughter of animals;

(g) Industrial starch manufacturing installations;

(h) Fish-meal and fish-oil factories;

(i) Sugar factories.

8. Textile, leather, wood and paper industries

(a) Industrial plants for the production of paper and board (projects not included in Annex I);

(b) Plants for the pretreatment (operations such as washing, bleaching, mercerization) or dyeing of fibres or textiles;

(c) Plants for the tanning of hides and skins;

(d) Cellulose-processing and production installations.

9. Rubber industry

Manufacture and treatment of elastomer-based products.

10. Infrastructure projects

(a) Industrial estate development projects;

(b) Urban development projects, including the construction of shopping centres and car parks;

(c) Construction of railways and intermodal transshipment facilities, and of intermodal terminals (projects not included in Annex I);

(d) Construction of airfields (projects not included in Annex I);

(e) Construction of roads, harbours and port installations, including fishing harbours (projects not included in Annex I);

(f) Inland-waterway construction not included in Annex I, canalization and flood-relief works;

(g) Dams and other installations designed to hold water or store it on a long-term basis (projects not included in Annex I);

(h) Tramways, elevated and underground railways, suspended lines or similar lines of a particular type, used exclusively or mainly for passenger transport;

(i) Oil and gas pipeline installations (projects not included in Annex I);

(j) Installations of long-distance aqueducts;

(k) Coastal work to combat erosion and maritime works capable of altering the coast through the construction, for example, of dykes, moles, jetties and other sea defence works, excluding the maintenance and reconstruction of such works;

(l) Groundwater abstraction and artificial groundwater recharge schemes not included in Annex I;

(m) Works for the transfer of water resources between river basins not included in Annex I.

11. Other projects

(a) Permanent racing and test tracks for motorized vehicles;

(b) Installations for the disposal of waste (projects not included in Annex I);

(c) Waste-water treatment plants (projects not included in Annex I);

(d) Sludge-deposition sites;

(e) Storage of scrap iron, including scrap vehicles;

(f) Test benches for engines, turbines or reactors;

(g) Installations for the manufacture of artificial mineral fibres;

(h) Installations for the recovery or destruction of explosive substances;

(i) Knackers' yards.

12. Tourism and leisure

(a) Ski-runs, ski-lifts and cable-cars and associated developments;

(b) Marinas;

(c) Holiday villages and hotel complexes outside urban areas and associated developments;

(d) Permanent camp sites and caravan sites;

(e) Theme parks.

13. – Any change or extension of projects listed in Annex I or Annex II, already authorized, executed or in the process of being executed, which may have significant adverse effects on the environment;
– Projects in Annex I, undertaken exclusively or mainly for the development and testing of new methods or products and not used for more than two years.

ANNEX III: Selection criteria referred to in Article 4 (3)

1. Characteristics of projects

The characteristics of projects must be considered having regard, in particular, to:
– the size of the project,
– the cumulation with other projects,
– the use of natural resources,
– the production of waste,
– pollution and nuisances,
– the risk of accidents, having regard in particular to substances or technologies used.

2. Location of projects

The environmental sensitivity of geographical areas likely to be affected by projects must be considered, having regard, in particular, to:
– the existing land use,
– the relative abundance, quality and regenerative capacity of natural resources in the area,
– the absorption capacity of the natural environment, paying particular attention to the following areas:
(a) wetlands;
(b) coastal zones;
(c) mountain and forest areas;
(d) nature reserves and parks;
(e) areas classified or protected under Member States' legislation; special protection areas designated by Member States pursuant to Directive 79/409/EEC and 92/43/EEC;
(f) areas in which the environmental quality standards laid down in Community legislation have already been exceeded;
(g) densely populated areas;
(h) landscapes of historical, cultural or archaeological significance.

3. Characteristics of the potential impact

The potential significant effects of projects must be considered in relation to criteria set out under 1 and 2 above, and having regard in particular to:
– the extent of the impact (geographical area and size of the affected population),

– the transfrontier nature of the impact,
– the magnitude and complexity of the impact,
– the probability of the impact,
– the duration, frequency and reversibility of the impact.

ANNEX IV: Information referred to in Article 5 (1)

1. Description of the project, including in particular:
– a description of the physical characteristics of the whole project and the land-use requirements during the construction and operational phases,
– a description of the main characteristics of the production processes, for instance, nature and quantity of the materials used,
– an estimate, by type and quantity, of expected residues and emissions (water, air and soil pollution, noise, vibration, light, heat, radiation, etc.) resulting from the operation of the proposed project.

2. An outline of the main alternatives studied by the developer and an indication of the main reasons for this choice, taking into account the environmental effects.

3. A description of the aspects of the environment likely to be significantly affected by the proposed project, including, in particular, population, fauna, flora, soil, water, air, climatic factors, material assets, including the architectural and archaeological heritage, landscape and the inter-relationship between the above factors.

4. A description (7) of the likely significant effects of the proposed project on the environment resulting from:
– the existence of the project,
– the use of natural resources,
– the emission of pollutants, the creation of nuisances and the elimination of waste, and the description by the developer of the forecasting methods used to assess the effects on the environment.

5. A description of the measures envisaged to prevent, reduce and where possible offset any significant adverse effects on the environment.

6. A non-technical summary of the information provided under the above headings.

7. An indication of any difficulties (technical deficiencies or lack of know-how) encountered by the developer in compiling the required information.

––––––

(7) This description should cover the direct effects and any indirect, secondary, cumulative, short, medium and long-term, permanent and temporary, positive and negative effects of the project.

APPENDIX 2
EU Member States' EIA systems

This appendix provides a brief overview of the differing EIA systems in each of the fifteen EU Member States. It is based largely on the bi-annual EIA Newsletters and EIA Leaflet Series of the EIA Centre, University of Manchester. Additional material has been obtained from the EC's five year review of Directive 85/337 (CEC 1993) and the United Nations Economic Commission for Europe (1991) report on EIA systems.

Austria

EIA was implemented in Austria by federal legislation, the Federal Act concerning EIA and Public Participation (the EIA Act), which was passed in September 1993 and came into operation on 1 July 1994. As well as establishing a system of EIA, the Act replaced the previous sectoral licensing procedures with a single, all-embracing, project licensing procedure.

Annex I of the EIA Act specifies those projects for which EIA is mandatory; this includes most of the project types in Annex I of Directive 85/337/EEC. Annex II of the Act deals with smaller-scale projects (basically those contained in Annex II of the Directive). For these Annex II projects, EIA is not required, although formal public consultation procedures must be followed. The EIA Act includes provisions covering scoping, EIS content, formal expert review of the EIS, public consultation and post-project auditing. Some of these provisions go beyond those required by Directive 85/337.

The exclusion of Annex II projects from the EIA procedures means that the annual number of EISs prepared in Austria is relatively small, although precise figures are not currently available.

Although there is no explicit requirement for the EIA of policies, plans or programmes, recent years have seen greater attention to environmental issues in the preparation of plans and programmes in areas such as waste and water management, traffic, energy and land-use planning. The Federal Ministry of the Environment has also commissioned research into SEA, the results of which were published in 1996.

Belgium

The responsibility for the implementation of EIA in Belgium rests with the country's three regions (Brussels, Flanders and Wallonia), with the exception of EIA for nuclear installations and the disposal or treatment of radioactive waste, which remains a Federal Government responsibility.

In Flanders, EIA was implemented by means of a number of Administrative Orders issued by the regional government. These Orders supplement existing legislation and incorporate EIA within existing consent and licensing procedures for industrial installations, certain infrastructure-related projects and development requiring building permits; the Orders became operational on 23 March 1989. Good practice guidelines on EIA procedures and methodology were issued by the regional government in 1997. The regional government has also commissioned a study into the development of a methodology for the EIA of policies, plans and programmes in Flanders.

In Wallonia, EIA was implemented by the regional government in September 1985, by the Decree on the Organisation of the Evaluation of Environmental Effects in the Walloon Region. This legislation was supplemented by a number of Administrative Orders, issued in July 1990, October 1991, March 1993 and July 1993. The legislation requires the preparation of an Initial Environmental Examination report, which is used in the screening of projects. Public involvement in the scoping stage of EIA is another feature of the system. Significant changes in EIA legislation are likely during the next few years, prompted by the regional government's desire to shorten and simplify the existing EIA procedures. Two new pieces of legislation are expected to replace the existing EIA Decree of September 1985, one of which will introduce the EIA of spatial plans. This will result in a number of projects being exempted from EIA if provision is already made for them in such plans.

In Brussels, EIA was implemented by the Ordinance of 30 July 1992 on the Environmental Assessment of Certain Projects in the Brussels Capital Region; this became operational on 1 December 1993. EIA within Brussels was incorporated into existing environmental and building permit procedures. Changes to the EIA procedures were introduced in June 1997, the Ordinance of 30 July 1992 being replaced by a new Ordinance on Environmental Permits and modifications being made to the related Ordinance on Organisation of Planning and Urban Development in the Brussels Capital Region. Under the new procedures, all projects requiring an environmental permit will be classified into one of four categories (IA, IB, II and III). Only those projects in category IA will be subject to a full EIA, while those in IB will require a more limited environmental impact study. Certain land-use plans will also be subject to EIA.

The number of EISs prepared in Belgium is relatively low, averaging fewer than 50 per annum in Flanders and Wallonia combined.

Denmark

Directive 85/337/EEC was implemented in July 1989, within the existing regional planning system. Provision for EIA is contained in the Planning Act (no. 388 of June 1991) and the Consolidated Planning Act (no. 746 of August 1994), as well as in a number of Executive Orders. The Planning Act (June 1991) includes a provision for public participation at the scoping stage of EIA, and an appeal case in 1993 revealed that comments made by the public at this stage must be dealt with in the EIA and reported in the EIS. EIA guidelines, including guidance on screening and on SEA, have been published by the Ministry of the Environment and Energy.

EIA procedures were strengthened in October 1994, increasing the number of project types for which EIA is mandatory (from 16 to 34) and establishing new screening procedures for Annex II projects. Despite these changes, the annual number of EISs in Denmark remains very low, totalling only 25 during 1997.

Provision for the SEA of government policies was contained in an Administrative Order, which came into force in October 1993. This requires an assessment of the environmental impacts of legislative bills and other government proposals. Work is also under way on the application of SEA to regional development plans.

Finland

EIA in Finland was implemented by means of an Act on EIA Procedure (468/94), as well as by amendments to existing legislation; this legislation became operational on 1 September 1994. The EIA Act contains provisions for the EIA of certain policies, programmes and plans. Amendments to the Building Act also include a provision for EIA within land-use planning. Guidelines on SEA, including the environmental assessment of government bills, are to be published by the Ministry of the Environment.

A total of 40 EIAs were undertaken during the first year after the implementation of EIA. Road projects accounted for almost half the EIAs undertaken, although this proportion has declined in more recent years.

France

EIA legislation in France dates from July 1976, with the adoption of the Environmental Protection Act and two Decrees which implement the Act. The French EIA system involves a two-tier procedure, with only certain projects subject to a full EIA and the remainder subject to a simplified procedure called a *notice d'impact*. Legislation passed in December 1983 established procedures for public consultation, including requirements for the publication of EISs for public inquiries, as well as for certain types of construction work, development plans and land-use plans.

The number of EIAs prepared in France is by far the highest in the EU, averaging between 5,000 and 6,000 per annum. Most EIAs are for Annex II projects, the low screening thresholds for such projects explaining the large number of EIAs undertaken.

Some progress has been made in the implementation of SEA in France. Examples include the publication of national and regional environmental plans for electricity transmission lines, the introduction of public participation on draft EISs for major transport projects, and a decree from the Ministry of the Environment (in February 1993), which states that EIAs for projects must also include an assessment of the programme to which the project is linked.

Germany

EIA was implemented by the federal EIA Act of 12 February 1990, and subsequent amendments. This Act provides for consequential changes to 16 existing Federal Acts. Amendments to the Federal Mining Act (June 1994) and to the Federal Land Use Planning Act (November 1994) and related Ordinances also require EIA. In March 1992, an Ordinance was passed which also requires EIA for industrial projects, and in November 1994 an Ordinance concerning nuclear installations was amended to introduce a requirement for EIA. General administrative provisions, which prescribe criteria and procedures for identifying, scoping, describing, assessing and summarizing environmental impacts, were implemented in September 1995.

Most of Germany's regions (Länder) have established their own EIA legislation or administrative provisions in response to the federal EIA Act; in some cases these are more stringent than the federal regulations. The need to speed up planning and permit procedures to allow for the rapid reconstruction of the former East Germany's infrastructure and economy has resulted in a number of changes to the EIA Act.

It is estimated that between 200 and 500 EIAs per annum are undertaken in Germany, although comprehensive information on the total numbers is not currently collected. Road projects and industrial plants, including modifications to existing plant, account for a high proportion of German EIAS.

Greece

The legal framework for EIA in Greece was created in 1986, by Constitutional Law 1650/86 for the Protection of the Natural Environment. However, this legislation was not fully implemented until October 1990 when a number of relevant Ministerial Decisions were issued; these set out the types of project requiring EIA (Category A and B projects, roughly equivalent to Annexes I and II projects in Directive 85/337/EEC). Also specified are the contents of EISs and the procedures for the publication of EISs and for public consultation. Circulars clarifying EIA and licensing procedures, and further specifying EIS contents for a range of project types, have also been published.

Project authorization involves two stages: the Initial Approval of Siting (for which a mini-EIS is prepared) and the Approval of Environmental Conditions. The first of these stages is not required for certain types of project. A Special Inspectors Body was established in 1994, one of whose responsibilities is monitoring the implementation of environmental conditions attached to project authorizations.

Ireland

The implementation of Directive 85/337/EEC in Ireland was in many respects similar to that in the UK, being achieved through a series of Regulations; these provide for the incorporation of EIA into existing development consent procedures. The most important Regulations are the European Communities (Environmental Impact Assessment) Regulations 1989 (SI 340) and the Local Government (Planning and Development) Regulations 1994 (SI 86). Separate Regulations cover motorways, fisheries, gas, air navigation and transport, petroleum and other minerals development, the foreshore, and arterial drainage. The range of projects subject to EIA incorporates almost all of those in Annexes I and II of the EC Directive, and an average of approximately 80 EISs are prepared annually. In response to concerns about the environmental impact of forestry development, an amendment to the EIA Regulations in 1996 reduced the screening threshold for afforestation projects from 200 to 70 hectares.

Some progress has been made in the implementation of SEA. For example, national operational programmes relating to transport, tourism and other areas incorporate an assessment of the programmes' likely environmental effects.

449

Italy

The implementation of Directive 85/337/EEC in Italy has been slow and problematic; in particular, legislation covering EIA of Annex II projects was not issued until April 1996. Provision for EIA was originally introduced by Article 6 of Law no. 349 (of July 1986), and by two subsequent Presidential Decrees (in August and December 1988). Law no. 349 envisaged future provisions for EIA for certain projects of "national importance" and for a determination of the environmental compatibility of such projects. These provisions were made in the Presidential Decrees of 1988. The first of these (no. 377 of August 1988) indicates the types of project subject to EIA – basically only those in Annex I of Directive 85/337/EEC, plus dams and other installations for the long-term storage of water above a certain capacity. No provisions were made for the EIA of Annex II projects. The second Decree (of December 1988) contains technical regulations governing the preparation of EISs and for judging the environmental compatibility of a project.

Between 1990 and 1992, further legislation extended the EIA provisions to a number of additional project types of "national importance", including electricity transmission lines and hydroelectric power plants. A number of regions also issued regional legislation on EIA, some of which takes into consideration Annex II projects.

Significant changes in the Italian EIA system took place during 1996, with the issuing of a Presidential Decree (of 12 April 1996) extending the application of EIA to Annex II projects. The Decree indicates those projects for which EIA is mandatory or discretionary, provides guidelines for the preparation of an EIA, provides for scoping to be requested either by the developer or the competent authority and outlines procedures on public consultation, EIS evaluation and inter-regional and trans-boundary issues.

The current annual number of EISs prepared in Italy is not known, but in the early years after the introduction of EIA (1989–91), an average of about 30 EISs were prepared per annum. Most of these were for Annex I projects, mainly industrial installations, dams and power stations.

Luxembourg

EIA in Luxembourg is carried out under a range of existing legislation and a Regulation of 4 March 1994 which implemented the provisions of Directive 85/337/EEC with regard to Annex I projects. Relevant legislation includes: a law of May 1990 concerning the control of dangerous, dirty and noxious installations; a law of August 1982 regarding the protection/conservation of nature and natural resources (this applies to projects located in areas not designated for urban and industrial development, for which an EIA can be requested by the Ministry of the Environment); and a law of August 1967 (modified in August 1986) concerning the creation of a communication network (this provides for the EIA of road projects).

The Netherlands

See Section 11.2.

Portugal

Portugal joined the EU in 1986, shortly after the publication of Directive 85/337/ EEC. However, as early as 1987, a number of EIAS – for roads and motorways – had been undertaken in accordance with the Directive, and the Portuguese Environmental Act of 1987 (Law 11/87) made provision for EIA. Despite this early progress, Directive 85/337/EEC was not implemented until the end of 1990, with the issuing of Decree Law 186/90 (EIA Process) and Regulatory Decree 38/90 (EIA Process). Decree Law 258/92, published in 1992, deals with the EIA of large commercial developments.

The annual number of EIAS currently undertaken in Portugal is not known, but it is estimated that by 1995 approximately 300 had been completed. EIAS of highway developments have been of particular importance.

The absence of provisions for SEA has highlighted the weaknesses of project level EIA, for example in dealing with the impacts of multiple tourism developments in particular regions.

Spain

EIA in Spain is required by Legislative Decree 1302/1986 (of June 1986) and Decree 1131/1988 (of September 1988). The need for EIA for certain projects is also set out in Act 25/1988 on highways and Act 4/1988 on the conservation of natural areas and wildlife. Under the Spanish legislation, all projects listed in Annex I of the EC Directive require EIA, but not all those projects in Annex II. Various legal provisions for EIA have also been enacted at the regional level, and several regions require EIA for all, or nearly all, Annex II projects. However, in certain cases such projects are subject only to a simplified EIA procedure.

Spanish EIA legislation is currently under review; draft amendments to the legislation (as at the end of 1997) proposed an increase in the number of project types for which EIA is mandatory, the establishment of clearer screening criteria, improved scoping and the early consideration of alternatives. The establishment of procedures for the SEA of plans and programmes is also proposed.

The number of EIAS undertaken in Spain appears to be relatively high and continues to grow, approximately 900 EISS being prepared between July 1988 and April 1995. Highways, mineral extraction, dams, waste-disposal installations, urban development, chemical industry schemes, forestry and coastal development have been of particular importance.

Sweden

EIA has been implemented in Sweden within existing legislation covering project authorization and land-use planning procedures. Some of this legislation, in particular the Environmental Protection Act (SFS 1969:387) already required a form of EIA to be undertaken before permission for a project could be granted. The legislation covers all the project categories listed in Annexes I and II of Directive 85/337/EEC, with the exception of rail projects (for which legislation is in preparation). No separate EIA procedure has been introduced in Sweden; projects requiring EIA are subject to the same public consultation procedures as all other projects. Little guidance has been provided to date on the need for EIA (for most project types, screening criteria or thresholds have not been specified) or on EIS contents.

Some progress with the implementation of SEA has been made, particularly in the areas of land-use, road and railway planning, and in the forestry and fisheries sectors.

The United Kingdom

See Chapter 3.

References

CEC (Commission of the European Communities) 1993. *Report from the Commission on the implementation of Directive 85/337/EEC on the assessment of the effects of certain public and private projects on the environment, and annex for the United Kingdom.* COM(93)28 final, Vol. 12.

EIA Centre 1995. *Environmental assessment in the European Union.* EIA *Newsletter* **10**, EIA Centre, University of Manchester, Summer.

EIA Centre 1996a. *Recent EIA developments within the European Union.* EIA *Newsletter* **13**, EIA Centre, University of Manchester, Winter.

EIA Centre 1996b. EIA *legislation and regulations in the EU.* EIA Leaflet Series no. 5, EIA Centre, University of Manchester, November.

EIA Centre 1997. *Recent EIA developments within the European Union.* EIA *Newsletter* **15**, EIA Centre, University of Manchester, Winter.

United Nations Economic Commission for Europe 1991. *Policies and systems of environmental impact assessment.* New York: United Nations.

APPENDIX 3

The Lee and Colley review package

The Lee Colley Method reviews ESS under four main topics, each of which is examined under a number of sub-headings:

(i)   Description of the development, the local environment, and the baseline conditions:
   • Description of the development
   • Site description
   • Residuals
   • Baseline conditions
(ii) Identification and evaluation of key impacts:
   • Identification of impacts
   • Prediction of impact magnitudes
   • Assessment of impact significance
(iii) Alternatives and mitigation:
   • Alternatives
   • Mitigation
   • Commitment to mitigation
(iv) Communication of results:
   • Presentation
   • Balance
   • Non-technical summary

In outline, the content and quality of the environmental statement is reviewed under each of the subheads, using a sliding scale of assessment symbols A–F:

Grade A   indicates that the work has generally been well performed with no important omissions.
Grade B   is generally satisfactory and complete with only minor omissions and inadequacies.
Grade C   is regarded as just satisfactory despite some omissions or inadequacies.
Grade D   indicates that parts are well attempted but, on the whole, just unsatisfactory because of omissions or inadequacies.

Grade E   is not satisfactory, revealing significant omissions or inadequacies.

Grade F   is very unsatisfactory with important task(s) poorly done or not attempted.

Having analysed each subhead, aggregated scores are given to the four review areas, and a final summary grade is attached to the whole statement.

APPENDIX 4

EIS review criteria (IAU, Oxford Brookes University)

EIS number:

project name:

reviewer name:

marking criteria
  (A–F) to summarise how well EIS fulfils criterion for *all* criteria
  A – *good*
  B – *generally satisfactory (minor omissions etc.)*
  C – *just satisfactory (despite omissions)*
  D – *just unsatisfactory (because of omissions etc.)*
  E – *not satisfactory (significant omissions etc.)*
  F – *poor*

1. DESCRIPTION OF THE DEVELOPMENT

Criterion	Performance against criteria	Comments
Principal features of the project		
1.1 Explains the purpose(s) and objectives of the development.		
1.2 Indicates the nature and status of the decision(s) for which the environmental information has been prepared.		
1.3 Gives the estimated duration of the construction, operational and, where appropriate, decommissioning phase, and the programme within these phases.		

	Criterion	Performance against criteria	Comments
1.4	Describes the proposed development, including its design and size or scale. Diagrams, plans or maps will usually be necessary for this purpose.		
1.5	Indicates the physical presence or appearance of the completed development within the receiving environment.		
1.6	Describes the methods of construction.		
1.7	Describes the nature and methods of production or other types of activity involved in the operation of the project.		
1.8	Describes any additional services (water, electricity, emergency services etc.) and developments required as a consequence of the project.		
1.9	Describes the project's potential for accidents, hazards and emergencies.		
Land requirements			
1.10	Defines the land area taken up by the development site and any associated arrangements, auxiliary facilities and landscaping areas and by the construction site(s), and shows their location clearly on a map. For a linear project, describes the land corridor, vertical and horizontal alignment and need for tunnelling and earthworks.		
1.11	Describes the uses to which this land will be put, and demarcates the different land use areas.		
1.12	Describes the reinstatement and after-use of landtake during construction.		

	Criterion	Performance against criteria	Comments
Project Inputs			
1.13	Describes the nature and quantities of materials needed during the construction and operational phases.		
1.14	Estimates the number of workers and visitors entering the project site during both construction and operation.		
1.15	Describes their access to the site and likely means of transport.		
1.16	Indicates the means of transporting materials and products to and from the site during construction and operation, and the number of movements involved.		
Residues and emissions			
1.17	Estimates the types and quantities of waste matter, energy (noise, vibration, light, heat, radiation etc.) and residual materials generated during construction and operation of the project, and rate at which these will be produced.		
1.18	Indicates how these wastes and residual materials are expected to be handled/treated prior to release/disposal, and the routes by which they will eventually be disposed of to the environment.		
1.19	Identifies any special or hazardous wastes (defined as . . .) which will be produced, and describes the methods for their disposal as regards their likely main environmental impacts.		

Criterion	Performance against criteria	Comments
1.20 Indicates the methods by which the quantities of residuals and wastes were estimated. Acknowledges any uncertainty, and gives ranges or confidence limits where appropriate.		

Overall mark:

2. DESCRIPTION OF THE ENVIRONMENT

Criterion	Performance against criteria	Comments
Description of the area occupied by and surrounding the project		
2.1 Indicates the area expected to be significantly affected by the various aspects of the project with the aid of suitable maps. Explains the time over which these impacts are likely to occur.		
2.2 Describes the land uses on the site(s) and in surrounding areas.		
2.3 Defines the affected environment broadly enough to include any potentially significant effects occurring away from the immediate areas of construction and operation. These may be caused by, for example, the dispersion of pollutants, infrastructural requirements of the project, traffic etc.		
Baseline conditions		
2.4 Identifies and describes the components of the affected environment potentially affected by the project.		

	Criterion	Performance against criteria	Comments
2.5	The methods used to investigate the affected environment are appropriate to the size and complexity of the assessment task. Uncertainty is indicated.		
2.6	Predicts the likely future environmental conditions in the absence of the project. Identifies variability in natural systems and human use.		
2.7	Uses existing technical data sources, including records and studies carried out for environmental agencies and for special interest groups.		
2.8	Reviews local, regional and national plans and policies, and other data collected as necessary to predict future environmental conditions. Where the proposal does not conform to these plans and policies, the departure is justified.		
2.9	Local, regional and national agencies holding information on baseline environmental conditions have been approached.		

Overall mark:

3. SCOPING, CONSULTATION, AND IMPACT IDENTIFICATION

	Criterion	Performance against criteria	Comments
	Scoping and consultation		
3.1	There has been a genuine attempt to contact the general public, relevant public agencies, relevant experts and special interest groups to appraise them of the project and its implication. Lists the groups approached.		

	Criterion	Performance against criteria	Comments
3.2	Statutory consultees have been contacted. Lists the consultees approached.		
3.3	Identifies valued environmental attributes on the basis of this consultation.		
3.4	Identifies all project activities with significant impacts on valued environmental attributes. Identifies and selects key impacts for more intense investigation. Describes and justifies the scoping methods used.		
3.5	Includes a copy or summary of the main comments from consultees and the public, and measures taken to respond to these comments.		
Impact identification			
3.6	Considers direct and indirect/secondary effects of constructing, operating and, where relevant, after-use or decommissioning of the project (including positive and negative effects). Considers whether effects will arise as a result of "consequential" development.		
3.7	Investigates the above types of impacts in so far as they affect: human beings, flora, fauna, soil, water, air, climate, landscape, interactions between the above, material assets, cultural heritage.		
3.8	Also noise, land use, historic heritage, communities.		
3.9	If any of the above are not of concern in relation to the specific project and its location, this is clearly stated.		

	Criterion	Performance against criteria	Comments
3.10	Identifies impacts using a systematic methodology such as project specific checklists, matrices, panels of experts, extensive consultations, etc. Describes the methods/approaches used and the rationale for using them.		
3.11	The investigation of each type of impact is appropriate to its importance for the decision, avoiding unnecessary information and concentrating on the key issues.		
3.12	Considers impacts which may not themselves be significant but which may contribute incrementally to a significant effect.		
3.13	Considers impacts which might arise from non-standard operating conditions, accidents and emergencies.		
3.14	If the nature of the project is such that accidents are possible which might cause severe damage within the surrounding environment, an assessment of the probability and likely consequences of such events is carried out and the main findings reported.		

Overall mark:

4. PREDICTION AND EVALUATION OF IMPACTS

Criterion	Performance against criteria	Comments
Prediction of magnitude of impacts		
4.1 Describes impacts in terms of the nature and magnitude of the change occurring and the nature, location, number, value, sensitivity of the affected receptors.		
4.2 Predicts the timescale over which the effects will occur, so that it is clear whether impacts are short, medium or long term, temporary or permanent, reversible or irreversible.		
4.3 Where possible, expresses impact predictions in quantitative terms. Qualitative descriptions, where necessary, are as fully defined as possible.		
4.4 Describes the likelihood of impacts occurring, and the level of uncertainty attached to the results.		
Methods and data		
4.5 The methods used to predict the nature, size and scale of impacts are described, and are appropriate to the size and importance of the projected disturbance.		
4.6 The data used to estimate the size and scale of the main impacts are sufficient for the task, clearly described, and their sources clearly identified. Any gaps in the data are indicated and accounted for.		

	Criterion	Performance against criteria	Comments
Evaluation of impact significance			
4.7	Discusses the significance of effects in terms of the impact on the local community (including distribution of impacts) and on the protection of environmental resources.		
4.8	Discusses the available standards, assumptions and value systems which can he used to assess significance.		
4.9	Where there are no generally accepted standards or criteria for the evaluation of significance, alternative approaches are discussed and, if so, a clear distinction is made between fact, assumption and professional judgement.		
4.10	Discusses the significance of effects taking into account the appropriate national and international standards or norms, where these are available. Otherwise the magnitude, location and duration of the effects are discussed in conjunction with the value, sensitivity and rarity of the resource.		
4.11	Differentiates project-generated impacts from other changes resulting from non-project activities and variables.		
4.12	Includes a clear indication of which impacts may be significant and which may not.		

Overall mark:

5. ALTERNATIVES

	Criterion	Performance against criteria	Comments
5.1	Considers the "no action" alternative, alternative processes, scales, layouts, designs and operating conditions where available at an early stage of project planning, and investigates their main environmental advantages and disadvantages.		
5.2	If unexpectedly severe adverse impacts are identified during the course of the investigation, which are difficult to mitigate, alternatives rejected in the earlier planning phases are re-appraised.		
5.3	Gives the reasons for selecting the proposed project, and the part environmental factors played in the selection.		
5.4	The alternatives are realistic and genuine.		
5.5	Compares the alternatives' main environmental impacts clearly and objectively with those of the proposed project and with the likely future environmental conditions without the project.		

Overall mark:

6. MITIGATION AND MONITORING

	Criterion	Performance against criteria	Comments
	Description of mitigation measure		
6.1	Considers the mitigation of all significant negative impacts and, where feasible, proposes specific mitigation measures to address each impact.		

	Criterion	Performance against criteria	Comments
6.2	Mitigation measures considered include modification of project design, construction and operation, the replacement of facilities/ resources, and the creation of new resources, as well as 'end-of-pipe' technologies for pollution control.		
6.3	Describes the reasons for choosing the particular type of mitigation, and the other options available.		
6.4	Explains the extent to which the mitigation methods will be effective. Where the effectiveness is uncertain, or where mitigation may not work, this is made clear and data are introduced to justify the acceptance of these assumptions.		
6.5	Indicates the significance of any residual or unmitigated impacts remaining after mitigation, and justifies why these impacts should not be mitigated.		
Commitment to mitigation and monitoring			
6.6	Gives details of how the mitigation measures will be implemented and function over the time span for which they are necessary.		
6.7	Proposes monitoring arrangements for all significant impacts, especially where uncertainty exists, to check the environmental impact resulting from the implementation of the project and their conformity with the predictions made.		
6.8	The scale of any proposed monitoring arrangements corresponds to the potential scale and significance of deviations from expected impacts.		

Criterion	Performance against criteria	Comments
Environmental effects of mitigation		
6.9 Investigates and describes any adverse environmental effects of mitigation measures.		
6.10 Considers the potential for conflict between the benefits of mitigation measures and their adverse impacts.		

Overall mark:

7. NON-TECHNICAL SUMMARY

Criterion	Performance against criteria	Comments
Non-technical summary		
7.1 There is a non-technical summary of the main findings of the study, which contains at least a brief description of the project and the environment, an account of the main mitigation measures to be undertaken by the developer, and a description of any remaining or residual impacts.		
7.2 The summary avoids technical terms, lists of data and detailed explanations of scientific reasoning.		
7.3 The summary presents the main findings of the assessment and covers all the main issues raised in the information.		
7.4 The summary includes a brief explanation of the overall approach to the assessment.		

Criterion	Performance against criteria	Comments
7.5 The summary indicates the confidence which can be placed in the results.		

Overall mark:

8. ORGANISATION AND PRESENTATION OF INFORMATION

Criterion	Performance against criteria	Comments
Organisation of the information		
8.1 Logically arranges the information in sections.		
8.2 Identifies the location of information in a table or list of contents.		
8.3 There are chapter or section summaries outlining the main findings of each phase of the investigation.		
8.4 When information from external sources has been introduced, a full reference to the source is included.		
Presentation of information		
8.5 Mentions the relevant EIA legislation, name of the developer, name of competent authority(ies), name of organisation preparing the EIS, and name, address and contact number of a contact person.		
8.6 Includes an introduction briefly describing the project, the aims of the assessment, and the methods used.		

	Criterion	Performance against criteria	Comments
8.7	The statement is presented as an integrated whole. Data presented in appendices are fully discussed in the main body of the text.		
8.8	Offers information and analysis to support all conclusions drawn.		
8.9	Presents information so as to be comprehensible to the non specialist. Uses maps, tables, graphical material and other devices as appropriate. Avoids unnecessarily technical or obscure language.		
8.10	Discusses all the important data and results in an integrated fashion.		
8.11	Avoids superfluous information (i.e. information not needed for the decision).		
8.12	Presents the information in a concise form with a consistent terminology and logical links between different sections.		
8.13	Gives prominence and emphasis to severe adverse impacts, substantial environmental benefits, and controversial issues.		
8.14	Defines technical terms, acronyms and initials.		
8.15	The information is objective, and does not lobby for any particular point of view. Adverse impacts are not disguised by euphemisms or platitudes.		
Difficulties compiling the information			
8.16	Indicates any gaps in the required data and explains the means used to deal with them in the assessment.		

Criterion	Performance against criteria	Comments
8.17 Acknowledges and explains any difficulties in assembling or analysing the data needed to predict impacts, and any basis for questioning assumptions, data or information.		

Overall mark:

COLLATION

1 Description of the development   ——
2 Description of the environment   ——
3 Scoping, consultation, and impact identification   ——
4 Prediction and evaluation of impacts   ——
5 Alternatives   ——
6 Mitigation and monitoring   ——
7 Non-technical summary   ——
8 Organisation and presentation of information   ——

Overall mark (A–F):   ——

APPENDIX 5

The EA process at the World Bank

The EA process is built into the Bank's project cycle as an integral part of project design. Different units of staff within the Bank have different responsibilities for the process. The main EA-related steps in the project cycle are described below.

Stage 1: Screening

To decide the nature and extent of the EA to be carried out, the Bank's environmental review process begins with environmental screening at the time a project is identified. In the screening, the Bank team determines the nature and magnitude of the proposed project's potential environmental and social impacts, and assigns the project to one of three environmental categories:

Category A: A full EA is required. Category A projects are those expected to have "adverse impacts that may be sensitive, irreversible, and diverse" (Operational Directive (OD) 4.01), with attributes such as direct pollutant discharges large enough to cause degradation of air, water or soil; large-scale physical disturbance of the site and/or surroundings; extraction, consumption, or conversion of substantial amounts of forest and other natural resources; measurable modification of hydrologic cycles; hazardous materials in more than incidental quantities; and involuntary displacement of people and other significant social disturbances.

Category B: Although a full EA is not required, some environmental analysis is necessary. Category B projects have impacts which are "less significant . . . not as sensitive, numerous, major or diverse. Few, if any of these impacts are irreversible, and remedial measures can be more easily designed" (OD 4.01). Typical Category B projects entail rehabilitation, maintenance or upgrading rather than new construction.

Category C: No EA or other environmental analysis is required. Category C projects have negligible or minimal direct disturbance on the physical setting. Typical Category C projects focus on education, family planning, health and human resource development.

Projects with multiple components are classified according to the component with the most significant adverse impact; if there is a Category A component, the full project is classified as A.

471

Stage 2: Scoping and Terms of Reference Development

Once a project is categorised, a scoping process is undertaken to identify key issues and develop the Terms of Reference (TOR) for the EA. It is essential to identify more precisely the likely environmental impacts and to define the project's area of influence at this stage. As part of this process, information about the project and its likely environmental effects is disseminated to local affected communities and NGOs, followed by consultations with representatives of the same groups. The main purpose of these consultations is to focus the EA on issues of concern at the local level.

Stage 3: Preparing the EA Report

When a project is classified as Category A, a full-scale EA is undertaken, resulting in an EA report. Category B projects are subject to a more limited EA, the nature and scope of which is determined on a case-by-case basis. The main components of a full EA report are the following:

Executive Summary
The Executive Summary should consist of a concise discussion of significant findings of the EA and recommended actions in the project.

Policy, Legal and Administrative Framework
Discussion of the policy, legal and administrative framework within which the EA is prepared. The environmental requirements of any co-financiers should be explained.

Project Description
In this section, staff should provide a concise description of the project's geographic, ecological, social and temporal context, including any off-site investments that may be required by the project, such as dedicated pipelines, access roads, power plants, water supply, housing and raw material and product storage materials.

Baseline Data
For EA purposes, baseline data includes an assessment of the study area's dimensions and a description of relevant physical, biological, and socioeconomic conditions, including any changes anticipated before the project begins, and current and proposed development activities within the project area, even if not directly connected to the project.

Impact Assessment
This section includes identification and assessment of the positive and negative impacts likely to result from the proposed project. Mitigation measures, and any residual negative impacts that cannot be mitigated, should be identified. Opportunities

for environmental enhancement should be explored. The extent and quality of available data, key data gaps, and uncertainties associated with predictions should be identified/estimated. Topics that do not require further attention should be specified.

Analysis of Alternatives

A key purpose of EA work is to assess investment alternatives from an environmental perspective. This is the more proactive side of EA – enhancing the design of a project through consideration of alternatives, as opposed to the more defensive task of reducing adverse impacts of a given design. The Bank's EA OD calls for the systematic comparison of the proposed investment design, site, technology, and operational alternatives in terms of their potential environmental impacts, capital and recurrent costs, suitability under local conditions, and institutional, training and monitoring requirements. For each alternative, the environmental costs and benefits should be quantified to the extent possible, economic values should be attached where feasible, and the basis for the selected alternative should be stated.

Mitigation or Management Plan

A mitigation plan consists of the set of measures to be taken during implementation and operation to eliminate, offset, or reduce adverse environmental impacts to acceptable levels. The plan identifies feasible and cost-effective measures and estimates their potential environmental impacts, capital and recurrent costs and institutional, training and monitoring requirements. The plan should provide details on proposed work programs and schedules to help ensure that the proposed environmental actions are in phase with construction and other project activities throughout implementation. The plan should consider compensatory measures if mitigation measures are not feasible or cost-effective.

Environmental Monitoring Plan

This plan specifies the type of monitoring, who will do it, how much it will cost, and what other inputs, such as training, are necessary.

Public Consultation

Consultation with affected communities is recognised as key to identifying environmental impacts and designing mitigation measures. The Bank's policy requires consultation with affected groups and local NGOs during at least two stages of the EA process: (1) at the scoping stage, shortly after the EA category has been assigned, and (2) once a draft EA report has been prepared. Consultation throughout EA preparation is also generally encouraged, particularly for projects that affect people's livelihood and for community-based projects. In projects with major social components, such as those requiring involuntary resettlement or affecting indigenous people, the consultation process should involve active public participation in the EA and project development process and the social and environmental issues should be closely linked.

Stage 4: EA Review and Project Appraisal

Once the draft EA report is complete, the borrower submits it to the Bank for review by environmental specialists. If found satisfactory, the Bank project team is authorised to proceed to appraisal of the project. On the appraisal mission, Bank staff review the EA's procedural and substantive elements with the borrower, resolve any outstanding issues, assess the adequacy of the institutions responsible for environmental management in light of the EA's findings, ensure that the mitigation plan is adequately budgeted, and determine if the EA's recommendations are properly addressed in project design and economic analysis.

Stage 5: Project Implementation

The borrower is responsible for implementing the project according to agreements derived from the EA process. The Bank supervises the implementation of environmental aspects as part of overall project supervision, using environmental specialists as necessary.

(*Source:* World Bank 1995)

Reference

World Bank 1995. *Environmental assessment: challenges and good practice*. Washington: World Bank.

APPENDIX 6
Key references

Principles and procedures

Bond, A. J. 1997. Environmental assessment and planning: a chronology of development in England and Wales. *Journal of Environmental Planning and Management* **40**(2) 261–71.

Clark, B. D. & R. G. H. Turnbull 1984. Proposals for environmental impact assessment procedures in the UK. In *Planning and ecology*, R. D. Roberts & T. M. Roberts (eds), 135–44. London: Chapman & Hall.

Commission of the European Communities 1985. On the assessment of the effects of certain public and private projects on the environment. *Official Journal* L175, 5.7.1985.

Commission of the European Communities 1992. *Towards sustainability: a European Community Programme of policy and action in relation to the environment and sustainable development*, vol. II. Brussels: Commission of the European Communities.

Commission of the European Communities 1997. Council Directive 97/11/EC of 3 March 1997 amending Directive 85/337/EEC on the assessment of certain public and private projects on the environment. *Official Journal*. L73/5.

Department of the Environment/Welsh Office 1988. DOE Circular 15/88 (Welsh Office Circular 23/88) *Environmental assessment*, 12 July 1988.

Department of the Environment/Welsh Office 1989. *Environmental assessment: a guide to the procedures*. London: HMSO.

Department of the Environment, Transport and the Regions 1998. Draft Town and Country Planning (Assessment of Environmental Effects) Regulations 1998. London: DETR.

HM Government 1994. *Sustainable development: the UK strategy*. London: HMSO.

Lee, N. & C. M. Wood 1984. Environmental impact assessment procedures within the European Economic Community. In *Planning and ecology*, R. D. Roberts & T. M. Roberts (eds), 128–34. London: Chapman & Hall.

Munn, R. E. 1979. *Environmental impact assessment: principles and procedures*. New York: John Wiley.

O'Riordan, T. & W. R. D. Sewell (eds) 1981. *Project appraisal and policy review*. Chichester: John Wiley.

Pearce, D. W. (ed) 1993. *Blueprint 3: measuring sustainable development.* London: Earthscan.

Pearce, D. W., A. Markandya, E. B. Barbier 1989. *Blueprint for a green economy.* London: Earthscan.

Redclift, M. 1987. *Sustainable development: exploring the contradictions.* London: Methuen.

Reid, D. 1995. *Sustainable development: an introductory guide.* London: Earthscan.

Tomlinson, P. 1986. Environmental assessment in the UK: implementation of the EEC Directive. *Town Planning Review* **57**(4), 458–86.

United Nations Economic Commission for Europe 1991. *Policies and systems of environmental impact assessment.* (Environmental Series). Geneva: United Nations Economic Commission for Europe.

United Nations World Commission on Environment and Development 1987. *Our common future.* Oxford: Oxford University Press.

Vanclay, F. & D. Bronstein (eds) 1995. *Environmental and social impact assessments.* London: Wiley.

Wathern, P. (ed) 1992. *Environmental impact assessment: theory and practice,* second edition. London: Routledge.

Westman, W. E. 1985. *Ecology, impact assessment and environmental planning.* New York: John Wiley.

Process

Atkinson, N. & R. Ainsworth 1992. Environmental assessment and the local authority: facing the European imperative. *Environmental Policy and Practice* **2**(2), 111–28.

Barde, J. P. & D. W. Pearce 1991. *Valuing the environment: six case studies.* London: Earthscan.

Bisset, R. 1983. A critical survey of methods for environmental impact assessment. In *An annotated reader in environmental planning and management,* T. O'Riordan & R. K. Turner (eds), 168–85. Oxford: Pergamon Press.

Bisset, R. 1989. Introduction to EIA Methods. Paper presented at the 10th International Seminar on Environmental Impact Assessment, University of Aberdeen, 9–22 July.

Bregman, J. I. & K. M. Mackenthun 1992. *Environmental impact statements.* Chelsea, Michigan: Lewis.

Buckley, R. 1991. Auditing the precision and accuracy of environmental impact predictions in Australia. *Environmental Impact Assessment Review* **11**, 1–23.

Canter, L. W. 1996. *Environmental impact assessment,* second edition. London: McGraw-Hill.

Clark, M. & J. Herington (eds) 1988. *The role of environmental impact assessment in the planning process.* London: Mansell Publishing Ltd.

Construction Industry Research and Information Association 1994. *Environmental assessment: a guide to the identification and mitigation of environmental issues in construction schemes.* CIRIA Special Publication 96. London: CIRIA.

Culhane, P. J. 1993. Post-EIS environmental auditing: a first step to making rational environmental assessment a reality. *The Environmental Professional,* **5**.

Department of the Environment 1991. *Policy appraisal and the environment.* London: HMSO.

Department of the Environment 1992. *Planning, pollution and waste management.* London: HMSO.

Department of the Environment 1994. *Evaluation of environmental information for planning projects – a good practice guide.* London: HMSO.

Department of the Environment 1994. *PPG 23 – planning and pollution control.* London: HMSO.

Department of the Environment 1995. *Preparation of environmental statements for planning projects that require environmental assessment: a good practice guide.* London: HMSO.

Department of the Environment, Transport and the Regions 1997. *Mitigation measures in environmental statements.* London: DETR.

Department of the Environment/Welsh Office 1989. *Environmental assessment: a guide to the procedures.* London: HMSO.

Department of Transport 1983. *Manual of environmental appraisal.* London: HMSO.

Department of Transport 1993. *Design manual for roads and bridges volume 11: environmental assessment.* London: HMSO.

Flynn, A. C. (ed) 1993. Special issue: Costing the countryside. *Journal of Environmental Planning and Management* **36**(1), 3–116.

Fortlage, C. A. 1990. *Environmental assessment: a practical guide.* Aldershot: Gower Technical.

Hansen, P. E. & S. E. Jorgensen (eds) 1991. *Introduction to environmental management.* New York: Elsevier.

Institute of Environmental Assessment and the Institute of Terrestrial Ecology 1995. *Guidelines for baseline ecological assessment.* London: E & FN Spon.

Institute of Environmental Assessment and the Landscape Institute 1995. *Guidelines for landscape and visual impact assessment.* London: E & FN Spon.

Jain, R. K., L. V. Urban, G. S. Stacey 1977. *Environmental impact analysis.* New York: Van Nostrand Reinhold.

Jain, R. K., L. V. Urban, G. S. Stacey, H. E. Balbach 1993. *Environmental assessment.* New York: McGraw-Hill.

Lee, N. 1989. *Environmental impact assessment – a training guide,* second edition. Occasional Paper 18, Department of Planning and Landscape, University of Manchester.

Lee, N. & R. Colley 1992. *Reviewing the quality of environmental statements,* second edition. Occasional Paper 24, Department of Planning and Landscape, University of Manchester.

Lichfield, N. 1996. *Community impact evaluation.* London: UCL Press.

McCormick, J. 1991. *British politics and the environment*. London: Earthscan.

Morris, P. & R. Therivel (eds) 1995. *Methods of environmental impact assessment*. London: UCL Press.

Munn, R. E. 1979. *Environmental impact assessment: principles and procedures*. New York: John Wiley.

Organization for Economic Co-operation and Development 1991. *Good practices for environmental impact assessment of development projects*. OECD/GD (91) 200. Paris: OECD.

O'Riordan, T. & R. K. Turner (eds) 1983. *An annotated reader in environmental planning and management*. Oxford: Pergamon Press.

Parkin, J. 1992. *Judging plans and projects*. Aldershot, England: Avebury.

Pearce, D. W. & A. Markandya 1989. *Environmental policy benefits: monetary valuation*. Paris: OECD.

Pearce, D. W., D. Whittington, S. Georgiou 1994. *Project and policy appraisal: integrating economics and environment*. Paris: OECD.

Petts, J. & G. Eduljee 1994. *Environmental impact assessment for waste treatment and disposal facilities*. Chichester: John Wiley & Sons.

Rau, J. G. & D. C. Wooten 1980. *Environmental impact analysis handbook*. New York: McGraw Hill.

Sadler, B. 1996. *Environmental assessment in a changing world: evaluating practice to improve performance*. International study of the effectiveness of environmental assessment. Canada: CEAA.

Salter, J. R. 1992. Environmental assessment: the challenge from Brussels. *Journal of Planning and Environment Law* (January), 14–20.

Salter, J. R. 1992. Environmental assessment: the need for transparency. *Journal of Planning and Environment Law* (March), 214–21.

Salter, J. R. 1992. Environmental assessment: the question of implementation. *Journal of Planning and Environment Law* (April), 313–18.

Suter II, G. W. 1993. *Ecological risk assessment*. Chelsea, Michigan: Lewis.

Tomlinson, P. 1989. Environmental statements: guidance for review and audit. *The Planner* **75**(28), 12–15.

Tomlinson, P. & S. F. Atkinson 1987. Environmental audits: proposed terminology. *Environmental Monitoring and Assessment* **8**, 187–98.

Turner, R. K., D. W. Pearce, I. Bateman 1994. *Environmental economics: an elementary introduction*. Hemel Hempstead: Harvester Wheatsheaf.

UNEP 1997. *Environmental impact assessment training resource manual*. Stevenage: SMI Distribution.

Wathern, P. (ed) 1992. *Environmental impact assessment: theory and practice*, second edition. London: Routledge.

Westman, W. E. 1985. *Ecology, impact assessment and environmental planning*. New York: John Wiley.

Winpenny, J. T. 1991. *Values for the environment: a guide to economic appraisal*. London: Overseas Development Institute/HMSO.

Wood, C. M. & N. Lee (eds) 1989. *Environmental impact assessment – five training case studies*, second edition. Occasional Paper 19, Department of Planning and Landscape, University of Manchester.

Practice

Canadian Environmental Assessment Agency 1996. *Environmental assessment in Canada: Achievements challenges and directions.* Canada: CEAA.

Coles, T. F., J. P. Tarling, K. Fuller 1992. *Practical experience of environmental assessment in the UK.* Lincolnshire: Institute of Environmental Assessment.

Council for the Protection of Rural England 1991. *The environmental assessment Directive – five years on.* Submission to the European Commission's five-year review of Directive 85/337/EEC on environmental assessment. London: CPRE.

Department of the Environment 1991. *Monitoring environmental assessment and planning.* Report by C. M. Wood & C. E. Jones, Department of Planning and Landscape, University of Manchester. London: HMSO.

Department of the Environment 1996. *Changes in the quality of environmental statements for planning projects: research report.* Report by the Impact Assessment Unit, School of Planning, Oxford Brookes University. London: HMSO.

Department of Transport 1992. *Assessing the environmental impact of road schemes.* Report of the Standing Advisory Committee on Trunk Road Assessment. London: HMSO.

English Nature 1994. *Nature conservation in environmental assessment.* Peterborough: English Nature.

Frost, R. & D. Wenham 1996. *Directory of environmental impact statements (July 1988–September 1995).* Impact Assessment Unit, School of Planning, Oxford Brookes University.

Glasson, J. & D. Heaney 1993. Socio-economic impacts: the poor relations in British environmental impact statements. *Journal of Environmental Planning and Management* **36**(3), 335–43.

Harvey, N. 1998. *EIA: procedures and prospects in Australia.* Melbourne: OUP.

Jones, C. E., N. Lee, C. M. Wood 1991. *UK environmental statements 1988–1990: an analysis.* Occasional Paper No. 29, Department of Planning and Landscape, University of Manchester.

Lambert, A. J. & C. M. Wood 1990. UK implementation of the European Directive on EIA – spirit or letter? *Town Planning Review* **61**(3), 247–62.

Lee, N. & D. Brown 1992. Quality control in environmental assessment. *Project Appraisal* **7**(1), 41–5.

Lee, N., F. Walsh, G. Reeder 1994. Assessing the performance of the EA process. *Project Appraisal* **9**(3), 161–72.

National Audit Office 1994. *Department of Transport: environmental factors in road planning and design.* London: HMSO.

Ortolano, L. 1993. Controls on project proponents and environmental impacts assessment effectiveness. *The Environmental Professional* **15**, 352–63.

Petts, J. & G. Eduljee 1994. *Environmental impact assessment for waste treatment and disposal facilities.* Chichester: John Wiley.

Royal Society for the Protection of Birds 1995. *Wildlife impact – the treatment of nature conservation in environmental assessment.* Sandy, Bedfordshire: RSPB.

Sadler, B. 1996. *Environmental assessment in a changing world: evaluating practice to improve performance.* International study of the effectiveness of environmental assessment. Canada: CEAA.

Sheate, W. R. 1994. *Making an impact: a guide to EIA law and policy.* London: Cameron May.

Sheate, W. R. & M. Sullivan 1993. *Campaigners' guide to road proposals.* London: Council for the Protection of Rural England/Transport 2000.

Sheial, J. 1991. *Power in trust: the environmental history of the Central Electricity Generating Board.* Oxford: Oxford University Press.

Thomas, I. 1996. *Environmental impact assessment in Australia.* Sydney: Federation Press.

Treweek, J. R., S. Thompson, N. Veitch, C. Japp 1993. Ecological assessment of proposed road developments: a review of environmental statements. *Journal of Environmental Planning and Management* **36**(3), 295–307.

UNEP 1997. *Environmental impact assessment training resource manual.* Stevenage: SMI Distribution.

United Nations Economic Commission for Europe 1991. *Policies and systems of environmental impact assessment.* Environmental Series 4. Geneva: United Nations Economic Commission for Europe.

Weston, J. (ed) 1997. *Planning and environmental impact assessment in practice.* Harlow: Longman.

Wood, C. M. 1994. Five years of British environmental assessment: an evaluation. In *Development and Planning 1994*, D. Cross, & C. Whitehead (eds). Department of Land Economy, University of Cambridge.

Wood, C. M. 1995. *Environmental impact assessment: a comparative review.* Harlow: Longman.

Wood, C. M. & C. E. Jones 1997. The impact of environmental assessment on local planning authorities. *Urban Studies* **34**(8), 1237–257.

Wood, C. M., N. Lee, C.E. Jones 1991. Environmental statements in the UK: the initial experience. *Project Appraisal* **6**, 187–94.

World Bank 1995. *Environmental assessment: challenges and good practice.* Washington DC: World Bank.

Prospects

Bedfordshire County Council/RSPB 1996. *A step by step guide to environmental appraisal.* Bedford: Bedfordshire County Council.

Department of the Environment 1991. *Policy appraisal and the environment: a guide for government departments.* London: HMSO.

Department of the Environment 1993. *The environmental appraisal of development plans: a good practice guide.* London: HMSO.

Department of the Environment 1994. *Environmental appraisal in government departments.* London: HMSO.

English Nature 1996. *Strategic environmental assessment and nature conservation.* Peterborough: English Nature.

Glasson, J. (ed) 1994. *Theme edition: Environmental impact assessment: the next steps.* Built Environment **20**(4), 277–344.

Grayson, L. (ed) 1992. *Environmental auditing: a guide to best practice in the UK and Europe.* London: British Library.

International Association for Impact Assessment (IAIA) 1994. Guidelines and principles for social impact assessment. *Impact Assessment* Vol **12**. Summer.

Jacobs, M. 1991. *Environmental auditing in local government: a guide and discussion paper.* Local Government Management Board.

Ledgerwood, G., E. Street, R. Therivel 1992. *The environmental audit and business strategy: a total quality approach.* London: *Financial Times*/Pitman.

Lee, N. & J. Hughes 1995. *Strategic environmental assessment: legislation and procedures in the Community.* Department of Planning and Landscape, University of Manchester.

Lee, N. & E. Walsh 1992. Strategic environmental assessment: special collection of articles. *Project Appraisal* **7**(3).

McClaren, D. 1993. *The future for environmental impact assessment: meeting the demands of sustainable development.* London: Friends of the Earth.

McDonald, G. T. & L. Brown 1995. Going beyond environmental impact assessment: environmental input to planning and design. *Environmental Impact Assessment Review* **15**, 483–95.

Ministry of Housing, Spatial Planning and the Environment 1996. *Strategic environmental assessment: status, challenges and future directions.* Report No.53. Zoetermeer, the Netherlands: VROM.

Ministry of Housing, Spatial Planning and the Environment 1996. *Strategic environmental assessment: environmental assessment of policies.* Report No.54. Zoetermeer, the Netherlands: VROM.

Pinfield, G. 1992. Strategic environmental assessment and land-use planning. *Project Appraisal* **7**(3), 157–63.

Project Appraisal 1996. Special Edition: environmental assessment and socio-economic appraisal in development. *Project Appraisal* Vol **11**(4).

Sadler, B. 1996. *Environmental assessment in a changing world: evaluating practice to improve performance.* International study of the effectiveness of environmental assessment. Canada: CEAA.

Therivel, R. 1995. Environmental appraisal of development plans: current status. *Planning Practice and Research* **10**(2), 223–34.

Therivel, R. & M. R. Partidario 1996. *The practice of strategic environmental assessment.* London: Earthscan.

Therivel, R., E. Wilson, S. Thompson, D. Heaney, D. Pritchard 1992. *Strategic environmental assessment.* London: RSPB/Earthscan.

Woolston, H. (ed) 1993. *Environmental auditing: an introduction and practical guide.* London: British Library.

World Bank 1995. *Environmental assessment: challenges and good practice.* Environment Department Papers No. 018. Washington: World Bank.

Useful journals and newsletters

Ambio: A Journal of the Human Environment. 8 issues per year. Published by the Royal Swedish Academy of Sciences, Stockholm, Sweden.

Australian Planner. Quarterly. Journal of the Royal Australian Planning Institute.

Built Environment. Quarterly. Published by Alexandrine Press, Oxford, UK.

Eco-Management and Auditing. 3 issues per year. Published by John Wiley and Sons Ltd., Chichester, West Sussex, UK and ERP Environment, Shipley, West Yorkshire, UK.

EIA Newsletter. Bi-annual newsletter from the EIA Centre, University of Manchester, UK.

The ENDS Report. Monthly. Published by Environmental Data Services Ltd., London, UK.

Environment Business News Briefing. Fortnightly newsletter. Published by Information for Industry Ltd., London, UK.

Environmental Assessment. Magazine of the Institute of Environmental Assessment and the Environmental Auditors Registration Association, Lincoln, UK.

Environmental Impact Assessment Review. 6 issues per year. Published by Elsevier Science Inc., New York, USA.

Environmental Law Bulletin. Quarterly. Published by Cameron McKenna, London, UK.

Environmental Monitoring and Assessment. 15 issues per year. Published by Kluwer Academic Publishers, Dordrecht, the Netherlands.

Environmental Policy and Law. 6 issues per year. Published by IOS Press, Amsterdam, the Netherlands.

The Environmental Professional. Journal of the National Association of Environmental Professionals, USA.

Environmental and Waste Management (formerly Environmental Policy and Practice). Quarterly. Published by EPP Publications, Richmond, Surrey, UK.

European Environment. 6 issues per year. Published by John Wiley and Sons Ltd., Chichester, West Sussex, UK and ERP Environment, Shipley, West Yorkshire, UK.

Impact Assessment. Quarterly. Journal of the International Association for Impact Assessment.

International Planning Studies. 3 issues per year. Published by Carfax Publishing Ltd., Abingdon, Oxfordshire, UK.

Journal of Environmental Assessment Policy and Management. 4 issues per year. Published by Imperial Colleges Press Ltd.

Journal of Environmental Law. Bi-annual. Published by Oxford University Press, Oxford, UK.

Journal of Environmental Management. Monthly. Published by Academic Press Ltd., London, UK.

Journal of Planning and Environment Law. Monthly. Published by Sweet and Maxwell Ltd., London, UK.

Journal of Planning and Environmental Management. 6 issues per year. Published by Carfax Publishing Ltd., Abingdon, Oxfordshire, UK.

Land Use Policy. Quarterly. Published by Elsevier Science Ltd., Exeter, UK.

Landscape Design. Monthly. Journal of the Landscape Institute, London, UK.

Landscape Research. 3 issues per year. Published by Carfax Publishing Ltd., Abingdon, Oxfordshire, UK.

Local Environment. 3 issues per year. Published by Carfax Publishing Ltd., Abingdon, Oxfordshire, UK.

Planning Practice and Research. Quarterly. Published by Carfax Publishing Ltd., Abingdon, Oxfordshire, UK.

Project Appraisal. Quarterly. Published by Beech Tree Publishing, Guildford, Surrey, UK.

Sustainable Development. Published by John Wiley and Sons Ltd., Chichester, West Sussex, UK and ERP Environment, Shipley, West Yorkshire, UK.

Town Planning Review. Quarterly. Published by Liverpool University Press, Liverpool, UK.

APPENDIX 7

Addresses of UK organizations with EIS collections

Crown Estate Office
10 Charlotte Street, Edinburgh EH2 4DR
Department of the Environment, Transport and the Regions
DETR Library, Room P3-006, 2 Marsham Street, London SW1P 3EB
Department of the Environment (Northern Ireland), Town and Country Planning Service
Commonwealth House, 35 Castle Street, Belfast BT1 1GU
Department of Transport
Room S4-13A, 2 Marsham Street, London SW1P 3EB
Department of Transport, Ports Division
Sunley House, 90–93 High Holborn, London SC1V 6LP
Forestry Authority
231 Corstorphine Road, Edinburgh EH12 7AT
Ministry of Agriculture, Fisheries and Food, Flood Defence Division
Eastbury House, 30–34 Albert Embankment, London SE1 7TL
Scottish Office, Environment Department
New St Andrews House, Edinburgh EH1 3SZ
Welsh Office, Planning Division
Cathays Park, Cardiff CF1 3NQ
EIA Centre, University of Manchester
Department of Planning and Landscape, University of Manchester, Manchester M13 9PL
Impacts Assessment Unit, Oxford Brookes University
School of Planning, Oxford Brookes University, Gipsy Lane, Headington, Oxford OX3 0BP
Institute of Environmental Assessment
Welton House, Limekiln Way, Lincoln LN2 4US

Author index

Subject index